T0305697

# Guidebook for Systems Applications in Astrobiology

This book addresses the timely subject of systems applications in astrobiology. It demonstrates how astrobiology – a multidisciplinary, interdisciplinary, and transdisciplinary field of science – can benefit from adopting the systems approach.

Astrobiology draws upon its founding sciences, such as astronomy, physics, chemistry, biochemistry, geology, and planetary sciences. However, astrobiologists can encounter difficulties working across these fields. The systems approach, we believe, is the best contemporary approach to consider astrobiology holistically. The approach is currently used in other fields, such as engineering, which uses systems analysis routinely.

Such an approach needs to be learned, both in principle and through examples, from the field. This book features chapters from experts across the field of astrobiology who have applied the systems approach. It will be a valuable guide for astrobiology students at the advanced undergraduate and graduate levels, in addition to researchers in the field, both in academia and the space industry.

**Key Features:**

- Offers a unique and novel approach to studying and understanding astrobiology
- Encourages astrobiologists to apply a holistic systems approach to their work, rather than being bogged down in details
- Imparts practical knowledge to readers which can be adopted in different research and job opportunities in the field of astrobiology

**Vera M. Kolb** obtained degrees in chemical engineering and organic chemistry from Belgrade University, Serbia, and earned her PhD in organic chemistry from Southern Illinois University, Carbondale, Illinois, United States. Following a 30-year career, she is Professor Emerita of Chemistry at the University of Wisconsin-Parkside, Kenosha, Wisconsin. During her first sabbatical leave with the NASA Specialized Center of Research and Training (NSCORT) in Astrobiology, she conducted research with Dr. Leslie Orgel at the Salk Institute and Prof. Stanley Miller at UC San Diego. Her second sabbatical was with Prof. Joseph Lambert at Northwestern University, where she studied sugar silicates and their potential astrobiological relevance. She is credited for authoring over 160 publications, in the fields of organic and medicinal chemistry, green chemistry, and astrobiology, including several books. Recently, she authored *Green Organic Chemistry and Its Interdisciplinary Applications* (CRC 2016). In the astrobiology field, she edited *Astrobiology: An Evolutionary Approach* (CRC 2015) and *Handbook of Astrobiology* (CRC 2019). She co-authored (with Benton C. Clark) *Astrobiology for a General Reader: A Questions and Answers Approach* (CSP 2020) and *Systems Approach to Astrobiology* (CRC 2023).

## Series in Astrobiology

Series Editors: Vera M. Kolb, University of Wisconsin-Parkside

This exciting new book series will systematically cover the latest topics in the rapidly developing field of astrobiology. Astrobiology is concerned primarily with the origin and evolution of life on Earth, and search for extra-terrestrial life, at the microbial level in our Solar System, and intelligent life elsewhere. This new field is interdisciplinary, since it draws upon other disciplines, such as chemistry, biology, physics, astronomy, geology, planetary geology, modern science, and the engineering involved in the exploration of our Solar System and beyond.

Recent books in the series:

**Astrobiology: An Evolutionary Approach**
*Edited By Vera M. Kolb*

**Handbook of Astrobiology**
*Edited By Vera M. Kolb*

**Systems Approach to Astrobiology**
*Vera M. Kolb and Benton C. Clark*

**Guidebook for Systems Applications in Astrobiology**
*Edited By Vera M. Kolb*

# Guidebook for Systems Applications in Astrobiology

Edited by Vera M. Kolb

CRC Press is an imprint of the
Taylor & Francis Group, an **informa** business

Designed cover image: Shutterstock_ 2061939557

First edition published 2024
by CRC Press
2385 NW Executive Center Drive, Suite 320, Boca Raton FL 33431

and by CRC Press
4 Park Square, Milton Park, Abingdon, Oxon, OX14 4RN

*CRC Press is an imprint of Taylor & Francis Group, LLC*



*Library of Congress Cataloging-in-Publication Data*
Names: Kolb, Vera M., editor.
Title: Guidebook for systems applications in astrobiology / edited by Vera M. Kolb.
Description: First edition. | Boca Raton, FL : CRC Press, 2024. | Includes bibliographical references and index. | Summary: "This book addresses the timely subject of systems applications in astrobiology. It demonstrates how astrobiology – a multidisciplinary, interdisciplinary, and transdisciplinary field of science – can benefit from adopting the systems approach. Astrobiology draws upon its founding sciences, such as astronomy, physics, chemistry, biochemistry, geology, and planetary sciences. However, astrobiologists can encounter difficulties working across these fields. The systems approach, we believe, is the best contemporary approach to consider astrobiology holistically. The approach is currently used in other fields, such as engineering, which uses systems analysis routinely. Such an approach needs to be learned, both in principle and through examples from the field. This book features chapters from experts across the field of astrobiology who have applied the systems approach. It will be a valuable guide for astrobiology students at the advanced undergraduate and graduate level, in addition to researchers in the field, both in academia and the space industry"— Provided by publisher.
Identifiers: LCCN 2023032497 | ISBN 9781032278216 (hardback) | ISBN 9781032278230 (paperback) | ISBN 9781003294276 (ebook)
Subjects: LCSH: Exobiology. | Exobiology—Simulation methods.
Classification: LCC QH326 .G85 2024 | DDC 576.8/39—dc23/eng/20230926
LC record available at https://lccn.loc.gov/2023032497

ISBN: 978-1-032-27821-6 (hbk)
ISBN: 978-1-032-27823-0 (pbk)
ISBN: 978-1-003-29427-6 (ebk)

DOI: 10.1201/9781003294276

Typeset in Times LT Std
by Apex CoVantage, LLC

*I dedicate this Guidebook to the memory of my dearests, who have passed away but who live in my heart: my father, Dr. Martin, my mother, Dobrila, and my brother, Vladimir.*

# Contents

# Preface

Recently, I co-authored a book, *Systems Approach to Astrobiology*, with Benton C. Clark, III. The book was published by CRC Press in May 2023. It filled the void on the subject. As I was working on this book, I saw a need for a follow-up book, the *Guidebook for Systems Applications in Astrobiology*, in which authors with a range of expertise on this topic would contribute their knowledge, insights, and in-depth coverage. I submitted the proposal to CRC to edit such a book and was happy that the proposal was approved. I am grateful to all the contributors to the Guidebook for making its publication possible and to Danny Kielty, from CRC, for facilitating the process.

This book guides the reader through the landscape of astrobiology with the focus on the systems approach applications. As different astrobiology topics are covered, a pattern emerges which shows how parts of the astrobiology systems interact among themselves to produce new and emerging functions. This Guidebook starts with an overview of astrobiology and a primer on the systems approach. The chapters which follow cover different systems applications in an astrobiological context. These applications include systems approach in prebiotic chemistry which led to life; assembly of membranous compartments as an essential step in the emergence and function of protocells; locating the cradle of life in the hydrothermal impact crater lakes on young Earth and Mars; systems geochemistry in an astrobiological context; asymmetric autocatalysis and the origins of homochirality; abiotic-to-biotic transition of chemical systems starting with thermodynamic inversion; systems biology; systems approach to microbiology; the role of astrobiology in systems thinking education; knowledge gaps in astrobiology and how systems approach can help bridge them; life's emergence by protocellular mutually catalytic networks of small molecules; and prototyping the colonizer cell by combining biology and systems engineering.

Although the list of the systems applications given in this Guidebook is not inclusive for all fields and subfields of astrobiology, enough examples are given, in my opinion, to show that astrobiology is making strides in adopting the system's approach in solving its complex problems. It is my hope that our selection will demonstrate to the readers the power of the systems approach and that it will motivate the astrobiologists to expand on the use of the systems approach throughout the field.

**Vera M. Kolb**
*July 2023*

# Acknowledgments for the Guidebook

I am grateful to the individual contributors to the Guidebook, who made this book possible. Thanks are expressed also to Danny Kielty, from CRC, who has helped with the editorial aspects of the Guidebook.

# Contributors

**Julia Brodsky**
Blue Marble Institute of Science
Rockville, Maryland, USA

**Sankar Chatterjee**
Museum of Texas Tech University
Lubbock, Texas, USA

**Benton C. Clark**
Space Science Institute
Boulder, Colorado, USA

**Bruce Damer**
UC Santa Cruz, Biota Institute
Boulder Creek, California, USA

**David Deamer**
University of California, Santa Cruz
Santa Cruz, California, USA

**Maheen Gull**
University of South Florida
Tampa, Florida, USA

**S. Jheeta**
Network of Researchers on the Chemical Evolution of Life, Leeds, UK

**Tsuneomi Kawasaki**
Tokyo University of Science
Tokyo, Japan

**Vera M. Kolb**
University of Wisconsin Parkside
Kenosha, Wisconsin, USA

**Vladimir N. Kompanichenko**
Institute for Complex Analysis of Regional Problems RAS
Birobidzhan, Russia

**Oleg R. Kotsyurbenko**
Yugra State University
Khanty-Mansyisk, Russia And Network of Researchers on the Chemical Evolution of Life, Leeds, UK

**Doron Lancet**
Department of Molecular Genetics The Weizmann Institute of Science
Rehovot, Israel

**Arimasa Matsumoto**
Faculty of Science, Nara Women's University
Nara, Japan

**Matthew A. Pasek**
University of South Florida
Tampa, Florida, USA

**Povilas Šimonis**
UCSC and FTMC
Boulder Creek, California, USA

**D.A. Skladnev**
Network of Researchers on the Chemical Evolution of Life
Leeds, UK
And
Research Center of Biotechnology of the Russian Academy of Sciences
Winogradsky Institute of Microbiology,
Moscow, Russia

**Kenso Soai**
Department of Applied Chemistry, Tokyo University of Science
Tokyo, Japan

**Iman Tavassoly**
Department of Pharmacological Sciences, Icahn School of Medicine at Mount Sinai
New York, New York, USA

**Roy Yaniv**
Department of Molecular Genetics
The Weizmann Institute of Science
Rehovot, Israel

# 1 Astrobiology
## *An Overview*

*Vera M. Kolb*

## 1.1 WHAT IS ASTROBIOLOGY?

### 1.1.1 List of Astrobiology Goals

The main astrobiology goal is to study the origins, evolution, distribution, and future of life on the Earth and in the universe (Des Marais et al., 2008; Des Marais, 2019). Within this main goal, astrobiology has numerous specific subgoals, which themselves have their own subgoals, nesting inside each other like Russian dolls. This is graphically depicted in Figure 1.1.

Examples of astrobiology subgoals include: The identification of the sources of organic compounds on the primordial Earth; the role of the environment in their production and accumulation; the origins of organic compounds in space, their diversity, and complexity; study of the origins of life on Earth, including laboratory modeling

**FIGURE 1.1** The picture of the Russian dolls, which is used here to show metaphorically the nested goals.

*Photo courtesy:* Shutterstock; www.shutterstock.com 1883780119

DOI: 10.1201/9781003294276-1

of the chemical pathways which can lead to the present-day biological constituents (such as DNA, RNA, proteins, and lipids); study of the chemical evolution that led to the emergence of life; study of the origins of self-organization and increase in complexity during chemical evolution; study of the co-evolution of life and its physical environment; the identification and study of the environments that are habitable for life on the Earth, our Solar Systems, and exoplanets; the identification of biosignatures for life detection; and the search for extraterrestrial life in our Solar System and elsewhere by space flight missions and remote sensing.

As the field of astrobiology grew further, it acquired additional goals for its study, ranging from heliophysics (often referred to as solar and space physics) to laboratory activities and field studies in terrestrial surface and marine environments.

We list here selected astrobiology topics of study, which are notoriously difficult to elucidate. They include defining life (Kolb, 2019b, 2019c; Popa, 2004, 2010, 2015; Gayon, 2010), its origins and the emergence on Earth (Oparin, 1924, 1938, 1968; Miller and Orgel, 1974; Miller, 2000; Zubay, 2000; Perry and Kolb, 2004; Peretó, 2005, 2012, 2019; Fry, 2019; Krishnamurthy, 2017, 2018); the nature of life forms which are in a gray area of life, such as viruses (Villarreal, 2004); reconstruction and back-engineering of a historic process of chemical evolution that led to life (Mason, 1991; Szostak, 2017); energetics of life and life's out-of-equilibrium status (Kompanichenko, 2017, and the references cited therein); search for the extraterrestrial life in our Solar System (Clark, 2019) and beyond (Drake equation: https://en.wikipedia.org/wiki/Drake_equation); and the habitability of various objects in space (Kasting, 2010) and its changes over a period of time, among others.

### 1.1.2 Astrobiology as a Field of Scientific Study

Astrobiology is a field of scientific study, rather than a scientific discipline. It draws upon its foundational disciplines, such as physics, chemistry, biology, geology, and astronomy. Astrobiology also relies on various subdisciplines and research areas, such as atmospheric science, oceanography, evolutionary science, paleontology, planetary science, biochemistry, molecular biology, microbiology, ecology, and the history and philosophy of science. In addition, astrobiology is increasingly linked to various space missions, some of which have a direct goal of searching for the evidence of extraterrestrial life. Table 1.1 lists selected foundational disciplines and subdisciplines of astrobiology.

These different disciplines, sub-disciplines, and their branches superficially look like a laundry list of scattered items. However, this is not the case. Astrobiology weaves together these items in a holistic way. The latter approach is described in general terms by Kasser (2013) and Reif (2010). It is said that astrobiology is a multidisciplinary, interdisciplinary, and transdisciplinary field of study (see Kolb, 2019b). For multi/inter/and trans/disciplinarity in general, see e.g., Repko (2012) and Choi and Pak (2006). We explain these interactions in astrobiology separately for each category.

Astrobiology is multidisciplinary since it simultaneously takes the perspective of multiple foundational disciplines or subdisciplines. According to Choi and Pak (2006), multidisciplinarity in general draws on knowledge from different disciplines but stays within their boundaries.

**TABLE 1.1**
**Selected Foundational Disciplines and Subdisciplines of Astrobiology in Alphabetical Order**

| |
|---|
| Analytical and instrumental chemistry |
| Astronomy |
| Atmospheric science |
| Biochemistry |
| Biology |
| Chemical evolution |
| Chemistry |
| Cosmic evolution |
| Cosmology |
| Cultural evolution |
| Ecology |
| Engineering |
| Evolutionary biology |
| Geology |
| Microbiology |
| Molecular biology |
| Oceanography |
| Paleontology |
| Philosophy and history of science |
| Physics |
| Planetary science |
| Robotics |
| Space engineering |

*Source*: Kolb (2019b).

The interdisciplinary nature of astrobiology is due to the extensive integration of its foundational disciplines and subdisciplines, their overlap, and the requirement for their synchronous examination. Choi and Pak (2006) give a general description of interdisciplinarity, in which the latter "analyzes, synthesizes and harmonizes links between disciplines into a coordinated and coherent whole".

However, astrobiology is also a transdisciplinary field of science. Transdisciplinarity means that a synergy of views of the foundational disciplines and their subdisciplines is achieved. This synergy leads to conclusions which are above and beyond any achieved in the individual disciplines and their subdisciplines, single or integrated with others (Repko, 2012). In other words, transdisciplinarity integrates the disciplines but transcends their traditional boundaries (Choi and Pak, 2006).

We can understand the idea behind the multidisciplinary (M), interdisciplinary (I), and transdisciplinary (T) approach to astrobiology. However, to apply these ideas and start thinking in the "M-I-T" way require a long practice. Mastering the M-I-T approach is difficult, and it requires constant updates and reevaluations. Thus, the M-I-T approach looks more like a goal to achieve rather than a straightforward process. There are no specific, prescribed steps which will help us navigate this process.

Another approach to astrobiology exists, a systems approach, which complements and enhances the M-I-T approach. It is described in Subsection **1.1.3**.

### 1.1.3 Systems Approach to Astrobiology

This subsection is a short rendition of the content of a book with the same title (Kolb and Clark, 2023).

Astrobiology is a systems-level science since its foundational disciplines, subdisciplines, and research areas ("parts") interact among themselves in such a way to create new capabilities and functions which are not achieved by any individual part. This is in line with the "mantra" about systems, which states that a system is more than a sum of its parts (e.g., Meadows, 2008). In the next subsection, we shall briefly expand on this catchword, and in Chapter 2 of this Guidebook, we shall give a primer on the systems approach.

## 1.2 A VERY BRIEF INTRODUCTION TO SYSTEMS THINKING AND SYSTEMS APPROACH

To understand astrobiology as a systems-level science, astrobiologists need to adopt the approach in which astrobiology is a system and its various disciplines and subdisciplines are its parts. Such systems approach to astrobiology was recommended by the National Academy of Sciences, Engineering, and Medicine (NASEM) in their recent report (NASEM, 2019). This report makes the point that while many modern science and technology fields now employ the systems approach with great success, astrobiology did not follow and is behind in this respect. To alleviate this problem, we have published a book titled *Systems Approach to Astrobiology* (Kolb and Clark, 2023).

We will give next a brief introduction to the systems approach for the purpose of understanding this chapter. A more in-depth coverage of the system's approach will be presented in Chapter 2.

A system is composed of its parts. These may be simple or complex. Depending on their complexity, the parts may also be referred to as components or subsystems. These parts interact with each other in such a way to produce networks, feedback loops, hierarchical organization, and other complex interactions. As the result of such interactions, a new property or function of the system emerges, which is a characteristic only for the system but not for any of its individual parts. Because of the emergent properties of the system, it is often stated that the system is more than the sum of its parts.

The system's approach has strong potential for the advancement of the field of astrobiology. These will be gradually introduced, presented, explained, and exemplified throughout this book.

However, the systems approach/analysis which is generally used in science and technology may not be directly applicable to astrobiology and needs to be modified. For example, engineering systems typically have a purpose, which is stated and designed before the engineering system is built. A set of these pre-stated, detailed specifications which allow the system to function to fulfill its purpose is a standard way engineering systems are built. In contrast, an *a priori* purpose of some astrobiology systems, particularly those which involve evolution, would not be feasible to formulate. The way around this problem is to substitute the purpose in these cases with an actualization of the functioning of the system.

## 1.3 CENTRAL IDEAS IN ASTROBIOLOGY

We have seen from the previous sections that astrobiology is a multidisciplinary, interdisciplinary, and transdisciplinary field of science, which recently has been approached as a systems-level science. Demand is placed on the astrobiologist to be a renaissance person and have command of all scientific knowledge which is relevant to astrobiology. This elusive goal is further undermined by the rapid growth of astrobiology. Notably, space missions to Mars, discovery of exoplanets, and deepening knowledge about the simplest forms of life come to mind.

An important question arises, namely how can astrobiologists, and especially beginning students, comprehend and gain expertise in this field.

Astrobiology shares this problem with the disciplines and subdisciplines upon which it is founded, e.g., chemistry, biology, geology, planetary science, physics, astronomy, and evolutionary science, which have grown so rapidly that it become difficult to obtain comprehensive expertise in all areas of study. This problem is especially acute for astrobiologists who need to gain knowledge across its foundational disciplines and subdisciplines and conquer the complex way these interact to address astrobiology problems.

There may be many ways to handle this problem, but we describe just one. It comes from chemistry, which is this author's foundational discipline and can serve as an exemplar of both the above problem and the proposed solution to it.

Chemistry is a scientific discipline which branched to many subdisciplines, which, in turn, branched even further. Chemistry's basic subdisciplines, such as general, inorganic, analytical, physical, organic, and biochemistry, forked further. New subfields and their branches such as nano chemistry, environmental chemistry, geochemistry, green chemistry, theoretical chemistry, computational chemistry, quantum chemistry, electrochemistry, chemical thermodynamics, spectroscopy, forensic chemistry, and industrial chemistry are mushrooming.

Because of this, a new approach toward teaching and learning chemistry has been proposed. It organizes chemistry knowledge around *central ideas* (which are sometimes also called the *big ideas*) (e.g., Talanquer, 2016 and the references therein). We give here just a few examples of these ideas: atoms and molecules are in a constant motion; elements form families (*cf.* periodic chart); shape is of utmost importance; molecules interact with each other; chemical changes have a time scale over which they occur; matter changes, forming products that have new chemical and physical properties. Without going any deeper into the chemistry applications, we now move on to the central ideas in astrobiology. We consider selected exemplars, which are shown in Table 1.2. These are described in detail and are well-referenced in the "Handbook of Astrobiology" (Kolb, 2019a).

**TABLE 1.2**
**Exemplars of Central Ideas in Astrobiology. These Are Described in Detail and are Well-Referenced in the *Handbook of Astrobiology***

1) The nature and definition of life
2) The origins of life on Earth and its emergence from abiotic matter
3) Life's gray areas: Are viruses alive?
4) Chemical evolution
5) Role of the environment for evolution and sustainability of life
6) Coevolution of metabolism, information, and membrane subsystems
7) Energetics of life and life's out-of-equilibrium status
8) The nature of the putative extraterrestrial life
9) Search for extraterrestrial intelligent life: Drake's equation
10) Habitability on the objects in our Solar System and exoplanets
11) Future of life on Earth and the possibility of our life's migration to more hospitable extraterrestrial settings

*Source*: Kolb (2019a)

These central ideas represent the focus of many astrobiologists' research studies. There may be many more such ideas, and we encourage the reader to formulate those which resonate with their research focus. Organizing one's research around the central ideas is also helpful for the scientific cooperations which are essential in astrobiology.

## REFERENCES

Chela-Flores, J. *The New Science of Astrobiology: From Genesis of the Living Cell to Evolution of Intelligent Behaviour in the Universe*. Kluwer Academic Publishers: Dordrecht, The Netherlands, **2001**.

Chela-Flores, J. *The Science of Astrobiology: A Personal View on Learning to Read the Book of Life*. Springer: Dordrecht, The Netherlands, **2011**.

Choi, BC; Pak, AW. Multidisciplinarity, interdisciplinarity and transdisciplinarity in health research, services, education and policy: 1. Definitions, objectives, and evidence of effectiveness. *Clin Invest Med*. **2006**, *29*(6), 351–364. PMID: 17330451.

Clark, BC. Searching for extraterrestrial life in our Solar System. In *Handbook of Astrobiology*, Kolb, VM. Ed. CRC Press: Boca Raton, FL, **2019**, pp. 801–817.

Des Marais, DJ. Astrobiology goals: NASA strategy and European roadmaps. In *Handbook of Astrobiology*, Kolb, VM. Ed. CRC Press: Boca Raton, FL, **2019**, pp. 15–26.

Des Marais, DJ; Nuth III, JA; Allamandola, LJ; Boss, AP; Farmer, JD; Hoehler, TM; Jakosky, BM; Meadows, VS; Pohorille, A; Runnegar, B; Spormann, AM. The NASA astrobiology roadmap. *Astrobiology* **2008**, *8*(4), 715–730. https://doi.org/10.1089/ast.2008.0819

Drake equation. https://en.wikipedia.org/wiki/Drake_equation

Fry, I. Philosophical aspects of the origin-of-life question: Neither by chance nor by design. In *Handbook of Astrobiology*, Kolb, VM. Ed. CRC Press: Boca Raton, FL, **2019a**, pp. 109–124.

Fry, I. The origin of life as an evolutionary process: Representative case studies. In *Handbook of Astrobiology*, Kolb, VM. Ed. CRC Press: Boca Raton, FL, **2019b**, pp. 437–462.

Gayon, G. Defining life: Synthesis and conclusions. *Orig Life Evol Biosph* **2010**, *40*, 231–244. https://doi.org/10.1007/s11084-010-9204-3

Kasser, J. *Holistic Thinking: Creating Innovative Solutions to Complex Problems*. The Right Requirement Publishers: Cranfield, **2013**.

Kasting, J. *How to Find a Habitable Planet*. Princeton University Press: Princeton, NJ, **2010**.

Kolb, VM, Ed. *Astrobiology, An Evolutionary Approach*. CRC Press: Boca Raton, FL, **2015**.

Kolb, VM, Ed. *Handbook of Astrobiology*. CRC Press: Boca Raton, FL, **2019a**.

Kolb, VM. Astrobiology: Definition, scope and a brief overview. In *Handbook of Astrobiology*, Kolb, VM. Ed. CRC Press: Boca Raton, FL, **2019b**, pp. 3–14.

Kolb, VM. Defining life: Multiple perspectives. In *Handbook of Astrobiology*, Kolb, VM. Ed. CRC Press: Boca Raton, FL, **2019c**, pp. 57–64.

Kolb, VM; Clark, BC. *Astrobiology for a General Reader: A Questions and Answers Approach*. Cambridge Scholars Publishing: New Castle Upon Tune, **2020**.

Kolb, VM; Clark, BC. *Systems Approach to Astrobiology*. CRC Press: Boca Raton, FL, **2023**.

Kompanichenko, VN. *Thermodynamic Inversion: Origin of Living Systems*. Springer International Publishing: Cham, Switzerland, **2017**.

Krishnamurthy, R. Giving rise to life: Transition from prebiotic chemistry to protobiology. *Acct Chem Res*. **2017**, *50*, 455–459. https://doi.org/10.1021/acs.accounts.6b00470

Krishnamurthy, R. Life's biological chemistry: A destiny or destination starting from prebiotic chemistry? *Chem Eur J*. **2018**, *24*(63), 16708–16715. https://doi.org/10.1002/chem.201801847

Longstaff, A. *Astrobiology: An Introduction*. CRC Press: Boca Raton, FL, **2015**.

Mason, SF. *Chemical Evolution, Origins of the Elements, Molecules and Living Systems*. Oxford University Press: Oxford, **1991**.

Meadows, DH. *Thinking in Systems: A Primer*. Chelsea Green Publishing: White River Junction, VT, **2008**.

Miller, SL. The endogenous synthesis of organic compounds. In *The Molecular Origins of Life: Assembling Pieces of the Puzzle*, Brack, A. Ed. Cambridge University Press: Cambridge, **2000**, pp. 59–85.

Miller, SL; Orgel, LE. *The Origins of Life on the Earth*. Prentice Hall: Englewood Cliffs, NJ, **1974**.

NASEM (National Academies of Sciences, Engineering, and Medicine). *An Astrobiology Strategy for the Search for Life in the Universe*. The National Academies Press, Washington, DC, **2019**.

Oparin, AI. *The Origin of Life*. Published in Russian in **1924**. English translation by A. Sygne published in *Origins of Life: The Central Concepts*, Deamer, DW; Fleischaker, GP. Eds. Jones and Bartlett: Boston, MA, **1994**, pp. 31–71.

Oparin, AI. *Origin of Life*. **1938**. Translated by S. Morgulis. Republication of the original publication by Macmillan Company by Dover Publications Inc., 2nd ed. Dover: New York, **1965**.

Oparin, AI. *Genesis and Evolutionary Development of Life*. Published in Russian in **1966**. English translation by Maass, E. Academic Press: New York, **1968**.

Peretó, J. Controversies on the origin of life. *Intern Microbiol*. **2005**, *8*, 23–31.

Peretó, J. Out of fuzzy chemistry: From prebiotic chemistry to metabolic networks. *Chem Soc Rev*. **2012**, *41*, 5394–5403. https://doi.org/10.1039/C2CS35054H

Peretó, J. Prebiotic chemistry that led to life. In *Handbook of Astrobiology*, Kolb, VM. Ed. CRC Press: Boca Raton, FL, **2019**, pp. 219–233.

Perry, RS; Kolb, VM. On the applicability of Darwinian principles to chemical evolution that led to life. *Int J Astrobiol*. **2004**, *3*(1), 45–53. https://doi.org/10.1017/S1473550404001892

Popa, R. *Between Necessity and Probability: Searching for the Definition and Origin of Life*. Springer-Verlag: Heidelberg, **2004**, pp. 197–205 (definitions of life are given in chronological order starting from 1855).

Popa, R. Necessity, futility and the possibility of defining life are all embedded in its origin as a punctuated-gradualism. *Orig Life Evol Biosph.* **2010**, *40*, 183–190. https://doi.org/10.1007/s11084-010-9198-x

Popa, R. Elusive definition of life: A survey of main ideas. In *Astrobiology: An Evolutionary Approach*, Kolb, VM. Ed. CRC Press: Boca Raton, FL, **2015**, pp. 325–348.

Reif, F. *Applying Cognitive Science to Education: Thinking and Learning in Scientific and Other Complex Domains*. The MIT Press: Cambridge, MA, **2010**.

Repko, AF. *Interdisciplinary Research: Process and Theory*, 2nd ed. Sage: Thousand Oaks, CA, **2012**, pp. 20–22, 73, 94–96.

Szostak, JW. The narrow road to the deep past: In search of the chemistry of the origin of life. *Angew Chem Intl Ed.* **2017**, *56*, 1107–11043. https://doi.org/10.1002/anie.201704048

Talanquer, V. Central ideas in chemistry: An alternative perspective. *J Chem Ed.* **2016**, *93*(1), 3–8. https://pubs.acs.org/doi/10.1021/acs.jchemed.5b00434

Villarreal, LP. Are viruses alive? *Sci Am.* **2004**, *291*(6), 100–105. Accessed July 10, 2021. www.jstor.org/stable/26060805

Zubay, G. 2000. *Origins of Life on the Earth and the Cosmos*, 2nd ed. Harcourt Academic Press: San Diego, CA, **2000**.

# 2 A Primer on the Systems Approach

*Vera M. Kolb*

## 2.1 INTRODUCTION, OBJECTIVES, AND BACKGROUND

In this chapter, we cover the basics of the systems approach. We do not focus on the applications which are specific for the fields in which the systems approach is commonly used, such as computer science, engineering, management, and economics. Instead, we extract from these applications the principles which we believe are useful for astrobiology. We also draw from the relevant literature on some recent systems applications in chemistry and education.

The systems approach, systems thinking, and systems analysis are often described in the same literature resources, since these topics are linked. Thus, the systems thinking is the foundation for the systems analysis; and the systems approach encompasses the systems thinking, analysis, and the systems applications.

The background of the systems approach is given in various fundamental books on this subject (e.g., Von Bertalanffy, 1993; Capra, 1996; Capra and Luisi, 2014; Chela-Flores, 2013; Lovelock, 1995; Meadows, 2008; Kasser, 2013; O'Connor and McDermott, 1997; Systems Thinking and Practice, 2016). Also, much useful material is presented in various journal articles, which we cite as we discuss them. The book *Systems Approach to Astrobiology* (Kolb and Clark, 2023) focuses on the systems approach to astrobiology. It also provides an extensive background on the systems approach which we draw upon in this chapter.

We start by delineating a system in the most basic terms. In Table 2.1, we present a selection of common definitions and attributes of a system.The parts of a system are system specific. For example, the parts of a chemical system comprise chemical

**TABLE 2.1**
**Some Common Definitions and Attributes of a System**

1) A system comprises parts that interact to form a unified whole
2) A system is more than the sum of its parts, since it exhibits novel properties and functions, which are not exhibited by the individual parts
3) A system possesses boundaries, organization, and is characterized by its function
4) Functioning of a system occurs in conjunction with its interaction with the environment
5) A system may be the subject of the arrow of time, such as an evolutionary system

*Sources*: Von Bertalanffy (1993), Capra (1996), Capra and Luisi (2014), Lovelock (1995), Meadows (2008), Kasser (2013), O'Connor and McDermott (1997), Systems Thinking and Practice (2016).

DOI: 10.1201/9781003294276-2

**FIGURE 2.1** A picture of the parts of the motorcycle, whose functioning is enabled by the interaction of these parts.

*Photo courtesy:* Shutterstock (www.shutterstock.com/image-photo/kyiv-ukraine-april-9-2020-product-2146325355)

atoms and compounds. In the biochemical or biological systems, the parts of the system are more complex and in some cases may be considered as subsystems of the main system. An example would be metabolism, which is one of the subsystems which allow a cell to function.

In Figure 2.1, we show an easy-to-understand example of a system which is composed of parts and whose function is intuitive.

## 2.2 SYSTEMS THINKING

In this section, we give definitions of systems thinking, which were compiled from various sources by Arnold and Wade (2015), who also provided the original references. We do not cite the latter here, in the interest of space. Among the definitions provided by these authors, we have selected and modified those that we believe are most important for astrobiology. They are presented in Table 2.2.

We have found the definition 1) from this Table 2.2 especially useful since it prompts us to focus on the functioning of the system from the point of view of both the whole and its parts. The definition 2) is critical for astrobiology. It tells us that we should not limit ourselves to our disciplinary perspective but should include other perspectives which are characteristic of other foundational disciplines of astrobiology. The definition 4) is also very important. It prompts us not only to look at the system within a "snapshot" in time (a static view) but also to seek the view over a period of time (a dynamic view). The latter is often more informative, especially for the systems which are undergoing evolution.

## 2.3 MORE ON THE DEFINITIONS OF THE SYSTEMS

In this section, we show a set of definitions of systems, which address complexity of the parts/systems and the system's purpose. The literature sources for these are Constable et al. (2019) and Meadows (2008). We also address the particular importance of these definitions for astrobiology. These definitions are shown in Table 2.3, together with our comments.

**TABLE 2.2**
**Definitions of Systems Thinking**

The systems thinking is the ability to:

1) See both the forest and the trees simultaneously
2) Utilize multiple perspectives
3) Examine relationships between the parts within the whole, including the feedback processes, and patterns of change both in the parts and in the whole
4) Consider not only static but also dynamic behavior of the interconnections and feedbacks within the system
5) Grasp and identify nonlinearity in the behavior of complex systems

*Source*: Selected and modified from Arnold and Wade (2015)

**TABLE 2.3**
**Definitions of Systems Which Focus on the Systems' Complexity and Purpose, Together with Our Comments**

1) "A system comprises a set of elements working together to form a complex whole that produces a function" (Meadows, 2008). This definition is important for astrobiology since it does not include the purpose of the system, which is common in, e.g., engineering. Astrobiology is concerned with the life in the universe, and thus life and its purpose need to be addressed. The purpose of life can be defined, but, depending on the way it is done, it might generate controversy, if construed as a divine purpose. The term purpose is often reserved for something which is preordained. The safest and least controversial way is to substitute purpose with function, as is done in the definition 1). The life's functions then would be reproduction and metabolism, for example.
2) "Complexity is a hallmark of systems" (Constable et al., 2019). The systems analysis needs to deal with the complexity of the system without oversimplifying the system and without dissecting it into small parts. The complexity of the system is one of its key characteristics and should be preserved in our considerations.
3) "Complexity differs from system to system. The nature of the complexities associated with each discipline . . . differ. Moreover, complexities between subsystems within a discipline also differ" (Constable et al., 2019). The important point here is that the understanding of a complex system in one discipline does not guarantee the understanding of a system in a different discipline.

*Sources*: Constable et al. (2019), Meadows (2008) for Entry 1.

**TABLE 2.4**
**Selected Statements about the Emergence in Different Systems, and Our Comments**

1) "Broadly, the concept of emergence in chemical systems refers to phenomena in which the structures, properties, or behavior of multicomponent systems exceed those predicted from knowledge of the individual components" (Constable et al., 2019). The authors also point out that some molecules have tendencies toward self-assembly that results in the membrane-like structures and other biologically important constructs, which are strongly related to the astrobiology goal of understanding the origins of life.
2) "Emergent behavior needs to be identified" (York and Orgil, 2020). This is an important point. There may be more than one type of the emergent behaviors in complex systems, such as the astrobiological ones. A continuous search for different types of emergent behaviors will steer us away from the simplistic approach in which we stop the search once we have identified just one type.
3) The interaction between the system's components can create emergent properties such as structures, patterns, and complex behavior (Bar-Yam, 2002; https://en.wikipedia.org/wiki/Complex_system). The emerging structures and patterns are important for astrobiology, for example, in the search for biosignatures of the past or present extraterrestrial life.

*Sources*: Constable et al. (2019), York and Orgill (2020), Bar-Yam (2002), https://en.wikipedia.org/wiki/Complex_system.

**TABLE 2.5**
**Some Complex Properties of the Systems**

1) Resilience: This is the tendency of a system to remain in a stable condition or return to it when there is a change in forces that act upon the system. This property is limited, and it depends on the magnitude or type of the force.
2) A tipping point is a threshold beyond which the complex system cannot return to the starting point.
3) A complex system has multiple interacting parts between which matter and/or energy can be exchanged.
4) A complex system can be open or closed: the latter does not allow the transfer of energy or matter across the boundary, while the former does.
5) Positive feedback in a complex system amplifies the rate of change, while the negative one does the opposite.
6) Complex systems can undergo nonlinear change over time; this means that the magnitude of the response is not linearly proportional to the magnitude of the input.

Source: Based on Bar-Yam (2002); https://en.wikipedia.org/wiki/Complex_system

Next, we address the topic of the emergence in a system. This is one of the critical features of a system and is especially important for astrobiology, since the latter studies life, its origins, and future. It is currently believed that life emerged from the abiotic matter by chemical evolution. The emergence means that a new property, behavior, or function of a system appears because of the interactions of the parts of the system.

**FIGURE 2.2** An example of resilience. It is intuitive that resilience has its limits.

*Photo courtesy:* Shutterstock (https://image.shutterstock.com/image-photo/beautiful-resilient-flower-growing-out-600w-1715067310.jpg)

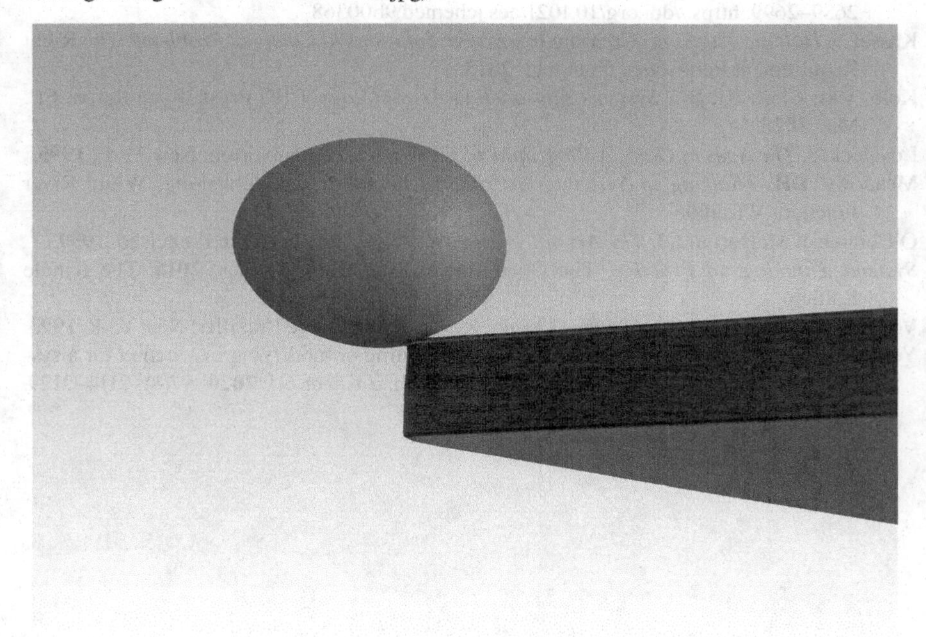

**FIGURE 2.3** An example of a system which is close to its tipping point. The point of no return is when the egg is pushed slightly to the left.

*Photo courtesy:* Shutterstock (https://image.shutterstock.com/image-photo/egg-on-edge-fine-balance-600w-334016795.jpg)

In Table 2.4, we show selected statements about the emergence in various systems and our comments about the relevance to astrobiology.

Next, we show some more complex properties of the systems, the examples of which are shown in Table 2.5.

In Figure 2.2, an easy-to-understand example of resilience is shown.

In Figure 2.3, we show the picture of a system which is close to its tipping point.

In conclusion, we have presented diverse definitions of systems and systems thinking, which are utilized in the systems analysis and approach, and have chosen those that are most relevant to astrobiology.

## REFERENCES

Arnold, RD; Wade, JP. A definition of systems thinking: A systems approach. *Procedia Comput Sci.* **2015**, *44*, 669–678. https://doi.org/10.1016/j.procs.2015.03.050

Bar-Yam, Y. General features of complex systems. *Enc Life Supp Syst.* **2002**, *1*, 1–57.

Capra, F. *The Web of Life.* Anchor Books, Doubleday: New York, **1996**.

Capra, F; Luisi, PL. *The Systems View of Life, A Unifying Vision.* Cambridge University Press: Cambridge, **2014**.

Chela-Flores J. From systems chemistry to systems astrobiology: Life in the universe as an emergent phenomenon. *Int J Astrobiol.* **2013**, *12*(1), 8–16.

Complex system. https://en.wikipedia.org/wiki/Complex_system (accessed on 9/12/2022).

Constable, DJC; Jimenéz-Gonzáles, C; Matlin, SA. Navigating complexity using systems thinking in chemistry with implications for chemistry education. *J Chem Ed.* **2019**, *96*, 2689–2699. https://doi.org/10.1021/acs.jchemed.9b00368

Kasser, J. *Holistic Thinking: Creating Innovative Solutions to Complex Problems.* The Right Requirement Publishers: Cranfield, **2013**.

Kolb, VM; Clark III, BC. *Systems Approach to Astrobiology.* CRC Press: Boca Raton, FL, May **2023**.

Lovelock, J. *The Ages of Gaia: A Biography of Our Living Earth.* Norton: New York, **1995**.

Meadows, DH. *Thinking in Systems: A Primer.* Chelsea Green Publishing: White River Junction, VT, **2008**.

O'Connor, J; McDermott, I. *The Art of Systems Thinking.* Thorsons: San Francisco, **1997**.

*Systems Thinkingand Practice.* The Open University: Milton Keynes, **2016**. The Kindle Edition.

Von Bertalanffy, L. *General Systems Theory*, Revised Ed. George Braziller: New York, **1993**.

York, S; Orgill, M. ChEMIST table: A tool for designing or modifying instruction for a systems thinking approach in chemistry education. *J Chem Ed.* **2020**, *97*(8), 2114–2129. https://doi.org/10.1021/acs.jchemed.0c00382

# 3 Systems Approach to Prebiotic Chemistry Which Led to Life

*Vera M. Kolb*

## 3.1 A BRIEF BACKGROUND ON SYSTEMS CHEMISTRY AS RELEVANT TO PREBIOTIC CHEMISTRY

The primer on systems approach is provided in Chapter 2 of this Guidebook. More extensive coverage is provided by Kolb and Clark (2023) and the references therein. Here, we provide a brief background about systems chemistry with the focus on the relevance to prebiotic chemistry. We start with the "mantra" of systems in general, according to which a system is composed of parts, which interact with each other in such a way to produce a new, emerging feature of the system. The latter is *not* exhibited by any of the system's individual parts. This mantra works well for the study of complex systems, which cannot be understood solely by looking at the properties of the individual systems parts.

The systems approach was applied with great success to various fields which study complex systems, such as engineering, computer science, biology, ecology, and economics. In contrast, systems chemistry has started and developed only relatively recently (e.g., Kindermann et al., 2005; Ludlow and Otto, 2008; von Kiedrowski, Otto, and Herdewijn, 2010; Ashkenasy et al., 2017; Ruiz-Mirazo, Briones, and de la Escosura, 2014; Strazewski, 2019a, 2019b). The scope of this field is broad, but we focus here only on the systems chemistry which is relevant to the prebiotic chemistry which led to life.

First, we provide various descriptions of the scope of systems chemistry, which are in line with our focus. In Table 3.1, we provide a selection of such descriptions.

The items in Table 3.1 are broad and inclusive. Some could serve as a roadmap for systems application to astrobiology. Currently, the pure systems approach is difficult since we still do not know enough about the parts of the system, and sometimes even less about the way these parts interact to enable the systems functions to emerge. However, a partial success is important, as long as we realize the limitations of our approach.

## 3.2 BACKGROUND AND OBJECTIVES OF THIS CHAPTER

This chapter gives selected recent examples of chemical systems, which are astrobiologically relevant and which utilize novel systems approaches. Systems chemistry in

DOI: 10.1201/9781003294276-3

**TABLE 3.1**
**Selection of Descriptions of the Scope of Systems Chemistry, Which Are Relevant to Prebiotic Chemistry that Led to Life.**

Systems chemistry:

1) Seeks the roots of Darwinian evolvability in chemical systems
2) Investigates prebiotic catalytic and autocatalytic systems
3) Seeks to understand the integration of metabolic, genetic, and membrane-forming subsystems into protocellular entities
4) Studies asymmetric autocatalysis and chiral symmetry breaking
5) Searches for structural and dynamic prerequisites which lead to chemical self-replication and self-reproduction
6) Studies self-organization as relevant to the origins of life
7) Studies far-from-equilibrium systems
8) Investigates energy dissipation in dynamic systems
9) Conceptualizes alternative genetic systems
10) Examines informational molecular systems, such as nucleic acid systems chemistry
11) Emphasizes the difference between kinetically controlled molecular networks and thermodynamically controlled ones; the former is associated with biological systems which operate far from equilibrium
12) Deals with challenging issues such as the creation of concurrent formation–destruction systems and continuous maintenance of a system that is far from equilibrium
13) Investigates organizational principles within molecular networks and the emergence of autocatalysis

*Source*: Modified from Kolb and Clark (2023); Ludlow and Otto (2008); Ashkenasy et al. (2017); and Ruiz-Mirazo, Briones, and de la Escosura (2014)

general (e.g., Ludlow and Otto, 2008; Ashkenasy et al., 2017) and its features which are relevant to astrobiology have been covered elsewhere (e.g., Strazewski, 2019a, 2019b; Islam and Powner, 2017; Ruiz-Mirazo, Briones, and de la Escosura, 2014; Szostak, 2009), most recently by Kolb and Clark (2023). Here, we focus on selected examples which are amenable to including *additional* systems approaches, which in some cases were not foreseen by the authors of these examples. The idea behind this approach is to show that the systems analysis is *dynamic* and that it evolves as we reanalyze the specific, previously published cases in light of new knowledge.

Prebiotic syntheses of the biologically relevant compounds, such as amino acids, sugars, nucleobases, and lipid constituents, have been reviewed previously (e.g., see Miller and Orgel, 1974, for the review of the early work, and Peretó, 2019, for the recent review). Thus, we do not review this material here. Instead, we cover selected prebiotic experimental *designs* for the syntheses of nucleosides and nucleotides, which are the components of the early genetic systems. These syntheses in the past were designed to mimic biological pathways, but in the laboratory, some gave poor results. Recently, progress has been made by adopting different experimental designs, some of which utilize the systems approach. We focus on these designs, rather than the experimental specifics, which are readily available in the literature we cite.

## 3.3 PREBIOTIC SYNTHESES OF NUCLEIC ACID COMPONENTS

Early prebiotic synthesis of nucleic acid components used a nucleobase as starting material, such as adenine, and reacted it with the sugar ribose or deoxyribose to form a nucleoside. The latter was then phosphorylated to form the nucleotide. This method seemed logical to pursue since some nucleobases could be easily obtained by prebiotic chemistry (from the oligomerization of HCN). The same is true for sugars (by the formose reaction). For the final phosphorylation step, various processes were developed to some degree early on (e.g., Miller and Orgel, 1974). Thus, it was believed that the nucleotide synthesis could be accomplished by chemical "stitching" of the constituent parts. The structure of a nucleotide with its constituent parts is shown in Figure 3.1. More such structures are readily available in any biochemistry textbook (e.g., Voet and Voet, 2011).

However, many syntheses of nucleosides which used the chemical stitching of nucleobases and sugars gave poor yields. Such poor results were ascribed to the "nucleosidation problem", as discussed in depth by Yadav, Kumar, and Krishnamurthy (2020). These authors attributed the failure of the "stitching" method to intrinsic chemical problems, such as the reduced nucleophilicity of the nucleobases and their insolubility in water which required heterogeneous reaction conditions and high temperatures. The latter led to decomposition and side reactions. Due to these problems, the chemical "stitching" approach eventually fell out of favor.

Novel approaches were undertaken, which were focused on the reactivities of simple prebiotic precursors and their potential to produce nucleosides and nucleotides, without a constraint of producing the intermediates which were mimicking the biological intermediate steps of such syntheses. These approaches were ultimately proven to be fruitful, as shown by the Carell and Sutherland's research groups

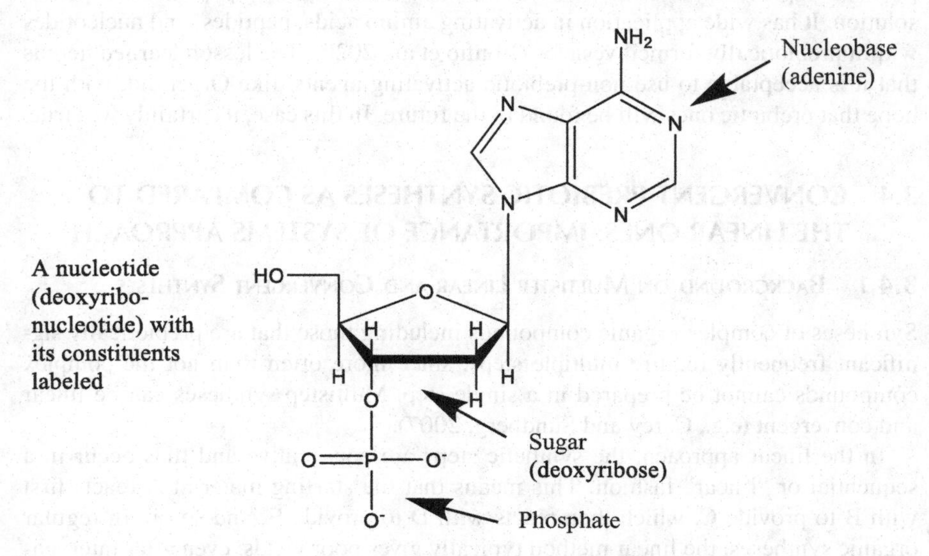


**FIGURE 3.1** The structure of a nucleotide which shows its constituent parts.

*Source:* Modified and redrawn from Voet and Voet (2011)

(Becker et al., 2018a, 2018b, 2019; e.g., Powner, Sutherland, and Szostak, 2011). The prebiotic pathways to RNA, a key chemical required for the emergence of life, were accomplished. Yadav, Kumar, and Krishnamurthy (2020) reviewed chemical principles of these new syntheses.

An important advancement of these syntheses was achieved by the systems approach. The principles of such an approach are covered in Chapter 2 of this book and Section 3.1 of this chapter. In this approach, the interactions of chemical components of the system give a new quality, such as autocatalysis, which did not exist in the separate parts. Another example is template-based syntheses. The parts of the system are the template, the chemicals which assemble on it, and other components which enable stitching together the chemicals on the template to give complex molecules. Cyclic reaction processes in which some chemicals are being fed into a cycle and some others removed as products represent another example.

The beginning of prebiotic systems chemistry was reviewed by Strazewski (Strazewski, 2019a, and the references cited therein). The examples covered include enzyme-free template-directed nucleic acid chain elongation by Orgel, enzyme-free autocatalytic ligations of oligonucleotides by von Kiedrowski, Ghadiri's autocatalytic ligation of activated peptide fragments, Szathmary's theoretical treatment on the selection of the faster replicating population by Szathmary, and Eigen's study of error threshold for replication fidelity.

However, not all chemical reactions that comprise these systems are prebiotically feasible. For example, Orgel used imidazole-based activated groups to achieve stitching of the compounds on the template, but these activated groups are not attainable under prebiotic conditions. Recently, progress has been made by using methyl isocyanide (MeNC), which is prebiotically feasible (Bonfio et al., 2020). This compound is capable of simultaneous activation of nucleotides and peptides in aqueous solution. It has wide application in activating amino acids, peptides, and nucleotides within prebiotically formed vesicles (Bonfio et al., 2020). The lesson learned here is that it is acceptable to use non-prebiotic activating agents, like Orgel did, with the hope that prebiotic ones will be found in the future. In this case, it certainly was true.

## 3.4 CONVERGENT PREBIOTIC SYNTHESES AS COMPARED TO THE LINEAR ONES: IMPORTANCE OF SYSTEMS APPROACH

### 3.4.1 Background on Multistep Linear and Convergent Syntheses

Syntheses of complex organic compounds including those that are prebiotically significant frequently require multiple steps, since more often than not the complex compounds cannot be prepared in a single step. Multistep syntheses can be linear and convergent (e.g., Carey and Sundberg, 2007).

In the linear approach, the synthetic steps are consecutive and thus occur in a sequential or "linear" fashion. This means that the starting material A reacts first with B to provide C, which then reacts with D to provide E, and so on. In regular organic syntheses, the linear method typically gives poor yields, even after interventions such as purification of intermediate products. In prebiotic experiments, the linear method is typically employed in one-pot syntheses. To make the prebiotic synthesis more successful, the input of the chemicals into the flask is carefully chosen. Only the compounds which are believed to give the desired results are fed into the reaction container. This is not realistic under prebiotic conditions, which do not include an outside intervention, and purification of the intermediate steps, although some general purification may occur e.g., via geo chromatography (e.g., Wing and Bada, 1991).

In regular organic syntheses (thus not prebiotic), the multistep convergent method is typically used for the preparation of complex organic molecules. This synthetic method involves separate syntheses of two or more parts of the complex molecule and, in yet another separate synthetic step, joining these parts to give the desired product. For example, one linear synthesis starts with A to give a product X (after several linear steps); another starts with B, to give a product Y (after several linear steps). Then, X and Y are joined together in a separate linear step to give the desired product Z. The convergent method gives in principle a better net yield than the linear sequence which gives the same end product.

### 3.4.2 Convergent Prebiotic Syntheses

Many convergent prebiotic syntheses have been performed in the past, but the method used was not explicitly classified as "convergent". For example, one component was made separately under the basic conditions, and the other one under the acidic conditions. This difference in the reaction conditions necessitated different locations for these reactions. Joining together the separately made components was then performed, under the conditions that are suitable for this reaction step, and at yet another location. This method in the past was constrained by the requirement that the intermediate components had to mimic biological pathways. Recently, the latter constraint was dropped which led to highly successful prebiotic syntheses. Thus, experimental pathways were designed on the basis of the intrinsic chemistry of the available prebiotic precursors, with the final product as a goal and no restrictions on the nature of the intermediates (Sutherland, 2016, 2017). Sutherland (2016) proposed prebiotic convergent syntheses in which the transport of the synthesized intermediate components from their initial synthetic locations to the final reaction site would occur by water streams and rivulets, and that the final reaction step would occur at the confluence of these.

We bring up some additional factors that need to be weighed. For example, the probability of the survival of the chemical components during the transport from their initial synthetic location to the final reaction site needs to be considered. Also, the probability of these units finding each other in space and time, so that they can react, needs to be considered. Further, when chemical components are transported via waterways, the details of possible chemical events during such a transfer must be addressed specifically. Thus, if the chemical components are water soluble, they could undergo unwanted in-water reactions in which they react with water or water-soluble components. If they are not water-soluble, they could still react with other chemicals which are present in the waterways via an "on-water" mechanism (Kolb, 2015, 2016, 2019), which sometimes is faster and more specific than the in-water reactions. The simplified mechanism is that water-insoluble compounds try to avoid water, since they are hydrophobic, and thus stick together among themselves. This brings them to a close proximity with each other, which facilitates their mutual reactions, which is often faster and more selective. Alternatively, the water-insoluble chemicals could stick together to make an oil slick in which they could react with each other via liquid–solid or solid–solid mechanisms. The astrobiological applications of the on-water, liquid–solid, and solid–solid reactions are well known (Kolb, 2012, 2015, 2016, 2019). A recent example of the synthesis of ribose in solid state has been published (e.g., Haas et al., 2020).

Therefore, we should not consider water just as an inert carrier of chemicals. Also, we should not assume that the rivulet banks are inert, since they usually contain various solid mineral catalysts, which could facilitate unwanted reactions, including decomposition.

The systems boundary for the convergent syntheses is difficult to define since it is dynamic. We can only consider snapshots of the transport of chemicals along the rivulets.

### 3.4.3 Linear One-Pot Prebiotic Syntheses

A single-pot prebiotic synthesis, in which all the reaction steps occur in approximately the same space and time, is expected to have a more favorable probability of producing the desired end-product. This point of view is in contrast with Sutherland's convergent synthetic approach, which we have covered in Section 3.4.2. The one-pot synthesis recently gained support by Szostak and co-workers who performed such a synthesis which led to template-directed RNA copying (Zhang et al., 2022). Other prebiotic syntheses of complex materials were achieved by one-pot method, such as for RNA nucleosides (Hud and Fialho, 2019) and nucleotides (Becker et al., 2019).

In one-pot synthesis, there is an advantage that the reaction vessel boundary is fixed and, thus, is not subject to dynamic fluctuations. Most previously performed prebiotic syntheses were of this type.

In a typical prebiotic experimental set-up for the one-pot synthesis, researchers all over the world use pretty much the same glass containers which are made of borosilicate. This choice was not questioned until recently. Criado-Reyes et al. (2021) reinvestigated the famous experiment by Stanley Miller (1953) in which he synthesized amino acids from $H_2O$, $NH_3$, $CH_4$, and $H_2$, by a spark discharge, in a specially designed apparatus, shown in Figure 3.2.

Criado-Reyers et al. questioned if the experimental outcome of the Miller reaction depends on the chemical nature of the reaction vessel. To answer their query, they used three different reaction vessels: 1) Teflon flask; 2) Teflon flask, inside which they placed some chips of borosilicate glass; and 3) borosilicate glass as a control. Their results showed that the nature of the reaction vessel plays an important role in the reaction outcome. When Teflon flask was used, the molecular diversity of the products was minimal. The latter increased when chips of borosilicate glass were added to the Teflon flask. The highest molecular diversity was observed when borosilicate flask was used, in which case the yield was also the highest.

Numerous prebiotic experiments use much simpler reaction setup, such as a single reaction flask, perhaps equipped with the condenser if heating in the volatile solvent is employed. In these experiments, borosilicate glass would also be used. We do

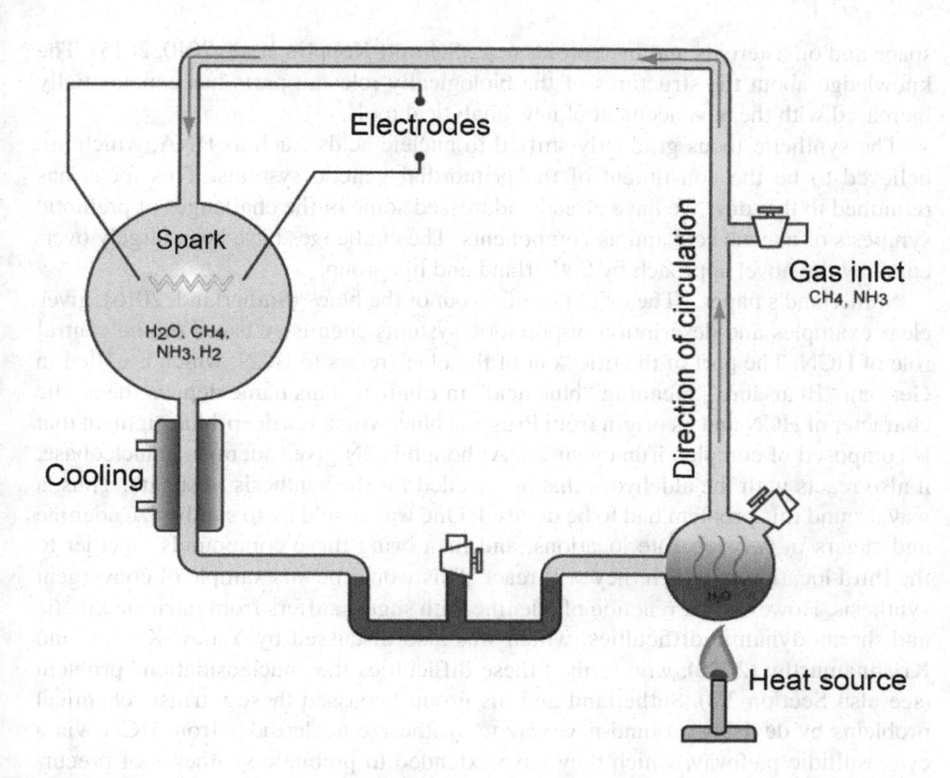


**FIGURE 3.2** A simplified scheme of Miller apparatus for the synthesis of amino acids by spark discharge from simple prebiotic precursors.

*Photo courtesy:* Wikimedia free depository

not know if the results would be different if the chemical nature of the flask were considered.

The lesson to be learned is that we should not assume that the boundary of the system is inert. Even if we do not know if it is inert or not, and perhaps do not have resources to investigate, we should at least leave the question open, rather than silently assuming that the boundary is inert.

## 3.5 MORE EXAMPLES OF PREBIOTIC SYSTEMS CHEMISTRY

In every systems approach, one needs to understand the structure and properties of the parts of the system and the interactions between these parts which are responsible for the emergence of the new properties and function of the system. Early prebiotic chemistry dealt mostly with the synthesis of parts which are biologically relevant as building blocks of life. Examples include amino acids, primitive proteinoids, sugars, primitive lipid-like compounds, and nucleic acid components. This topic was extensively reviewed (e.g., Miller and Orgel, 1974; Peretó, 2012, 2019). A huge number of biologically relevant compounds were found on meteorites, such as Murchison, which shows that such compounds can be made under unassisted reaction conditions in the space and on asteroids and meteoroids (e.g., Schmitt-Kopplin et al., 2010, 2015). The knowledge about the structures of the biologically relevant parts has exponentially increased with the advancement of new analytical tools.

The synthetic focus gradually shifted to nucleic acids, such as RNA, which are believed to be the constituent of the primordial genetic systems. This focus has remained to this day. We have already addressed some of the challenges of prebiotic synthesis of nucleic acid and its components. The challenges have been largely overcome by the novel approach by Sutherland and his group.

Sutherland's paper, "The origin of life – out of the blue" (Sutherland, 2016), gives clear examples and description of prebiotic systems chemistry based on the central role of HCN. The part of the title "out of the blue" refers to HCN, which is called in German "Blausäure", meaning "blue acid" in English. This name depicts the acidic character of HCN and its origin from Prussian blue, which is a deep-blue pigment that is composed of complex iron cyanides. Although HCN gives adenine, a nucleobase, it also reacts with the aldehydes that are needed for the synthesis of sugars. Thus, a way around this problem had to be devised. One way would be to synthesize adenine and sugars in two separate locations, and then bring these compounds together to the third location in which they will react. This would be an example of convergent synthesis. However, the reaction of adenine with sugars suffers from intrinsic kinetic and thermodynamic difficulties, which was also discussed by Yadav, Kumar, and Krishnamurthy (2020), who termed these difficulties the "nucleosidation" problem (see also Section 3.3). Sutherland and his group bypassed these intrinsic chemical problems by devising a brand-new way to synthesize nucleosides from HCN, via a cyanosulfidic pathway, which they have extended to prebiotic syntheses of precursors of other important prebiotic building blocks such as amino acids, ribonucleotides, and lipids. Many of these syntheses utilize systems approach (e.g., Patel et al., 2015; Sutherland, 2016, 2017; Powner and Sutherland, 2011; Powner, Sutherland, and Szostak, 2011; Ranjan and Sasselov, 2016; Ranjan et al., 2018; Islam and Powner, 2017; Ritson et al., 2018; Sasselov, Grotzinger, and Sutherland, 2020; Bonfio et al., 2020). These pathways include HCN and its selected derivatives, hydrogen sulfide as the reductant, UV light, and Cu(I)–Cu(II) photo-redux cycling. In some specific applications, these pathways also may utilize Fe, P, Ca, and wet–dry cycles. The systems approach is augmented by including plausible geological, atmospheric, and meteorite-impact conditions. The latter is important as prebiotic source of phosphate. One example is the phosphide mineral schreibersite, $(Fe,Ni)_3P$ (e.g., Lang, Lago, and Pasek, 2019, and the references therein).

A review by Islam and Powner's (2017), titled "Prebiotic systems chemistry: Complexity overcoming cluster", provides the details of the synthesis of RNA, lipid, and protein precursors by cyanosulfidic protometabolism and synthesis of activated ribonucleotides by bypassing ribose and nucleobases, among other innovations. Ritson (2021) provides experimental support for the cyanosulfidic origin of the Krebs cycle. This cycle is central for the present-day metabolism (e.g., Voet and Voet, 2011).

Some researchers are skeptical of the cyanosulfidic approach. For example, Harrison and Lane (2018) believe that prebiotic nucleotide synthesis should be guided by the biochemical pathways. Their belief is due to a lack of resemblance of

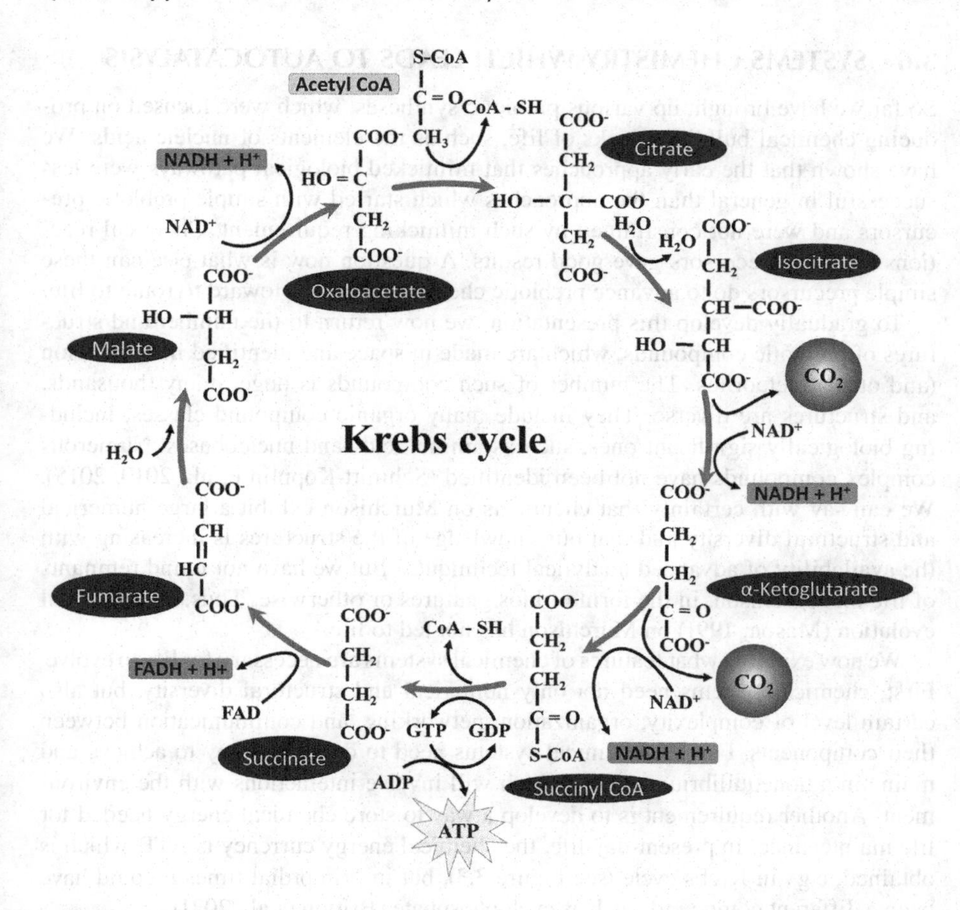


**FIGURE 3.3** Krebs cycle.
*Photo courtesy:* Shutterstock

biological pathways to cyanosulfidic protometabolism, and the difficulty in extrapolating backwards from LUCA (Last Universal Common Ancestor) to prebiotic chemistry, among other factors. They make the following statement:

> The demonstration that activated nucleotides can be formed from cyanide . . . proves it can be done, and it eliminates some of the mystique. But it does not prove this is the only way to do it. Life itself hints that this way was not the way it happened. If we want to understand the origin of life, we would be foolish to ignore life as a guide.

Also: "Perhaps the biggest problem is that the chemistry involved in these clever syntheses does not narrow the gap between prebiotic chemistry and biochemistry – it does not resemble extant biochemistry in terms of substrates, reaction pathways, catalysts or energy coupling".

## 3.6 SYSTEMS CHEMISTRY WHICH LEADS TO AUTOCATALYSIS

So far we have brought up various prebiotic syntheses, which were focused on producing chemical building blocks of life, such as the elements of nucleic acids. We have shown that the early approaches that mimicked biological pathways were less successful in general than the approaches which started with simple prebiotic precursors and were not constrained by such mimicking requirement. Chemical reactions of these precursors gave good results. A question now is what else can these simple precursors do to advance prebiotic chemistry systems toward to route to life.

To gradually develop this presentation, we now return to the number and structures of prebiotic compounds, which are made in space and identified in Murchison (and other) meteorites. The number of such compounds is huge, many thousands, and structures are diverse. They include many organic compound classes, including biologically significant ones, such as amino acids and nucleobases. Numerous complex compounds have not been identified (Schmitt-Kopplin et al., 2010, 2015). We can say with certainty that chemicals on Murchison exhibit a large numerical and structural diversity and that our knowledge of the structures is increasing with the availability of advanced analytical techniques. But we have not found remnants of life on Murchison, in the form of biosignatures or otherwise. Thus, the chemical evolution (Mason, 1991) on Murchison has not led to life.

We now examine what features of chemical systems are necessary for life to evolve. First, chemical systems need not only numerical and structural diversity, but also certain level of complexity, organization, networking, and communication between their components. Further, chemical systems need to develop a way to achieve and maintain a nonequilibrium status, which will involve interactions with the environment. Another requirement is to develop a way to store chemical energy needed for life maintenance. In present-day life, the chemical energy currency is ATP, which is obtained, e.g., in Krebs cycle (see Figure 3.3), but in primordial times it could have been a different compound, such as cyclophosphate (Britvin et al., 2021).

One important step in the evolution of chemical systems is the emergence of autocatalysis. Autocatalysis is observed in formose reaction, an important prebiotic reaction, by which sugars are produced under the basic conditions. Figure 3.4 shows a simplified version of the formose reaction.

Figure 3.4 shows an autocatalytic cycle in which glycolaldehyde (2-C unit) (bottom left structure) reacts with formaldehyde (1-C unit). A series of tautomerizations (double-headed arrows) and formaldehyde additions lead to a tetrose (4-C unit) which falls apart to two molecules of glycolaldehyde (C-2 unit), which completes the cycle when formaldehyde is added.

A more detailed scheme which includes the formation of a pentose (5-C units) is given by Miljanić (2017) in his paper titled "Small-Molecule Systems Chemistry". We show a modified redrawn version in Figure 3.5.

Figure 3.5 shows the formation of higher, 5-C sugars. This is still not a complete picture of the formose reaction, but it shows, like Figure 3.4, the autocatalytic nature of the formose reaction and various steps, including tautomerizations.

The classic formose reaction starts with formaldehyde and glycolaldehyde in aqueous base, which results in typically calcium hydroxide. Recently, formose reaction was reinvestigated under dry conditions, which resulted in a higher selectivity toward unbranched

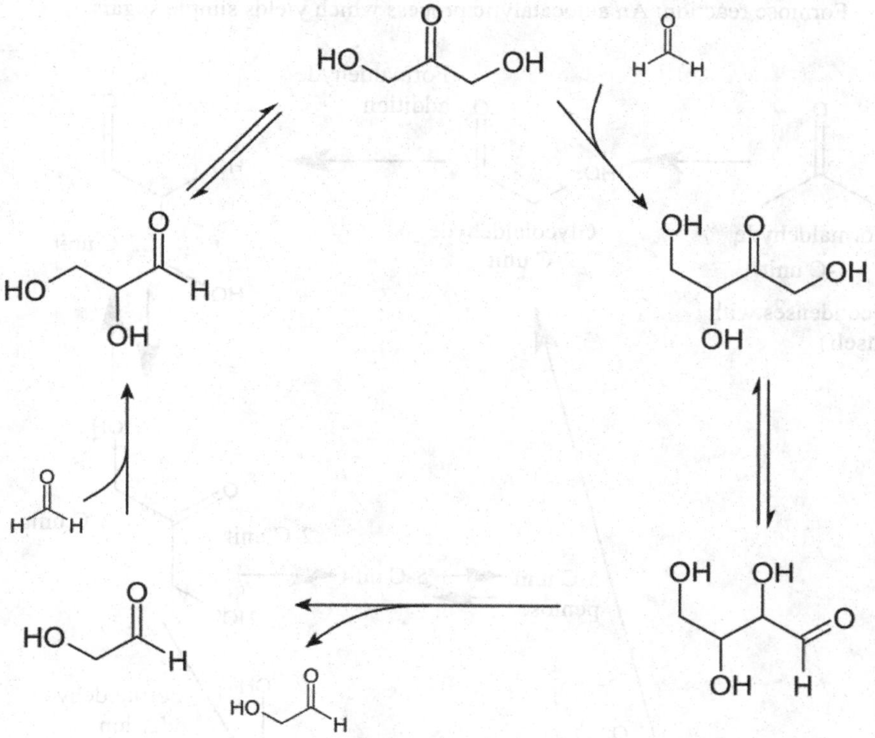

**FIGURE 3.4** A simplified version of the formose reaction.
*Source:* From Wikimedia, no permission required

monosaccharides. When mechanical force and a variety of basic minerals were used, the synthesis of monosaccharides was greatly accelerated (Haas et al., 2020; Lamour et al., 2019). In the experiments, the authors used a ball mill, but in the prebiotic geochemical environment, such a force could have been generated by meteorite impacts of lithospheric activity. Haas et al. (2020) systematically investigated the role of the different basic catalysts for the formose reaction in the solid state. The basic catalysts included minerals such as portlandite, talc, montmorillonite 1 and 2, brucite, apatite, calcite, schreibersite, and others, among those listed by Hazen (2021). There are two lessons to be learned from this work. The first one is that it does pay to reinvestigate important reactions, despite the fact that they are old and studied for many years. All it takes is a new approach, which in this case was a change from aqueous to dry reaction conditions and an application of mechanical force. The latter would represent a case of application of mechanochemistry (Friščić, Mottillo, and Titi, 2020) to prebiotic chemistry.

Formose reaction starts with a *single* prebiotic compound, namely formaldehyde. A question arises if one can start from a *mixture* of simple prebiotic compounds and expect autocatalysis to emerge from it. The answer is yes, as shown by the recent work by Wołos and co-workers (Wołos et al., 2020).

These authors considered a mixture of simple prebiotic molecules, namely $CH_4$, $NH_3$, $H_2O$, HCN, $N_2$, and $H_2S$. They first performed computer simulations of chemically

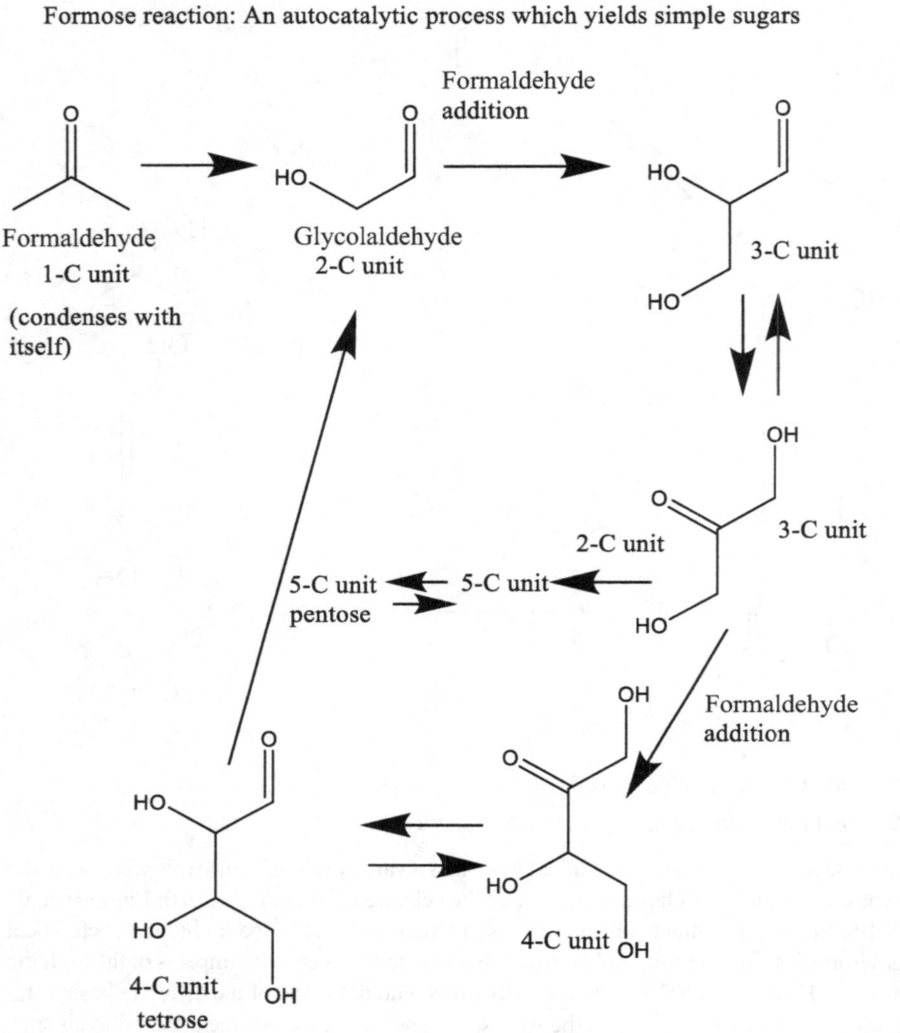

Formose reaction: An autocatalytic process which yields simple sugars

**FIGURE 3.5** A more detailed scheme for the formose reaction, which includes also pentoses.
*Source:* Modified and redrawn from Miljanić (2017)

feasible prebiotic reactions, starting from these chemicals. These simulations predicted numerous reaction products and networks of molecules, which contained not only reported syntheses of biologically relevant compounds, but also new, unreported synthetic routes. The authors experimentally validated several of these new synthetic pathways, such as those of acetaldehyde; diglycine; and various acids, such as malic, fumaric, citric, and uric. In addition, they discovered the following types of chemical emergence. In the first type, the molecules within the network were found to act as catalysts for subsequent reaction types. In the second type, they discovered the emergence of chemical systems which consisted of self-regenerating cycles.

The latter formed within a few synthetic generations. The self-regenerative cycle of iminodiacetic acid was confirmed experimentally. In the third type, the authors identified the network of chemical pathways to surfactants, both peptide-based and long-chain carboxylic acids. This is significant since surfactants are important components of prebiotic compartments.

A few words about the computer simulations which authors used are given here. These simulations were customized for the purpose of prebiotic organic syntheses and were developed by using 614 reaction rules which were grouped within 72 reaction classes, the knowledge of the underlying reaction mechanisms, and the reaction conditions, all of which were reported in the literature. One shortcoming was that the reaction rules did not include stereochemistry.

## 3.7 BUILDING CHEMICAL NOVELTY IN PREBIOTIC CHEMICAL SYSTEMS

Building chemical novelty in the prebiotic chemical systems is important for chemical evolution. We consider two methods. The first one explores the process in which the boundary of the system expands toward the environment and is based on the principle of adjacent possible by Kauffman (2013). The second one includes the mass transfer from the environment into the system.

Most traditional prebiotic organic syntheses are planned and designed to prepare the desired compounds. When these syntheses are put to practice, unexpected products are often formed alongside the desired products. The unexpected products which are not planned for are often considered undesired and are not examined further for their prebiotic potential. While this approach is practical, considering analytical difficulties in separating and identifying the undesired products, it is not conducive for recognizing novel compounds which may be prebiotically relevant. Further, chemicals which are considered undesired at some point of time may become desired later. Next, we describe one such scenario.

Starting materials that give the desired product become depleted with time, and the synthesis of the desired product ceases. This would be an example of a synthesis which is sound in principle but becomes unsustainable. However, the desired product could be made from a fresh supply of the needed chemicals or from some other chemicals which were not present in the system, perhaps using a different chemical route. This view is based on a variation of Kauffman's "adjacent possible" reaction principle (Kauffman, 2013). This principle involves further exploration of the chemical reaction space. Some chemicals which are present in the system reach the boundary of the system, which is in contact with the environment. The latter is the source of new chemicals. The reactions at the boundary cannot be predicted, since the nature of the chemicals which are present there is unknown. New products are obtained at the boundary, some of which may be our original desired products, or some novel products which are prebiotically relevant. When the synthesis of these ceases because of the depletion of the necessary starting components within this newly expanded system, the system can explore a new boundary, which is further into the environment. This process creates an opportunity for novel and unanticipated reactions and products.

The second method for introducing novelty into prebiotic chemistry is to allow the mass transfer between the system and the environment. Prebiotic experiments are typically performed in a reaction vessel, containing starting materials and a solvent and a supply of prebiotically available energy source such as heat, UV light, and spark discharge. However, the reaction vessel model for a prebiotic reaction system is limited since only the energy exchange is allowed with the system but not the mass exchange with the environment. The environment, both atmospheric and geological, needs to be somehow included in our experimental design. For example, the geological environment contains various metals, minerals, and clays that may catalyze or otherwise influence the reaction. To simulate the input of such materials from the environment to our reaction system, we can add some of them to the reaction vessel. By doing this, we would bring a part of the environment to the reaction vessel, so to speak. Such reaction design is expected to result in complex reaction mixtures. Out of these mixtures, the important compounds can be separated by natural (unassisted) purification such as geochromatography (e.g., Wing and Bada, 1991), crystallization, sublimation, and surface adsorption (e.g., Islam and Powner, 2017).

In the conclusion of this section, we need to apply new ways of systems thinking to the prebiotic reactions systems to enable the discovery of novel prebiotic pathways and compounds. We considered moving the boundaries of the system into the environment and allowing mass input from the environment to the system.

## REFERENCES

Ashkenasy, G; Hermans, TM; Otto, S; Taylor, AF. Systems chemistry. *Chem Soc Rev.* **2017**, *46*(9), 2543–2554. https://doi.org/10.1039/C7CS00117G

Becker, S; Feldmann, J; Wiedemann, S; Okamura, H; Schneider, C; Iwan, K; Crisp, A; Rossa, M; Amatov, T; Carell, T. Unified prebiotically plausible synthesis of pyrimidine and purine RNA ribonucleotides. *Science.* **2019**, *366*(6461), 76–82. https://doi.org/10.1126/science.aax2747

Becker, S; Schneider, C; Crisp, A; Carell, T. Non-canonical nucleosides and chemistry of the emergence of life. *Nat Commun.* **2018a**, *9*(1), 1–4. https://doi.org/10.1038/s41467-018-07222-w

Becker, S; Schneider, C; Okamura, H; Crisp, A; Amatov, T; Dejmek, M; Carell, T. Wet-dry cycles enable the parallel origin of canonical and non-canonical nucleosides by continuous synthesis. *Nat Commun.* **2018b**, *9*, 163. https://doi.org/10.1038/s41467-017-02639-1

Bonfio, C; Russell, DA; Green, NJ; Mariani, A; Sutherland, JD. Activation chemistry drives the emergence of functionalised protocells. *Chem Sci.* **2020**, *11*(39), 10688–10697. https://doi.org/10.1039/D0SC04506C

Britvin, SN; Murashko, MN; Vapnik, Y; Vlasenko, NS; Krzhizhanovskaya, MG; Vereshchagin, OS; Bocharov, VN; Lozhkin, MS. Cyclophosphates, a new class of native phosphorus compounds, and some insights into prebiotic phosphorylation on early Earth. *Geology.* **2021**, *49*(4), 382–386. https://d oi.org/10.1130/G48203.1

Carey, FA; Sundberg, RJ. *Advanced Organic Chemistry, Part B: Reactions and Synthesis.* 5th ed. Springer: New York, **2007**, pp. 1164–1166.

Criado-Reyes, J; Bizzarri, BM; Garcia-Ruiz, JM; Saladino, R; Di Mauro, E. The role of borosilicate glass in Miller-Urey experiment. *Sci Rep.* **2021**, *11*, 21009. https://doi.org/10.1038/s41598-021-00235-4

Friščić, T; Mottillo, C; Titi, HM. Mechanochemistry for synthesis. *Angew Chem.* **2020**, *132*(3), 1030–1041. http://doi.org/10.1002/anie.201906755

Haas, M; Lamour, S; Christ, SB; Trapp, O. Mineral-mediated carbohydrate synthesis by mechanical forces in a primordial geochemical setting. *Commun Chem.* **2020**, *3*(1), 140. https://doi.org/10.1038/s42004-020-00387-w

Harrison, SA; Lane, N. Life as a guide to prebiotic nucleotide synthesis. *Nat Commun.* **2018**, *9*(1), 1–4. https://doi.org/10.1038/s41467-018-07220-y

Hazen, RM; Morrison, SM. Mineralogical environments of the Hadean eon: Rare elements were ubiquitous in surface sites of rock-forming minerals. In *Prebiotic Chemistry and the Origin of Life. Advances in Astrobiology and Biogeophysics*, Neubeck, A; McMahon, S. Eds. Springer: Cham, Switzerland, **2021**, pp. 43–61. https://doi.org/10.1007/978-3-030-81039-9_2

Hud, NV; Fialho, DM. RNA nucleosides built in one prebiotic pot. *Science*. **2019**, *366*, 32–33. https://doi.org/10.1126/science.aaz1130

Islam, S; Powner, MW. Prebiotic systems chemistry: Complexity overcoming clutter. *Chem*. **2017**, *2*(4), 470–501. https://doi.org/10.1016/j.chempr.2017.03.001

Kauffman, SA. Evolution beyond Newton, Darwin, and entailing law: The origin of complexity in the evolving biosphere. In *Complexity and the Arrow of Time*, Lineweaver, CG; Davies, PCW; Ruse, M. Eds. Cambridge University Press: Cambridge, **2013**.

Kindermann, M; Stahl, I; Reimold, M; Pankau, WM; von Kiedrowski, G. Systems chemistry: Kinetic and computational analysis of a nearly exponential organic replicator. *Angew Chem*. **2005**, *117*(41), 6908–6913. https://doi.org/10.1002/ange.200501527

Kolb, VM. On the applicability of solventless and solid state reactions to the meteoritic chemistry. *Int J Astrobiol*. **2012**, *11*, 43–50. https://doi.org/10.1017/S1473550411000310

Kolb, VM. Prebiotic chemistry in water and in the solid state. In *Astrobiology: An Evolutionary Approach*, Kolb, VM. Ed. CRC Press: Boca Raton, FL, **2015**, pp. 199–216.

Kolb, VM. *Green Organic Chemistry and its Interdisciplinary Applications*. CRC Press: Boca Raton, FL, **2016**.

Kolb, VM. Prebiotic reactions in water, "on water," in superheated water, solventless, and in the solid state. In *Handbook of Astrobiology*, Kolb, VM. Ed. CRC Press: Boca Raton, FL, **2019**, pp. 331–340.

Kolb, VM; Clark, BC. *Systems Approach to Astrobiology*. CRC Press: Boca Raton, FL, **2023**.

Lamour, S; Pallmann, S; Haas, M; Trapp, O. Prebiotic sugar formation under nonaqueous conditions and mechanochemical acceleration. *Life*. **2019**, *9*(2), 52. https://doi.org/10.3390/life9020052

Lang, C; Lago, J; Pasek, M. Phosphorylation on the early Earth: The role of phosphorus in biochemistry and its bioavailability. In *Handbook of Astrobiology*, Kolb, VM. Ed. CRC Press: Boca Raton, FL, **2019**, pp. 361–370.

Ludlow, RF; Otto, S. Systems chemistry. *Chem Soc Rev*. **2008**, *37*(1), 101–108. https://doi.org/10.1039/B611921M

Mason, SF. *Chemical Evolution, Origins of the Elements, Molecules and Living Systems*. Oxford University Press: Oxford, **1991**.

Miljanić, OŠ. Small-molecule systems chemistry. *Chem*. **2017**, *2*(4), 502–524. https://doi.org/10.1016/j.chempr.2017.03.002

Miller, SL. A production of amino acids under possible primitive earth conditions. *Science*. **1953**, *117*, 528–529.

Miller, SL; Orgel, LE. *The Origins of Life on the Earth*. Prentice Hall: Englewood Cliffs, NJ, **1974**.

Patel, BH; Percivalle, C; Ritson, DJ; Duffy, CD; Sutherland, JD. Common origins of RNA, protein and lipid precursors in a cyanosulfidic protometabolism. *Nat Chem*. **2015**, *7*, 301–307.

Peretó, J. Out of fuzzy chemistry: From prebiotic chemistry to metabolic networks. *Chem Soc Rev*. **2012**, *41*, 5394–5403. https://doi.org/10.1039/C2CS35054H

Peretó, J. Prebiotic chemistry that led to life. In *Handbook of Astrobiology*, Kolb, VM. Ed. CRC Press: Boca Raton, FL, **2019**, pp. 219–233.

Powner, MW; Sutherland, JD. Prebiotic chemistry: A new modus operandi. *Phil Trans Royal Soc B: Biol Sci*. **2011**, *366*(1580), 2870–2877. https://doi.org/10.1098/rstb.2011.0134

Powner, MW; Sutherland, JD; Szostak, JW. The origins of nucleotides. *Synlett*. **2011**, *14*, 1956–1964. https://doi.org/10.1055/s-0030-1261177

Ranjan, S; Sasselov, DD. Influence of the UV environment on the synthesis of prebiotic molecules. *Astrobiology*. **2016**, *16*(1), 68–88. https://doi.org/10.1089/ast.2015.1359

Ranjan, S; Todd, Z; Sutherland, JD; Sasselov, DD. Sulfidic anion concentrations on early Earth for surficial origins-of-life chemistry. *Astrobiology*. **2018**, *18*(8), 1023–1040.

Ritson, DJ. A cyanosulfidic origin of the Krebs cycle. *Sci Adv*. **2021**, *7*(33), eabh3981. https://doi.org/10.1126/sciadv.abh3981

Ritson, DJ; Battilocchio, C; Ley, SV; Sutherland, JD. Mimicking the surface and prebiotic chemistry of early Earth using flow chemistry. *Nat Comm*. **2018**, *9*(1), 1–10. https://doi.org/10.1038/s41467-018-04147-2

Ruiz-Mirazo, K; Briones, C; de la Escosura, A. Prebiotic systems chemistry: New perspectives for the origins of life. *Chem Rev*. **2014**, *114*, 285–366. https://doi.org/10.1021/cr2004844

Sasselov, DD; Grotzinger, JP; Sutherland, JD. The origin of life as a planetary phenomenon. *Sci Adv*. **2020**, *6*(6), eaax3419. https://doi.org/10.1126/sciadv.aax3419

Schmitt-Kopplin, P; Gabelica, Z; Gougeon, RD; Fekete, A; Kanawati, B; Harir, M; Gebefuegi, I; Eckel, G; Hertkorn, N. High molecular diversity of extraterrestrial organic matter in Murchison meteorite revealed 40 years after its fall. *Proc Natl Acad Sci USA*. **2010**, *107*, 2763–2768. https://doi.org/10.1073/pnas.0912157107

Schmitt-Kopplin, P; Harir, M; Kanawati, B; Gougeon, R; Moritz, F; Hertkorn, N; Clary, S; Gebefügi, I; Gabelica, Z. Analysis of extraterrestrial organic matter in Murchison meteorite: A progress report. In *Astrobiology: An Evolutionary Approach*, Kolb, VM. Ed. CRC Press: Boca Raton, FL, **2015**, pp. 63–82.

Strazewski, P. The beginning of systems chemistry. *Life*. **2019a**, *9*(1), 11. https://doi.org/10.3390/life9010011

Strazewski, P. The essence of systems chemistry. *Life*. **2019b**, *9*(3), 60. https://doi.org/10.3390/life9030060

Sutherland, JD. The origin of life – out of the blue. *Angew Chem Int Ed*. **2016**, *55*(1), 104–121. https://doi.org/10.1002/anie.201506585

Sutherland, JD. Studies on the origin of life-the end of the beginning. *Nat Rev Chem*. **2017**, *1*, 12. https://doi.org/10.1038/s41570-016-0012

Szostak, JW. Systems chemistry on early Earth. *Nature*. **2009**, *459*, 171–172. https://doi.org/10.1038/459171a

Voet, D; Voet, JG. *Biochemistry*, 4th ed. Wiley: Hoboken, NJ, **2011**.

Von Kiedrowski, G; Otto, S; Herdewijn, P. Welcome home, systems chemists! *J Syst Chem*. **2010**, *1*(1). https://doi.org/10.1186/1759-2208-1-1

Wing, MR; Bada, JL. Geochromatography on the parent body of the carbonaceous chondrite. *Geochim Cosmochim Acta*. **1991**, *55*(10), 2937–2942. https://doi.org/10.1016/0016-7037(91)90458-H

Wołos, A; Roszak, R; Żądło-Dobrowolska, A; Beker, W; Mikulak-Klucznik, B; Spólnik, G; Dygas, M; Szymkuć, S; Grzybowski, BA. Synthetic connectivity, emergence, and self-regeneration in the network of prebiotic chemistry. *Science*. **2020**, *369*(6511). https://doi.org/10.1126/science.aaw1955

Yadav, M; Kumar, R; Krishnamurthy, R. Chemistry of abiotic nucleotide synthesis. *Chem Rev*. **2020**, *120*(11), 4766–4805. https://doi.org/10.1021/acs.chemrev.9b00546

Zhang, SJ; Duzdevich, D; Ding, D; Szostak, JW. Freeze-thaw cycles enable a prebiotically plausible and continuous pathway from nucleotide activation to nonenzymatic RNA copying. *Proc Natl Acad Sci*. **2022**, *119*(17), e2116429119. https://doi.org/10.1073/pnas.2116429119

# 4 Assembly of Membranous Compartments

## *An Essential Step in the Emergence and Function of Protocellular Systems*

*David Deamer, Bruce Damer and Povilas Šimonis*

## 4.1 INTRODUCTION AND BACKGROUND

Although significant progress has been made, the origin of life remains unexplained because of multiple gaps in our knowledge. It is worth listing these gaps in the introduction as examples of questions to be considered in this review.

1. Which conditions on the early Earth were conducive to the processes leading to the origin of life?
2. What were the primary sources of organic compounds?
3. Which sources of energy were available to drive mixtures of simple compounds away from equilibrium toward increasing complexity?
4. The organic solutes were present in aqueous solutions of ionic solutes exposed to mineral surfaces. How did the solutes and surfaces affect chemical reactions such as polymerization?
5. How did condensation reactions synthesize polymers like peptides and oligonucleotides?
6. How did the polymers evolve functions required for life's origin?
7. What were the initial chemical reactions leading to a primitive metabolism?
8. How did membranous compartments assemble and encapsulate polymers to form protocells?
9. How did protocells undergo selection and evolution?
10. When did homochirality emerge as a property of life and what was the mechanism?
11. What pigment systems began to capture light energy and transduce it into chemical energy?

DOI: 10.1201/9781003294276-4

12. How are vesicles, protocells, progenotes, and progenitor conditions related in early evolutionary progression?

This review will not address all these questions in detail. Instead, we will focus on six examples – prebiotic environment, sources of organic compounds, energy sources, condensation reactions, encapsulation of polymer products, and the initial steps toward primitive life. We are working to establish model protocells which contain potentially functional polymers encapsulated in membranous boundaries. These are microscopic, self-assembled systems, and their populations can undergo selection and evolution toward increasingly complex structures and functions. Our aim is to understand how molecular systems can emerge within the complex environment of the prebiotic Earth, then use this insight to guide experimental approaches that expand our understanding of life's origin.

## 4.2 WHICH CONDITIONS ON THE EARLY EARTH WERE MOST CONDUCIVE TO THE PROCESSES LEADING TO THE ORIGIN OF LIFE?

In a recent paper, we proposed that a new word – urability – would be useful to describe planetary conditions conducive to the origin of life. It is related to the word habitability which defines habitable planets in terms of astrophysical parameters that allow liquid water to exist for extended periods of time. The assumption is that microorganisms could survive and even thrive on a habitable planet. Although a planet may be habitable, it does not necessarily follow that life inevitably begins on such planets. The conditions required for the origin of life on habitable planets like the early Earth and Mars are much more complex than the conditions for habitability. This complexity of geochemical factors is illustrated in the following list (Deamer et al. 2022):

**Geochemical factors**

- Anoxic atmosphere perhaps with admixtures of reactive gasses such as HCN (hydrogen cyanide) and HCHO (formaldehyde)
- Liquid water within temperature ranges conducive to sustained prebiotic reactions
- Ionic concentrations ranging from fresh water to salty seas
- Acidity or alkalinity of aqueous solutions
- A continuous source of key organic compounds made available by local synthesis or exogenous delivery
- Synthesis or delivery of specific compounds that are capable of serving as monomers, including amino acids, nucleobases, monosaccharides, and phosphate
- Amphiphilic compounds available for assembly into vesicular boundary membranes
- Sources of energy to drive reactions: Chemical energy, redox potentials, light energy, wet–dry cycles

- Processes that concentrate dilute solutions of reactants sufficiently to react
- Conditions that capture energy to enable polymerization reactions such that populations of polymers of a sufficient length emerge to support catalytic and information storage functions
- Mixtures of organic compounds capable of being incorporated into systems related to autocatalysis and primitive metabolism
- Selective processes that lead toward homochirality
- Encapsulation processes to enclose sets of polymers and other molecules into populations of protocells

**Combinatorial factors**

- Cycling of sets of encapsulated polymers through dynamic environmental stresses to drive the first steps of evolution by combinatorial selection

Although the origin of life has been viewed as fundamentally a chemical process, we will argue here that biophysical properties also play an essential role, such as spontaneous self-assembly of amphiphilic compounds into membranous compartments. Basic biophysical properties also put constraints on the conditions that would be conducive to life's origin. These include the effects of temperature, concentration of potential reactants, and interactions between ionic solutes and organic compounds. These are important properties of liquid water in which interactions between organic compounds and energy sources led to the origin of life.

## 4.3 PROPERTIES OF LIQUID WATER

It is a given that liquid water is essential not only for all living systems but also for the origin of life. The presence of liquid water over extended periods of time is a fundamental factor of planetary urability. There are several physical properties of water that dramatically affect its ability to support the emergence of primitive forms of cellular life: Temperature, pH, ionic solutes, and the ability to evaporate. Over the past 50 years, it has become clear that water on Earth and presumably other habitable planets can exist in two general conditions. The first is exemplified by salty seawater which composes >99% of the Earth's water. Typical concentrations of ionic solutes in seawater are 0.58 M NaCl, 53 mM $MgCl_2$, and 10 mM $CaCl_2$. The second kind of water is present in hydrothermal fields associated with volcanic land masses. These are supplied with distilled water that evaporates from the oceans and falls as precipitation. In contrast to salty seawater, the concentrations of ionic solutes in hydrothermal pools are relatively low, in the micromolar to millimolar range (Deamer et al. 2019).

Because nearly all of the Earth's water is salty seawater, it has been generally assumed that life must have begun in the ocean. However, bulk seawater is a virtual desert in terms of chemical energy and nutrients. Hydrothermal vents called Black Smokers were discovered in 1977 by Jack Corliss during a dive in the deep submersible Alvin (Deamer et al. 2019). The vents were an obvious source of energy, and multiple forms of prokaryotic and eukaryotic life already take advantage of the available chemical nutrients. It was soon proposed that Black Smokers may have also served as an environment where life could begin (Baross and Hoffman 1985; Martin and Russell 2003).

A few years later, another version of hydrothermal vents was discovered (Kelly et al. 2005). These are referred to as alkaline vents because they are produced by serpentinization, a chemical reaction between seawater and olivine minerals in the sea floor. Several researchers realized that the moderate temperature ranges of alkaline vents (60–90°C) might be preferable to the extreme temperatures of Black Smokers which can exceed 400°C. Alkaline vents produce alkaline fluid with abundant reducing power in the form of dissolved hydrogen. Furthermore, when the alkaline effluent within the mineral matrix encounters relatively acidic seawater, a pH gradient could develop across mineral membranes, which offers additional chemical energy in the form of a primitive version of chemiosmosis (Lane 2017). These ideas have evolved into elaborate schemes in which the potential chemical reactions possible within alkaline vents could become a primitive version of metabolism that could support the origin of life.

There have been several attempts to test the hydrothermal vent idea using laboratory simulations (Herschy et al. 2014; Sojo et al. 2016; Barge et al. 2015), but it is very difficult to mimic the pressure, temperature, and mineral composition of actual vents. Furthermore, experiments at vents would be extremely expensive because they would involve diving in submersible vehicles like the Alvin and injection of labeled carbon dioxide into the vent minerals followed by capture of any reaction products in the vent fluid being emitted. Although these ideas are attractive in terms of potential chemical energy, from a biophysical perspective, there are significant limitations (Jackson 2016). The most significant is the thermodynamic problem that condensation reactions required for polymerization of monomers cannot occur spontaneously in bulk seawater. Instead, just as in metabolism in life today, the monomers must be chemically activated, and no obvious activation mechanism has yet been demonstrated in the vent environment.

Freshwater associated with hydrothermal fields offers an alternative to hydrothermal vents and seawater. Because the water supplied to hydrothermal fields is distilled by evaporation, the ionic concentrations are very low, ranging from micromolar to millimolar (Deamer et al. 2019) (Figure 4.1). It is important to note here that divalent cations in seawater are 53 mM $Mg^{2+}$ and 10 mM $Ca^{2+}$. Divalent cations tend to bind strongly to divalent anions in solution. Examples include a variety of relatively insoluble minerals such as limestone (calcium carbonate), gypsum (calcium sulfate), and apatite (calcium phosphate). This tendency can cause significant problems for solutions of organic compounds. For instance, calcium can bind to carboxyl groups and thereby disrupt the ability of amphiphilic molecules such as fatty acids to assemble into boundary membranes. It also removes phosphate from solution when calcium and phosphate form relatively insoluble apatite. Phosphate is essential to life today, and at some point, there must have been sufficient phosphate in solution for it to be incorporated into the initial reactions of a primitive metabolism.

Apel et al. (2002) pointed out that for fatty acids to assemble into membranes, the pH must be titrated to near the pK in order to have a mixture of protonated and anionic head groups required for stability. Maurer and Nguyen (2016) has noted that

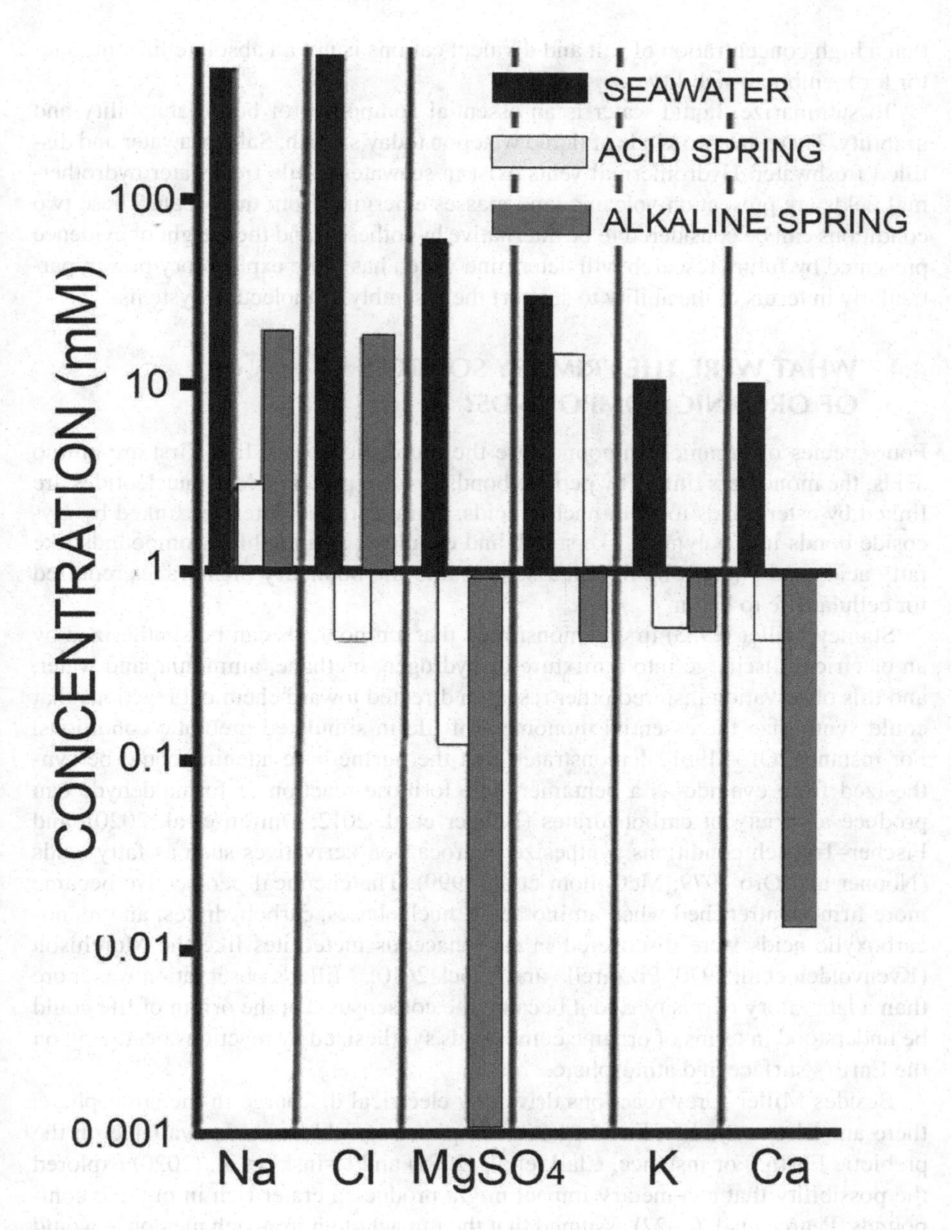


**FIGURE 4.1** Ionic solutes in seawater and freshwater hot springs.
*Source:* The authors

mixtures of fatty acid chain lengths are more stable than pure fatty acids. This was first demonstrated by Monnard et al. (2002) and later by Jordan et al. (2019) who prepared mixtures of amphiphiles that formed stable membranes in hot seawater, simulating temperature ranges on alkaline hydrothermal vents. These results show that a high concentration of salt and divalent cations is not an absolute limiting factor for membrane stability.

To summarize, liquid water is an essential component of both habitability and urability. There are two kinds of liquid water on today's Earth: Salty seawater and distilled freshwater. Hydrothermal vents exist in seawater, while freshwater hydrothermal fields are present on volcanic land masses emerging from the ocean. These two conditions can be considered to be alternative hypotheses, and the weight of evidence presented by future research will determine which has more explanatory power, particularly in terms of the ability to support the assembly of molecular systems.

## 4.4 WHAT WERE THE PRIMARY SOURCES OF ORGANIC COMPOUNDS?

Four species of organic compounds are the foundation of all life. First are amino acids, the monomers linked by peptide bonds to form proteins. Mononucleotides are linked by ester bonds to form nucleic acids, while carbohydrates are linked by glycoside bonds into polymers like starch and cellulose. Amphiphilic compounds like fatty acids and phospholipids can assemble into the boundary membranes required for cellular life to begin.

Stanley Miller (1953) first demonstrated that amino acids can be synthesized by an electrical discharge into a mixture of hydrogen, methane, ammonia, and water, and this observation inspired other research directed toward chemical reactions that could synthesize the essential monomers of life in simulated prebiotic conditions. For instance, Oro (1961) demonstrated that the purine base adenine could be synthesized from cyanide as a pentamer. The formose reaction of formaldehyde can produce a variety of carbohydrates (Benner et al. 2012; Omran et al. 2020), and Fischer–Tropsch conditions synthesize hydrocarbon derivatives such as fatty acids (Nooner and Oro 1979; McCollom et al. 1999). The chemical perspective became more firmly entrenched when amino acids, nucleobases, carbohydrates, and monocarboxylic acids were discovered in carbonaceous meteorites like the Murchison (Kvenvolden et al. 1970; Pizzarello and Shock 2010). Miller's observation was more than a laboratory curiosity, and it became the consensus that the origin of life could be understood in terms of organic compounds synthesized by reactions occurring on the Earth's surface and atmosphere.

Besides Miller–Urey reactions driven by electrical discharge in the atmosphere, there are other ways in which organic compounds could be made available on the prebiotic Earth. For instance, Clark et al. (2021) and Osinski et al. (2020) explored the possibility that a cometary impact might produce a crater rich in organic compounds. Pearce et al. (2022) assumed that the impact of an iron-rich meteorite would add significant hydrogen cyanide to the atmosphere, which could undergo a variety of known reactions to produce amino acids and nucleobases.

Although it is inferred from laboratory simulations that biologically relevant small organic molecules could be synthesized in this way, there is no direct evidence that such reactions were a major source of organic compounds on the early Earth. An alternative source is organics delivered by interplanetary dust particles (IDPs), meteorites, and comets, and direct evidence is the fact that this is still occurring today. A cometary source was first suggested by Oro (1961). Anders (1989) proposed that interplanetary dust particles (IDPs) were likely to be a major source. This was confirmed by Chyba and Sagan (1992) who calculated the rate at which organics were delivered by these sources. Four billion years ago, IDPs were by far the most abundant source ($10^8$ kg/year) followed by comets ($10^5$ kg/year) and carbonaceous meteorites ($10^3$ kg/year). The calculation for IDPs was anchored by the delivery rate today, which has been estimated to be $3.2 \times 10^5$ kg/year.

A useful perspective on these numbers is to divide the rates by the surface area of the Earth ($5.1 \times 10^8$ km$^2$), which would be 0.2 kg of organic material per square kilometer per year. Most of this would fall into the sea and become unavailable for further processing, but volcanic land masses resembling Hawaii and Iceland today were common even though plate tectonics had not yet produced continental land masses (Van Kranendonk et al. 2021). If so, 200 kg would accumulate per km$^2$ in a 1,000 years. This would be further concentrated when precipitation flushed the IDPs and organic material into hydrothermal fields analogous to those in Yellowstone National Park where they would become available as potential reactants.

What would be the composition of the organic material? Carbonaceous meteorites provide a clue (Figure 4.2). Most of the organic carbon in such meteorites is an insoluble kerogen-like polymer composed of polycyclic aromatic hydrocarbons, but a smaller fraction is a mixture of thousands of organic compounds

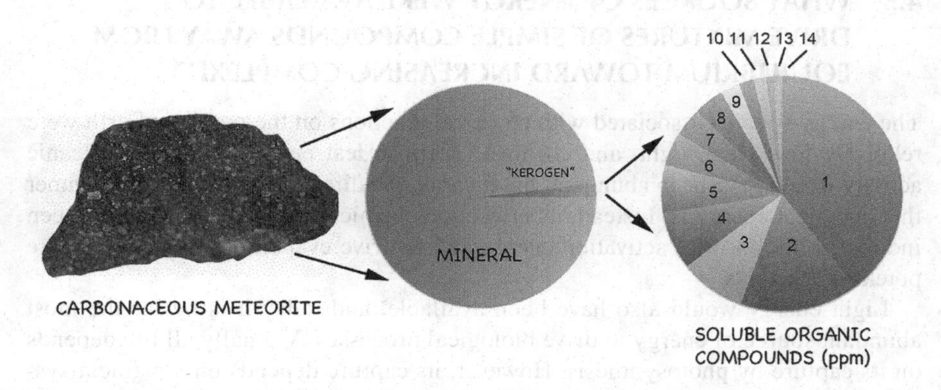


**FIGURE 4.2** Organic compounds in carbonaceous meteorites. Organic carbon composes a few percent of the mass of such meteorites, mostly in the form of an insoluble kerogen-like polymer. A second fraction is soluble in water or organic solvents, and the pie chart shows the composition in the Murchison meteorite. The most abundant compounds by mass are monocarboxylic acids shown in blue (1–300 ppm); then going clockwise around the chart in order of abundance are polar hydrocarbons (2–120 ppm), sulfonic acids (3–67 ppm), amino acids (4–60 ppm), dicarboximides (5–50 ppm), aliphatic hydrocarbons (6–35 ppm), dicarboxylic acids (7–30 ppm), polyols (8–30 ppm), aromatic hydrocarbons (9–20 ppm), hydroxy acids (10–15 ppm), amines (11–13 ppm), N-heterocycles (12–7 ppm), pyridine carboxylic acids (13–7 ppm), and phosphonic acids (14–2 ppm). Because carbonaceous meteorites are not uniform in composition, the measured values are typical averages of a small sample specimen, rather than being characteristic of all such meteorites.

*Source:* Prepared using data from Pizzarello et al. (2006)

including amino acids, nucleobases, simple carbohydrates, and monocarboxylic acids ranging up to 12 carbons in length. Assuming that this composition is an approximation of the organic suite that might be delivered by IDPs, all four main species required for assembly into living systems would be available as solutes in hydrothermal fields.

To summarize, a source of organic compounds is an essential factor of urability. Although it is uncertain which of the two possible sources is primary, it is important to understand that this may not be essential to know with certainty. The organic compounds of both sources will be mixed and undergo physical and chemical processing before they become involved in the process by which they assemble into protocells. Another point to consider is that the mixtures would be undergoing continuous degradation and replenishment. One example of this was described by Pearce et al. (2017) who used a numerical approach to study what would happen over time after the nucleobase adenine was delivered to small hydrothermal pools by a carbonaceous meteorite. The results are instructive. The primary loss of adenine from the pools was by seepage into underlying porous minerals rather than chemical or photochemical decomposition. At first, the rate of loss is balanced by infall, but at some point, the declining rate of delivery was such that adenine would no longer be available as a potential reactant.

## 4.5 WHAT SOURCES OF ENERGY WERE AVAILABLE TO DRIVE MIXTURES OF SIMPLE COMPOUNDS AWAY FROM EQUILIBRIUM TOWARD INCREASING COMPLEXITY?

The energy sources associated with chemical reactions on the prebiotic Earth were relatively few: Heat, light, and chemical energy. Heat energy related to volcanic activity would have been abundant, but it cannot be directly captured in the manner that chemical energy is. Instead, its effects on chemical reactions would have been indirect in the form of activation energy and to drive evaporation and concentrate potential reactants.

Light energy would also have been available, and today it is by far the most abundant source of energy to drive biological processes. Virtually all life depends on its capture by photosynthesis. However, its capture depends on a pigment system that begins with the pigment chlorophyll. When photons strike chlorophyll molecules in the membranes of photosynthetic bacteria and higher plants, the electronic structure jumps to an excited state that releases electrons to an electron transport chain embedded in the membranes. The electron ends up on $NADPH_2$ which is then used in the dark reactions to reduce $CO_2$ to carbohydrates and other reduced compounds.

But what pigment was available on the early Earth? It was not chlorophyll, which is a highly evolved pigment that is synthesized by a series of enzyme-catalyzed metabolic reactions. An interesting possibility is that polycyclic aromatic compounds could serve as primitive pigments and components of electron transport systems. Examples include pyrene, fluoranthene, and anthracene derivatives. These absorb light in the UV range and are highly fluorescent. Significantly, the excited state of pyrene can donate electrons to an acceptor like benzoquinone (Escabi-Perez et al. 1979), and quinone compounds isolated from carbonaceous meteorites have been demonstrated to generate pH gradients in lipid vesicles (Milshteyn et al. 2019. Further research on the photochemical properties of PAH is likely to be fruitful.

A final source of energy must be mentioned. A dilute solution of potential reactants is at its highest entropy state, but if the solution evaporates and concentrates the reactants, the entropy decreases. This follows from the fundamental equation of thermodynamics: $\Delta G = \Delta H - T\Delta S$. If there is a favorable change in $\Delta S$, the free energy ($\Delta G$) available to drive a reaction increases. An example of this process is crystallization. As dilute, disordered solutes become concentrated, they can form orderly crystals which are at the lowest possible entropy. Furthermore, if the solutes are reactants, a reaction that is improbable in dilute solution becomes highly probable. This thermodynamic fact leads to another source of chemical energy – wet–dry cycles – which is discussed in the next section.

## 4.6 HOW DID CONDENSATION REACTIONS SYNTHESIZE POLYMERS LIKE OLIGONUCLEOTIDES?

Condensation reactions link biological monomers such as mononucleotides with nucleic acid polymers. It has been known for many years that in a DNA duplex strand, hydrogen bonding between purines and pyrimidines stabilizes the double helix structure. What is less well known is that the nucleotide pairs are also stabilized by base stacking and that stacking is an even more important factor. Base stacking does not only occur in duplex strands, but also when concentrated mononucleotides ultimately undergo crystallization. This point was demonstrated by Himbert et al. (2016) who used X-ray diffraction to study a 1:1 mixture of AMP and UMP that had been dried into a film by evaporation. A strong pattern emerged that resulted from the 3.4 A distance between stacked base pairs. In other words, as the solution dried, the bases assembled into linear crystals referred to as pre-polymers. This organizing effect would markedly reduce the entropy of the system in a favorable direction.

The final reaction that linked the monomers was the reduction in water activity as the solution evaporated to dryness (Ross and Deamer 2016). At some point, all the bulk phase water was gone, but condensation reactions occurring along the pre-polymer could release water as a leaving group to form phosphodiester bonds. One way to think about this reaction is to consider the opposite reaction of hydrolysis in which water molecules are added to an ester bond. The hydrolysis reaction is spontaneous, but the reverse reaction of condensation can also become spontaneous when water activity is sufficiently decreased.

This hypothetical possibility is supported by multiple laboratory simulations in which mononucleotides and amino acids were subjected to wet–dry cycles. The polymers were detected by electrophoresis of labeled products (Rajamani et al. 2008) and by nanopore analysis (DeGuzman et al. 2014). The polymers have also been imaged by atomic force microscopy (Hassenkam and Deamer 2022).

## 4.7 BALANCE BETWEEN DEGRADATION AND SYNTHESIS: IMPORTANCE OF CYCLES AND KINETIC TRAPS

In most chemical reactions, a reaction mixture away from equilibrium is allowed to proceed toward an equilibrium in which products accumulate. It is rare that reactions are exposed to multiple cycles, but this is precisely what becomes possible in hydrothermal fields on volcanic land masses. The pools are periodically filled with freshwater distilled from the salty ocean and delivered as precipitation, and then it evaporates quickly because the pools are heated by underlying magma. The pools are cycled continuously, and this drives a process that has not yet been carefully studied. In the dry phase of a cycle, condensation reactions produce polymers as described before. In the wet phase, the polymers are subjected to hydrolysis. One might expect that the products of condensation reactions are simply degraded, but this doesn't return to equilibrium for two reasons. First, the energetically uphill synthesis of polymers occurs rapidly when evaporation is completed, but, upon rehydration, the downhill, thermodynamically favored reaction of hydrolysis is much slower, so the polymers accumulate even though they may be far from equilibrium. Second, the polymers have the potential to fold into secondary and tertiary structures that are protected from hydrolysis. An example is DNA in which the duplex structure is much less labile to hydrolysis than single-stranded DNA.

## 4.8 HOW DID MEMBRANOUS COMPARTMENTS ASSEMBLE AND ENCAPSULATE POLYMERS TO FORM PROTOCELLS?

The physical properties of lipid vesicles relevant to the origin of life have been recently reviewed by Imai et al. (2022). Here, we will focus on a second property: Encapsulation of polymers. Encapsulation of drugs in lipid compartments is now an important process in the pharmaceutical industry. An example is the messenger RNA of the COVID virus which is delivered by a lipid carrier. However, the industrial encapsulation process is highly technical. How could protocells assemble spontaneously on the prebiotic Earth with encapsulated polymers like nucleic acids?

The answer is surprisingly simple. When a mixture of membrane-forming lipids and biopolymers such as nucleic acids is exposed to a single wet–dry cycle, the vesicles fuse into a multilamellar matrix with polymers trapped between lipid layers (Toppozini et al. 2013). Upon rehydration, the lipid layers swell as water enters the dry matrix and vesicles containing the polymers referred to as protocells dominate the mixture (Shew and Deamer 1985). This process is illustrated in Figure 4.3. Vesicles were prepared from a 2:1 mixture by weight of soy lecithin and yeast RNA. The preparation was then exposed to a single wet–dry–wet cycle at 80°C. A phase microscopy image of the vesicles is shown on the left at 400× magnification, and the same preparation on the right was stained with the fluorescent dye acridine orange. The encapsulated RNA adsorbs the dye and can be seen filling the vesicles.

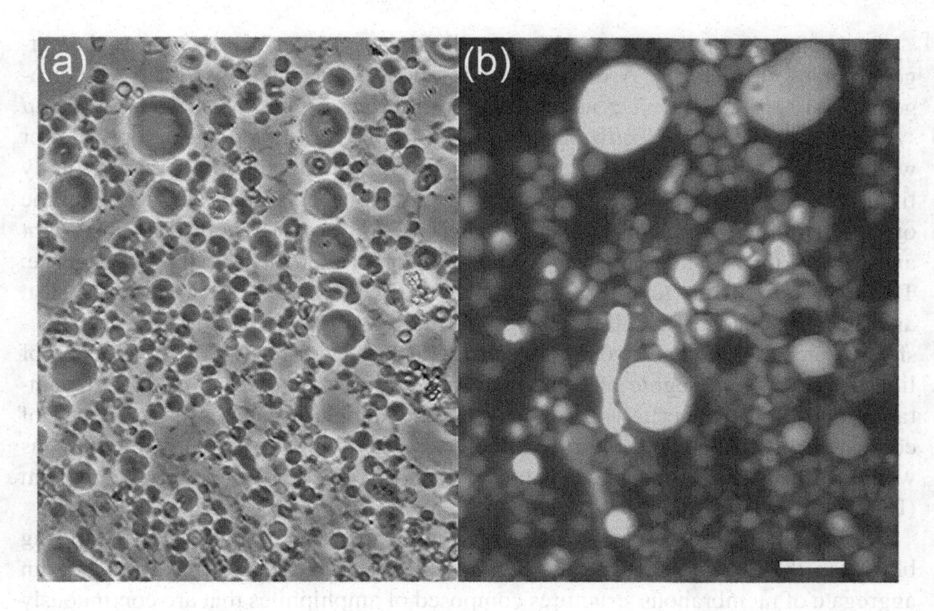


**FIGURE 4.3** Micrographs of vesicles produced from a 2:1 mixture by weight of soy lecithin and yeast RNA exposed to a single wet–dry–wet cycle at 80°C. The vesicles on the left (a) were photographed at 400× using phase microscopy. The same preparation on the right (b) was stained with the fluorescent dye acridine orange to reveal the encapsulated RNA. Bar = 20 µm.

*Source:* The authors

## 4.9 HOW CAN POPULATIONS OF POLYMERS AND PROTOCELLS BE CONTINUOUSLY GENERATED AND SUBJECT TO SELECTION?

Having established how protocell compartments can assemble on the early Earth, we can now consider a broader question: *Which processes can continuously generate large populations of protocells while supporting their evolution by selection?* A single protocell or small group of protocells with similar contents would be insufficient in terms of molecular evolution because protocells must be subjected to selection and amplification in large populations having diverse internal cargos. In addition, a single initial population of fragile protocells would not progress very far because disruptive forces would soon cull their numbers. Therefore, a system must be present which can generate a continuous flux of protocell populations, analogous to a factory assembly line that fabricates a stream of random mechanical devices. Extending the factory metaphor to a closed-loop evolutionary system, each device would be tested for a specific function. Those that perform well enough are selected and then fed (along with the blueprints for their construction) back into the assembly line to guide future generations with ever more novel capabilities.

Such a factory environment might be called a "progenitor" for the eventual emergence of useful devices. A dictionary definition (Merriam-Webster n.d.) for progenitor is: *A physical object which can be considered both a "precursor" to and "originator" of a biologically ancestral form.* In this scientific context, a progenitor would be a medium or substrate from which protocell populations and eventually living cells can emerge. In fact, this term has already been applied in the field of the origin of life by proponents of the submarine alkaline vent hypothesis. Cartwright and Russell (2019) and Simon Duval and co-workers (2019) described "Fougerite: The not so simple progenitor of the first cells." Fougerite is also called green rust, and, in a later paper, Russell (2021) referred to it as "first seed of life . . . a 'makeshift' protocell". In other words, for a mineral to have a role in the emergence of life, it must have a degree of complexity beyond its crystal structure and elemental composition. Fougerite is given as such an example because it has a matrix of compartments that could support gradients of concentration, pH and temperature, as well as mineral surfaces that can catalyze synthetic reactions related to metabolism (Russell 2018).

In our view, a progenitor must self-assemble from available organic building blocks in the environment, so an alternative to a mineral substrate would be an aggregate of membranous structures composed of amphiphiles that are continuously provided from natural sources. If these become concentrated in small hydrothermal pools subjected to periodic cycles of wetting and drying, the amphiphiles will self-assemble into vesicles and then settle into a "bathtub ring" around the mineral edge as the water line rises and recedes. During evaporation to dryness, the membranous vesicles fuse into multilamellar sheets that capture solutes from the pool by sandwiching them between lipid lamellae. At this point, reduced water activity and elevated temperatures provide an energy source for condensation reactions within the progenitor, which allows the synthesis of ester and peptide bonds. The result is a primitive chemical factory capable of producing a continuous stream of short polymers composed of random sequences of amino acids and nucleotides.

As described earlier, the polymers can be encapsulated in microscopic lipid vesicles and form protocells, each having a different composition from all the rest. The progenitor provides a sort of "proto-niche" (Damer 2019; Odling-Smee et al., 2024; Olding-Smee, Laland, and Feldman, 2013) that protects organic compounds from degradative forces such as hydrolysis. The dynamic fluidity of membranes also offers intrinsic mechanisms for concentration, transport, and combinatorial selection. At this stage, the boundaries of a lipid vesicle population can enter into a new type of interaction with their entrapped polymers. One such interaction is that polymers provide a stabilizing influence to otherwise fragile boundary membranes. This stabilizing effect introduces *beneficial mutualism* (Bronstein 2015) which is universal to life: The machinery of the cell is produced within and protected by its membranous boundary, and the machinery in turn works to stabilize the boundary structures. Cytoskeletal proteins such as spectrin and actin have a similar function in cellular life today.

The next phase in the factory-like production cycle of a progenitor begins when the dry, multilamellar matrix is rehydrated. The influx of water penetrates the membranes and causes budding to form membranous compartments. Some of these compartments will be empty while others encapsulate polymers and trapped solutes from the lamellar matrix of the progenitor. We refer to the compartments as protocells, and these are initial units subject to selection and evolution. Can a fragile protocell remain intact within the bulk of a small pond while being subjected to shear forces, temperature and pH changes, and hydrolytic and other selective factors? Experiments both in laboratory and field analogs in hot spring conditions (Deamer et al. 2019) have demonstrated that populations of protocells do survive some of these stresses, delivering their cargoes back to the lipid aggregate phase. To support the amplification of selected populations of encapsulated polymers through multiple wet–dry cycles, the stabilizing effect of polymers initiates a closed-loop cycle which sets up a *kinetic trap* in which the rate of polymer synthesis exceeds the rate of their degradation through hydrolysis.

The processes we described here are not speculation but are supported by laboratory results and field observations. Figure 4.4 summarizes the mechanism and experimental evidence for polymer synthesis and encapsulation within lipid vesicles to form protocells. The progenitor is a collective environment which initially assembles from aggregates of amphiphiles and monomers in fresh-water hot spring pools. In the dry phase, the vesicles aggregate during evaporation and dry on mineral surfaces

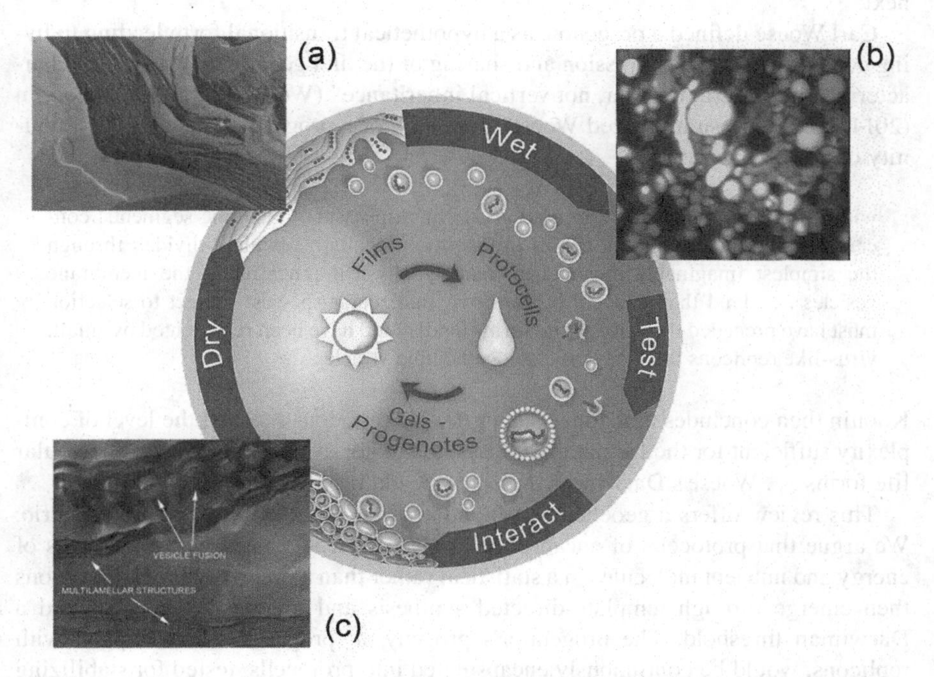


**FIGURE 4.4** The three-phase kinetic trap within which populations of polymers can be synthesized, become encapsulated and subjected to selection. (a) Synthesis and selection between dried lipid films, (b) inclusion within budding compartments during rehydration and testing for stability during aqueous immersion, (c) returning as selected populations of polymers for fusing, mixing, and further interaction within a gel phase and subsequent dry lamellar phase.

*Source:* The authors

to form multilamellar matrices. The inset (a) is a freeze-fracture electron micrograph that reveals the multiple lipid bilayers as stacks. If monomers such as nucleotides and amino acids are present, they undergo condensation reactions within the layers to form polymers. In the transition to the wet phase, vesicles bud off when the multilamellar matrix is rehydrated. The vesicles have encapsulated nucleic acids labelled with a fluorescent dye (depicted in the microscopic image b) and are now defined as protocells containing newly synthesized polymers. The protocells are tested for stability during the aqueous phase, and surviving compartments return to form a moist "gel" during evaporation. The inset (c) is another freeze-fracture micrograph showing vesicular compartments fusing into a dry, multilamellar matrix.

A "progenote" is formed and is sustained within the three-phase kinetic trap as a population of polymers amplifies and attains sufficient number and length to support the emergence through the selection of catalytic and informational functions. Such emergence was demonstrated by Holliger and coworkers through in-vitro evolution of ribozymes which were capable of triplet polymerase activity (Attwater et al. 2018). Our employment of the term progenote, suggested in the figure to originate during the gel phase, was first introduced by Woese and Fox (1977) and will be detailed next.

Carl Woese defined a progenote as a hypothetical transitional form leading to living cells in which the expression and sharing of the first genetic material were characterized by "lateral transfer, not vertical inheritance" (Woese 1998, 2002). Koonin (2014) succinctly summarized Woese' definition of a progenote as being a community of primitive entities

> with imprecise, "statistical" translation, and multiple, small genomic segments, conceivably present in multiple copies in each (proto)cell . . . [which] divided through the simplest imaginable mechanism, namely physical pinching of the membrane vesicles . . . [and that as such] "organisms" (molecular species) subject to selection must have preceded the cellular life forms [and] could have been represented by small, virus-like replicons that populated abiogenic lipid vesicles.

Koonin then concludes that "once such growing replicons reached the level of complexity sufficient for the formation of cells, the major evolution transition to cellular life forms . . . Woese's Darwinian threshold would be crossed".

This review offers a geochemical foundation for Woese and Koonin's scenario. We argue that protocells of encapsulated polymers would interact with sources of energy and nutrient molecules in a statistical rather than a precise manner. Replicons then emerge through template-directed synthesis and gradually evolve toward a Darwinian threshold. The progenitor's progeny, a 'progenote soup' replete with replicons, would be continuously encapsulated into protocells, tested for stabilizing properties, returned, mixed, and tested again for catalytic and metabolic capacities, and then sandwiched within drying and concentrating lamellae to participate in new rounds of polymer formation.

This progenitor proto-niche (Odling-Smee et al., 2024; Olding-Smee, Laland, and Feldman, 2013) self-assembles from membranous vesicles (Figure 4.5, lower left). It is continuously growing and changing, driven by cycles of dehydration and rehydration supplemented with additional amphiphiles and other organic compounds from exogenous sources to sustain the generation and mutation of its encapsulated progenote (center). A progenote is a stream of vast numbers of interacting polymers which can undergo a form of individual and group selection while transiently occupying lamellae and protocells. Progenitor material can also be distributed among nearby interconnected pools, thereby providing exchanges of molecular innovations as they evolve within widely varying resource and selection environments (Damer 2016). Through a lengthy process of starts, failures, and exchanges, protocells gradually capture more functions and express an identity clearer from the progenote substrate and ultimately evolve toward self-reproducing, living cells (upper right).

The transition to self-reproducing microbial communities is a vast evolutionary chasm crossed by populations of protocells depending on the stacking up of a towering structure of evolved molecular tools, each discovered solely through selective forces. One key molecular tool arose when wet–dry cycling as a sole source of energy for polymer formation is replaced by energy-driven enzymes which can enable condensation reactions while in a continuously hydrated state. Another molecular innovation enabled the first expression of genes, a proto-ribosomal set of steps which could translate code stored in nucleotide sequences into functional chains of amino acids. The division of protocells into viable daughter cells is a somewhat arbitrary

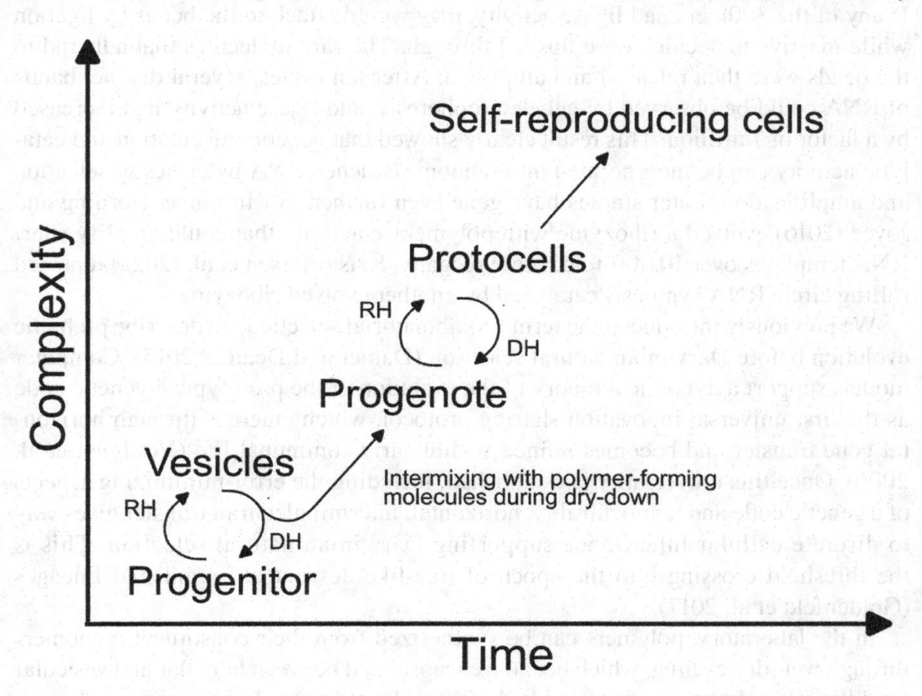


**FIGURE 4.5** Illustrating the formation of a progenitor, the emergence of a progenote, and the generation of protocell populations into cycles of selection and evolution toward living cells. *Abbreviations*: RH, rehydration; DH, dehydration.

*Source:* The authors

boundary signifying the start of life. These early living cells initiated the phase transition to more specialized forms within microbial communities, not only passing genes vertically to their progeny but also continuing to depend upon the *progenote prerogative* of the sharing of genetic material and other resources horizontally.

## 4.10 HOW DID POLYMERS EVOLVE FUNCTIONS REQUIRED FOR LIFE'S ORIGIN?

Long before LUCA (Last Universal Common Ancestor), cycling populations of encapsulated polymers would have been the basic units of selection. Over time, additional polymer functions would become incorporated and shared within the progenote. Polymers perform the central roles as the stepping-stones in prebiotic evolution because they are capable of catalysis and storing the first forms of genetic information. When cycled through selection and amplification processes, populations of polymers such as nucleic acids can evolve these functions. This was first demonstrated by Bartel and Szostak (1993) who prepared random sequences of RNA ~300 nucleotides in length and tested whether ligase activity could emerge when the system was exposed to cycles of selection and amplification. Selection was performed by putting the RNA through a column that had small strands of RNA attached to beads. If any of the 300mers had ligase activity, they would attach to the beads by ligation while inactive molecules were flushed through. The rare molecules that adhered to the beads were then released and amplified. After ten cycles, several distinct bands of RNA could be observed by gel electrophoresis, and ligase activity had increased by a factor of 7 million. This result clearly showed that genetic information and catalytic activity can be incorporated into random sequence RNA by cycles of selection and amplification. Later studies have gone even further. For instance, Horning and Joyce (2016) evolved a ribozyme with polymerase activity that could amplify short RNA templates over 10,000-fold. More recently, Kristoffersen et al. (2022) reported rolling circle RNA synthesis catalyzed by another evolved ribozyme.

We previously introduced the term "combinatorial selection" to describe prebiotic evolution before Darwinian natural selection (Damer and Deamer 2015). Computer models support a dynamical theory of the evolution of the prototypical genetic code as the first universal innovation-sharing protocol, which emerges through horizontal gene transfer and becomes refined within early communal life (Vetsigian et al. 2006). Once this universality is established, including the error-minimizing aspects of a genetic code and homochirality, horizontal, intermingled trait transfer gives way to discrete cellular inheritance supporting Darwinian natural selection. This is the threshold crossing into the epoch of tree-like descent of organismal lineages (Goldenfeld et al. 2017).

In the laboratory, polymers can be synthesized from their constituent monomers through wet–dry cycling, which become encapsulated between lamellar and vesicular lipid boundary layers in simulated hydrothermal conditions. Because such polymers are random sequences, this represents a *de novo* system in which functions are discovered in populations of random sequence polymers tested by selective stresses. An example of a simple selection process is that encapsulated polymers have a stabilizing influence on surrounding membranes. Vesicles containing polymers are more robust against degradative environmental stresses such as flocculation (Black et al. 2013). Polymer-hosting vesicles could also be more resistant to dissociation from shear forces, elevated temperatures, varying pH, or osmotic pressure related to concentration gradients across membranes. Membrane enclosures themselves have been shown to reduce hydrolysis rates of encapsulated polymers (Woods et al. 2023). Thus, stability is the first selective adaptation to be demonstrated in protocellular evolution.

Synthetic biology methods can also be employed in which specific polymer sequences are designed, synthesized, and introduced into protocell populations. Testing can then focus on predicted functions related to membrane stability, protocell growth, and protocell-to-protocell interactions. Several studies have demonstrated that this approach is effective (Chen et al. 2004; Adamala and Szostak 2013a, 2013b; Adamala et al. 2016); Mayer et al. 2018).

Recent investigations carried out by Gözen and coworkers (Köksal et al. 2019; Spustova et al. 2021; Gozen et al. 2022) reported novel properties of lipid aggregates. Such aggregates can form compartments of varying sizes and properties, including "tethers" which could operate as concentrating mechanisms and distribution webs. Such complex and dynamic morphologies will undoubtedly have an impact on polymer populations, reaction rates and intermixing, and resulting products of combinatorial selection, and this is a fruitful area for future investigation.

Another recent research direction is the study of membrane systems in which monomers are not simply free-floating in solution but are adsorbed on the membrane surface (Black et al. 2013). This co-localization effectively concentrates monomers while permitting the increased probability that they will react to form or extend polymers. In addition, polymers co-localized with membranes will become mobilized at different rates, promoting polymer-to-polymer encounters (Cornell et al. 2019; Xue et al. 2020). Such interactions can also occur between membrane-bound and free-floating interlamellar polymers. Figure 4.6 shows a cartoon of this scenario that illustrates how monomers and polymers, diffusing between fluid lipid lamellae, provide a specialized environment with the potential to promote combinatorial reactions that cannot occur in the bulk phase aqueous medium.

We emphasize here that all of the polymers begin as random sequences of their monomers, and functions emerge when certain sequences happen to have a function that allows a protocell to resist environmental stresses and to initiate metabolism, growth, and reproduction (Xavier and Kauffman 2022). The logic of this progression is simplistic with many gaps, but in order to find experimental approaches, we can build on the fact that biological membranes in cellular life today depend on three kinds of proteins that have specific functions. The first are the cytoskeletal proteins that stabilize fragile membranes composed of lipid bilayers. Without specifying the monomer composition of prebiotic polymers, it seems reasonable that some of them would have a similar stabilizing effect on self-assembled membranes of protocells. A second essential function is that there must be a way for solutes to be transported across membrane barriers. Living cells have protein pores that allow selective transport of ions and nutrients, and protocells would also require other polymers to serve this function. At some point, certain polymers would happen to have sequences that can function as catalysts. If the polymer is RNA, these would be ribozymes, but certain peptides can also have weak catalytic activity. With a primitive metabolism in

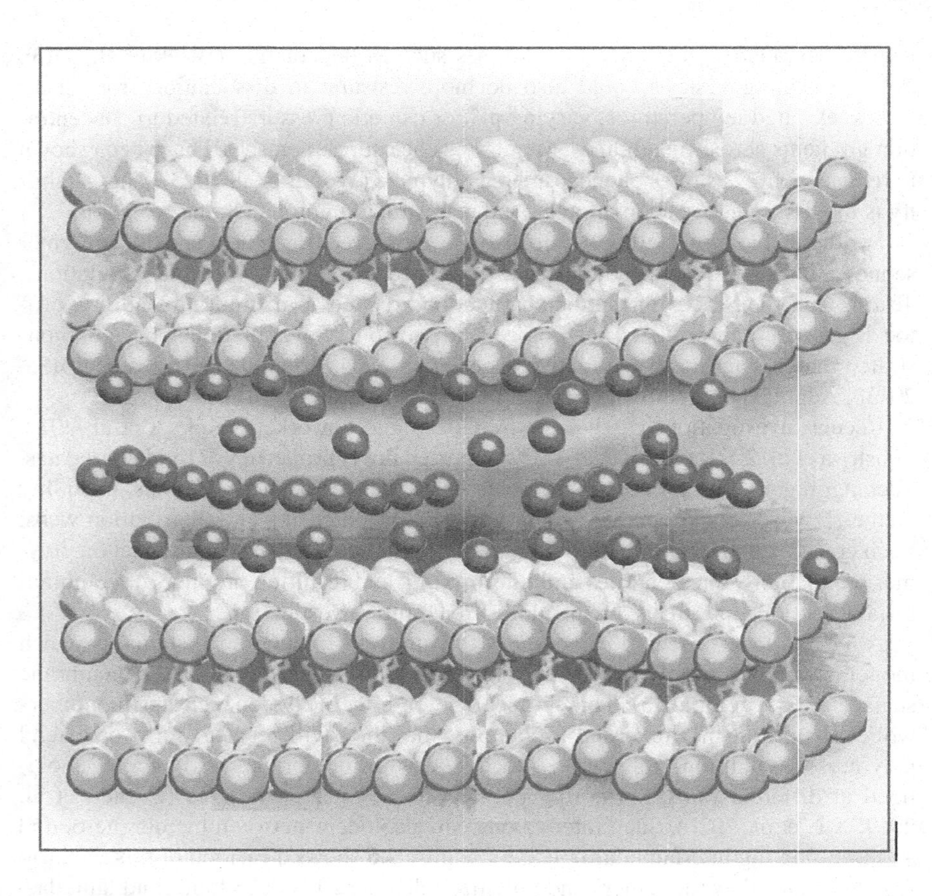

**FIGURE 4.6** A nano-scale conceptual view of a membranous lamellar "sandwich" showing two polymers and multiple monomers mixing within the volume between lipid lamellae. The increased concentration of the solutes in a relatively non-polar matrix promotes condensation reactions between monomers, which allow polymers to be synthesized.

*Source:* The authors

place, a supply of products could give rise to closed-loop cycles in which polymers catalyze their own replication. To support replication, a duplex polymer must emerge to encode and store informational sequences, guiding the future supply of specific functional polymers.

Each of these chemical circuits, especially autocatalytic sets (Hordijk et al. 2019), must come under the influence of rate-controlling feedback polymers. It is suggested that much of this chemical evolution would take place in the lamellar setting within the protected, concentrated progenote mixture. The reason for this is that free-floating protocell interiors and the surrounding bulk of a pool would be very dilute, and energy sources would also be severely limited. During the dilute stage of protocellular encapsulation, stabilization, pore formation, and some metabolic processes would gradually come into play. Only much later when protocells had acquired more identity in which they can not only sequester and replicate sets of polymers but also are able to independently capture sources of energy do they begin to live more independently from the progenitor medium. The transition to living cells occurs when protocells become capable of duplicating complete sets of interacting polymers and undergo division into viable daughter cells. Such a transition might not initially be possible for individual protocells in a high-stress dilute bulk solution, but the supportive aggregate of a progenitor medium could provide significant protection and resources to enable the first primitive form of cellular fission and the origin of species of primitive cellular life (Damer and Deamer 2020).

Conceiving even a vastly simplified description of essential polymer structures and functions required for life to emerge on the early Earth makes clear what a daunting task researchers face. It also illustrates why it might have required as much as a half a billion years from the time that liquid water became available to the emergence of the earliest microbial life we call LUCA. But understanding how life can begin is not an impossible task. It took chemists and biochemists half a century to discover the structure and function of DNA, and in the time since that discovery, scientists have managed to dissect the machinery of living cells and make each component function independently in the laboratory. Origins of life research is progressing in a similar way, with small advances in our knowledge of possible key chemical prebiotic stepping-stones. We are therefore optimistic that, at some point, the advances will be integrated into a more complete viable path to the first microbial communities, just as we now know precisely how nucleic acids serve as the core of all life today.

## REFERENCES

Adamala, K., Engelhart, A.E., Szostak, J.W. 2016. Collaboration between primitive cell membranes and soluble catalysts. *Nature Commun* 7:11041. https://doi.org/10.1038/ncomms11041

Adamala, K., Szostak, J.W. 2013a. Competition between model protocells driven by an encapsulated catalyst. *Nat Chem* 5:495–501.

Adamala, K., Szostak, J.W. 2013b. Nonenzymatic template-directed RNA synthesis inside model protocells. *Science* 342:1098–1100.

Anders, E. 1989. Pre-biotic organic matter from comets and asteroids. *Nature* 342:255–257.

Apel, C., Mautner, M., Deamer, D.W. 2002. Self-assembled vesicles of monocarboxylic acids and alcohols: Conditions for stability and for the encapsulation of biopolymers. *BBA Biomembranes* 1559:1–9.

Attwater, J., Raguram, A., Morgunov, A.S., Gianni, E. and Holliger, P. 2018. Ribozyme-catalysed RNA synthesis using triplet building blocks. *Elife* 7:e35255.

Barge, L.M., et al. 2015. From chemical gardens to chemobrionics. *Chem Rev* 115:8652–8703.

Baross, J.A., Hoffman, S.E. 1985. Submarine hydrothermal vents and associated gradient environments as sites for the origin and evolution of life. *Orig Life Evol Biospheres* 15:327–345.

Bartel, D.B., Szostak, J.W. 1993. Isolation of new ribozymes from a large pool of random sequences. *Science* 261:1411–1418.

Black, R.A., Blosser, M.C., Stottrup, B.L., Tavaklev, R., Deamer, D.W., Keller, S.L. 2013. Nucleobases bind to and stabilize aggregates of a prebiotic amphiphile, providing a viable mechanism for the emergence of protocells. *Proc Natl Acad Sci USA* 110:13272–13276.

Bronstein, J.L., ed. 2015. *Mutualism*. Oxford University Press.

Cartwright, J.H., Russell, M.J. 2019. The origin of life: The submarine alkaline vent theory at 30. *Interface Focus* 9:20190104.

Chen, I.A., Roberts, R.W., Szostak, J.W. 2004. The emergence of competition between model protocells. *Science* 305:1474–1476.

Chyba, C., Sagan, C. 1992. Endogenous production, exogenous delivery and impact-shock synthesis of organic molecules: An inventory for the origins of life. *Nature* 355:125–132.

Clark, B.C., Kolb, V.M., Steele, A., House, C.H., Lanza, N.L., Gasda, P.J., VanBommel, S., Martinez-Frias, J., et al. 2021. Origin of life on Mars: Suitability and opportunities. *Life* 11:539.

Corliss, J.B., et al. 1979. Submarine thermal springs on the Galápagos rift. *Science* 203:1073–1083.

Cornell, C.E., Black, R.A., Xue, M., Litz, H.E., Ramsay, A., Gordon, M., Mileant, A., Cohen, Z.R., Williams, J.A., Lee, K.K., Drobny, G.P., Keller, S.L. 2019. Prebiotic amino acids bind to and stabilize prebiotic fatty acid membranes. *Proc Natl Acad Sci USA* 35:17239–17244.

Damer, B.F. 2016. A field trip to the Archaean in search of Darwin's warm little pond. *Life* 6:21.

Damer, B.F. 2019. *The Hot Spring Hypothesis for the Origin of Life and the Extended Evolutionary Synthesis. Essay for Extended Evolutionary Synthesis Project.* Available online: http://extendedevolutionarysynthesis.com/the-hot-spring-hypothesis-for-the-origin-of-life-and-the-extended-evolutionary-synthesis/ (accessed 7 February 2023).

Damer, B.F., Deamer, D.W. 2015. Coupled phases and combinatorial selection in fluctuating hydrothermal pools: A scenario to guide experimental approaches to the origin of cellular life. *Life* 5:872–887.

Damer, B.F., Deamer, D.W. 2020. The hot spring hypothesis for an origin of life. *Astrobiol* 20:429–452.

Deamer, D.W., Cary, F., Damer, B. 2022. Urability: A property of planetary bodies that can support an origin of life. *Astrobiology* 22:889–900.

Deamer, D.W., Damer, B., Kompanichenko, V. 2019. Hydrothermal chemistry and the origin of cellular life. *Astrobiology* 19:1523–1537.

De Guzman, V., Shenasa, H., Vercoutere, W., Deamer, D. 2014. Generation of oligonucleotides under hydrothermal conditions by non-enzymatic polymerization. *J Mol Evol* 78:251–262.

Duval, S., Baymann, F., Schoepp-Cothenet, B., Trolard, F., Bourrié, G., Grauby, O., Branscomb, E., Russell, M.J., Nitschke, W. 2019. Fougerite: The not so simple progenitor of the first cells. *Interface Focus* 6:20190063.

Escabi-Perez, J.R., Romero, A., Lukac, S., Fendler, J.H. 1979. Aspects of artificial photosynthesis. Photoionization and electron transfer in dihexadecyl phosphate vesicles. *J Amer Chem Soc* 101:2231–2233.

Goldenfeld, N., Biancalani, T., Jafarpour, F. 2017. Universal biology and the statistical mechanics of early life. *Phil Trans R Soc A* 375:20160341.

Gözen, I., Köksal, E.S., Põldsalu, I., Xue, L., Spustova, K., Pedrueza-Villalmanzo, E., Jesorka, A. 2022. Protocells: Milestones and recent advances. *Small* 18:2106624.

Hassenkam, T., Deamer, D. 2022. Visualizing RNA polymers produced by hot wet-dry cycling. *Sci Rep* 12:10098.

Herschy, B., Whicher, A., Camprubi, E., Watson, C., Dartnell, L., Ward, J., Evans, J.R.G., Lane, N. 2014. An origin-of-life reactor to simulate alkaline hydrothermal vents. *J Mol Evol* 79:213–227.

Himbert, S., Chapman, M., Deamer, D.W., Rheinstadter, M.C. 2016. Organization of nucleotides in different environments and the formation of pre-polymers. *Sci Rep* 6. https://doi.org/10.1038/srep31285

Hordijk, W., Steel, M., Kauffman, S.A. 2019. Molecular diversity required for the formation of autocatalytic sets. *Life* 9:23.

Horning, D.P., Joyce, G.F. 2016. Amplification of RNA by an RNA polymerase ribozyme. *Proc Natl Acad Sci USA* 113:9786–9791.

Imai, M., Sakuma, Y., Minoru Kurisu, M., Walde, P. 2022. From vesicles toward protocells and minimal cells. *Soft Matter* 18:4823–4849.

Jackson, J.B. 2016. Natural pH gradients in hydrothermal alkali vents were unlikely to have played a role in the origin of life. *J Mol Evol* 83:1–11.

Jordan, S.F., Nee, E., Lane, N. 2019. Isoprenoids enhance the stability of fatty acid membranes at the emergence of life potentially leading to an early lipid divide. *Interface Focus* 9:20190067.

Kelly, D.S., et al. 2005. A serpentinite-hosted ecosystem: The Lost City hydrothermal field. *Science* 307:1428–1434.

Köksal, E.S., Liese, S., Kantarci, I., Olsson, R., Carlson, A., Gozen, I. 2019. Nanotube-mediated path to protocell formation. *ACS Nano* 13:6867–6868.

Koonin, E.V. 2014. Carl Woese's vision of cellular evolution and the domains of life. *RNA Biol* 11(3).

Kristoffersen, E.L., Burman, M., Noy, A., Holliger, P. 2022. Rolling circle RNA synthesis catalyzed by RNA. *Elife* 11:e75186.

Kvenvolden, K., Lawless, J., Pering, K., Peterson, E., Flores, J., Ponnamperuma, C., Moore, C. 1970. Evidence for extraterrestrial amino-acids and hydrocarbons in the Murchison meteorite. *Nature* 228:923–926.

Lane, N. 2017. Proton gradients at the origin of life. *BioEssays* 39(6). https://doi.org/10.1002/bies.201600217

Martin, W., Russell, M.J. 2003. On the origins of cells: A hypothesis for the evolutionary transitions from abiotic geochemistry to chemoautotrophic prokaryotes, and from prokaryotes to nucleated cells. *Phil Trans R Soc Lond B* 35859–35885.

Maurer, S.E., Nguyen, G. 2016. Prebiotic vesicle formation and the necessity of salts. *Orig Life Evol Biosph* 46:215–222.

Mayer, C., Schreiber, U., Davila, M.J., Schmitz, O.J., Bronja, A., Meyer, M., Klein, J., Meckelmann, S.W. 2018. Molecular evolution in a peptide-vesicle system. *Life, 8*:16.

McCollom, T., Ritter, G., Simoneit, B.R.T. 1999. Lipid synthesis under hydrothermal conditions by Fischer–Tropsch-Type reactions. *Orig Life Evol Biosph* 29:153–166.

Merriam-Webster. n.d. Progenitor. In *Merriam-Webster.com Dictionary*. Retrieved September 9, 2023, Available online: https://www.merriam-webster.com/dictionary/progenitor

Miller, S.L. 1953. A production of amino acids under possible primitive Earth conditions. *Science* 117:528–529.

Monnard, P.-A., Apel, C.L., Kanavarioti, A., Deamer, D.W. 2002. Influence of ionic inorganic solutes on self-assembly and polymerization processes related to early forms of life: Implications for a prebiotic aqueous medium. *Astrobiology* 2:139–152.

Nooner, D.W., Oro, J. 1979. Synthesis of fatty acids by a closed system Fischer-Tropsch process. *Adv Chem* 178:159–171.

Odling-Smee, J., et al. 2024. *How Life Contributes to its Own Evolution*? MIT Press. In press.

Odling-Smee, J., Laland, K., Feldman, M. 2013. *Niche Construction: The Neglected Process in Evolution*. Princeton University Press.

Omran, A., Pasek, M., Menor-Salvan, C., Springsteen, G. 2020. The messy alkaline formose reaction and its link to metabolism. *Life* 10:120.

Oró, J. 1961a. Comets and the formation of biochemical compounds on the primitive Earth. *Nature* 190:389–390.

Oró, J. 1961b. Mechanism of synthesis of adenine from hydrogen cyanide under possible primitive Earth conditions. *Nature* 191:1193–1194.

Osinski, G.R., Cockell, C.S., Pontefract, A., Sapers, H.M. 2020. The role of meteorite impacts in the origin of life. *Astrobiology* 20:1121–1149.

Pearce, B.K.D., Molaverdikhani, K., Pudritz, R.E., Henning, T., Cerrillo, K.E. 2017. Origin of the RNA world: The fate of nucleobases in warm little ponds. *Proc Natl Acad Sci USA* 114:11327–11332.

Pearce, B.K.D., Molaverdikhani, K., Pudritz, R.E., Henning, T., Cerrillo, K.E. 2022. Toward RNA life on early Earth: From atmospheric HCN to biomolecule production in warm little ponds. *Astrophys J* 932:9.

Pizzarello, S., Cooper, G.W., Flynn, G.J. 2006. The nature and distribution of the organic material in carbonaceous chondrites and interplanetary dust particles. In *Meteorites and the Early Solar System II*, 625–651. The University of Arizona Press.

Pizzarello, S., Shock, E. 2010. The organic composition of carbonaceous meteorites: The evolutionary story ahead of biochemistry. *Cold Spring Harb Perspect Biol* 2:a002105.

Rajamani, S., Vlassov, A., Benner, S., Coombs, A., Olasagasti, F., Deamer, D. 2008. Lipid-assisted synthesis of RNA-like polymers from mononucleotides. *Orig Life Evol Biosph* 38:57–74.

Ross, D., Deamer, D.W. 2016. Dry/wet cycling and the thermodynamics and kinetics of prebiotic polymer synthesis. *Life* 6:28.

Russell, M.J. 2018. Green rust: The simple organizing 'seed' of all life? *Life* 8:35.

Russell, M.J. 2021. The 'water problem' (sic), the illusory pond and life's submarine emergence – a review. *Life* 11:429.

Shew, R., Deamer, D.W. 1985. A novel method for encapsulation of macromolecules in liposomes. *Biochim Biophys Acta* 816:1–8.

Sojo, V., Herschy, B., Whicher, A., Camprubí, E., Lane, N. 2016. The origin of life in alkaline hydrothermal vents. *Astrobiology* 16:181–197.

Spustova, K., Köksal, E.S., Ainla, A., Gözen, I. 2021. Subcompartmentalization and pseudo-division of model protocells. *Small* 17:2005320.

Toppozini, L., Dies, H., Deamer, D.W., Rheinstädter, M.C. 2013. Adenosine monophosphate forms ordered arrays in multilamellar lipid matrices: Insights into assembly of nucleic acid for primitive life. *PLOS ONE* 8:e62810.

Van Kranendonk, M.J., Baumgartner, R., Djokic, T., Ota, T., Steller, L., Garbe, U., Nakamura, E. 2021. Elements for the origin of life on land: A deep-time perspective from the Pilbara Craton of Western Australia. *Astrobiology* 21:39–59.

Vetsigian, K., Woese, C., Goldenfeld, N. 2006. Collective evolution and the genetic code. *PNAS* 103(28):10696–10701.

Woese, C.R. 1998. The universal ancestor. *Proc Natl Acad Sci USA* 12:6854–6859.

Woese, C.R. 2002. On the evolution of cells. *Proc Natl Acad Sci USA* 99:8742–8747.

Woese, C.R., Fox, G.E. 1977. The concept of cellular evolution. *J Mol Evol* 10:1–6.

Woods, B., Thompson, K.C., Szita, N., Chen, S., Milanesi, L., Tomas, S. 2023. Confinement effect on hydrolysis in small lipid vesicles. *Chem Sci* 14:2616–2623.

Xavier, J.C., Kauffman, S. 2022. Small-molecule autocatalytic networks are universal metabolic fossils. *Phil Trans R Soc A* 380:20210244.

Xue, M., Black, R.A., Cornell, C.E., Drobny, G.P., Keller, S.L. 2020. A step toward molecular evolution of RNA: Ribose binds to prebiotic fatty acid membranes, and nucleosides bind better than individual bases do. *ChemBioChem* 21:2764–2767.

# 5 Locating the Cradle of Life
## *The Hydrothermal Impact Crater Lakes on Young Earth and Mars*

*Sankar Chatterjee*

## 5.1 INTRODUCTION

### 5.2 INTERPLANETARY NURSERIES OF LIFE'S BIOMOLECULES

The CHONPS (carbon, hydrogen, oxygen, nitrogen, phosphorous, and sulfur) elements and biomolecules of life were synthesized in interplanetary space during the cataclysmic explosion of a nearby supernova, many times more massive than our Sun. The explosion of the dying star dispersed the stardust material into the solar nebula that collapsed and coalesced to form our solar system. After the Sun formed, the gas giants Jupiter and Saturn followed, and the rocky planets attained most of their present size. Today's asteroids and comets are planetesimals that escaped participation in planet formation (Boss and Keiser 2010). Meteorites preserve a record of the elements, isotopes, and organic compounds derived from the stardust of the Solar System's formation. The building blocks of life started in the interplanetary icy grains composed of gas and dust and distributed in comets and asteroids (Delsemme 1988). Asteroid fragments, such as carbonaceous chondrites that fall to Earth from time to time along with continuous rain of interstellar dust, offer a glimpse of the interplanetary nurseries of the building blocks of life. Complex biomolecules such as lipid membranes, amino acids, nucleobases, phosphorous, and ribose sugars have been detected in carbonaceous meteorites (Bernstein et al. 1999). Moreover, these compounds are abundant in interplanetary space. Complex carbon-rich organic molecules, precursors to life, have been found everywhere in space – in comets, carbonaceous asteroids, and interstellar dust. The Stardust mission detected these building blocks in interstellar dust particles, which provide a fundamental constraint for testing the foregoing theory of life's cosmic seeding by the supernova explosion (Brownlee et al. 2003).

Carbonaceous chondrites are so primitive in the asteroid belt that they contain traces of interstellar dust, including organic compounds that survived the thermal processing of solar nebulae. The presolar minerals and organic compounds were probably formed during the supernova explosion. Of all chondrites, carbonaceous chondrites formed farthest from the Sun in the asteroid belt by the coalescence of

DOI: 10.1201/9781003294276-5

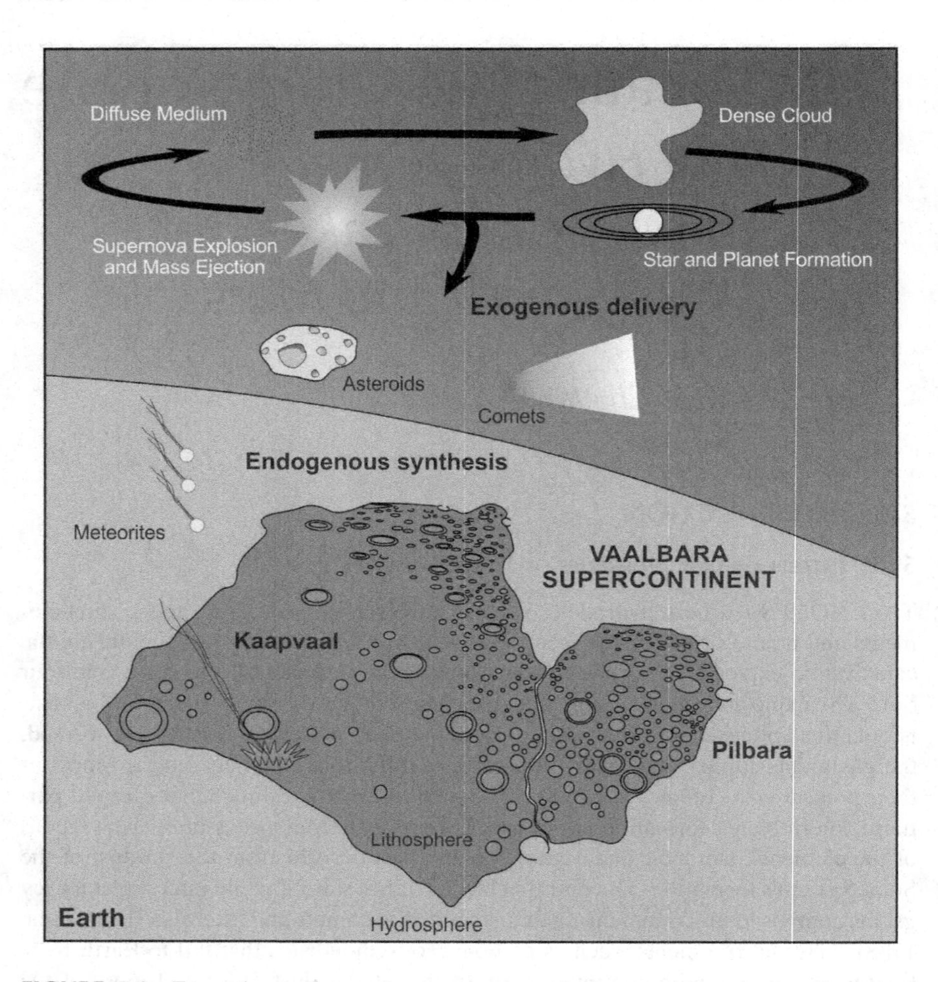


**FIGURE 5.1** From the outflows of a supernova explosion, reactions, and radiations of the interstellar medium, through nebular and Solar System chemistry, cosmic building blocks were delivered to young Earth by meteorite impacts during the tail end of the Heavy Bombardment period (exogenous delivery). However, the abiogenesis occurred in the hydrothermal crater lakes on the ancient Archean supercontinent, such as Vaalbara (endogenous production), leading to the first cells.

stardust material with the highest proportion of volatile compounds and water and played critical roles in abiogenesis. The most celebrated carbonaceous chondrite is the Murchison meteorite that has yielded essential components of living cells, including pyrimidine and purine nucleobases; sugars and phosphorous for making nucleotides; 70 amino acids; and membrane-forming compounds such as lipids, long-chain monocarboxylic acids, and carbohydrates (Deamer 2011, 2019).

Chyba and Sagan proposed that the prebiotic building blocks of life on Earth came from interstellar space delivered exogenously by meteorites. But life itself was an endogenous production synthesized in the womb of our planet (Chyba and Sagan 1992). The icy cold vacuum of interplanetary space is not conducive to abiogenesis. On the other hand, because of intense bombardment on young Earth, the original organic compounds were depleted and supplemented by cosmic ingredients to promote life synthesis. Life also requires water, an essential solvent, which was delivered by carbonaceous asteroids during the late heavy bombardment period (Alexander et al. 2012). Water can remain liquid on our Goldilocks planet but not in space. Thus, the origin of life is an extraordinary event of two worlds: the biomolecules of life and ice were delivered from interplanetary space by meteorites and interplanetary dust particles, but life emerged and diversified exclusively on Earth, making it the only life-sustaining planet on the solar system (Figure 5.1). Here, we try to reconstruct the plausible site where this momentous genesis began in young Earth.

## 5.3 HADEAN EARTH

The intense impact environment of the Hadean (~4.5–4 Ga) Earth, an eon of extreme geological violence, largely determined the initial physical and chemical states of the planet (Ryder 2003). For much of the Hadean, cosmic impacts pummeled Earth and its sister worlds in the inner Solar System. Impact evidence from lunar samples, meteorites, and the pockmarked surfaces of the inner planets paints a picture of a violent environment during the Hadean eon. Asteroids were battering the young Earth and the other inner planets, a ceaseless violent cosmic warfare that continued for the first 600 million years. The impacting objects were mostly asteroids in solar orbit between Mars and Jupiter. Comets from the distant Kuiper Belt appear to have been a small fraction of the cataclysmic projectiles (Kring and Cohen 2002; Marchi et al. 2014). Impact evidence from lunar samples, meteorites, and the pockmarked surfaces of the inner planets paints a picture of a violent environment during the Hadean eon. The heat generated by ceaseless impacts caused large-scale melting of the Earth, forming an early magma ocean. Swirling currents churned the interior of the Earth, and the materials in the hot magma were eventually sorted into three layers: The core, mantle, and crust (Marchi et al. 2014).

The Hadean impacting record is reconstructed from clues from the Moon, Mercury, Mars, meteorites, and computer simulations. It has been suggested that the lack of Hadean rocks on Earth is attributable to their destruction by meteorite bombardment. Impactors that bombarded Earth and Moon are thought to have been dominated by asteroids. During the Hadean, the basaltic crust began to form from the magma ocean and was then covered by the global sea (Mojzsis et al. 2001). But it was not a tranquil sea. The impact continued throughout Hadean and early Eoarchean times, between 4.4 and 3.8 billion years ago (Figure 5.2). Each successive, large-impact event (caused by asteroids larger than 100 km in diameter) would have partially or even wholly vaporized the oceans, sterilizing Earth's surface in the process (Arndt and Nisbet 2012). Most likely, the Haden Ocean was boiling by recurrent impacts, and the vapor condensed, creating torrential rain. Abramov and Mojzsis (2009), on the other hand, contradicted this view and suggested that the habitable zone was not fully sterilized and championed microbial habitat during the Hadean Earth. A molecular clock model suggests that LUCA may have existed before the late heavy bombardment (>4.1 Ga), during the most violent period in our planet's history (see Table 5.1), soon after the moon-forming impact (Betts et al. 2018).

In contrast to violent Hadean Earth, a habitable Hadean world with liquid water is emerging from recent geological evidence. Studies of detrital zircons from Australia (Harrison et al. 2008) and India (Miller et al. 2018) suggest that Earth began a pattern of crustal formation, erosion, and sediment recycling as early as 4.35 billion years ago during the Hadean. The survival of these detrital zircons over time indicates that continental granitic crust exited in Australia and India by 4.4 Ga and possibly as early as 4.5 Ga. Overall, the late Hadean Earth most probably consisted of a thick basaltic crust covered by a global ocean punctuated by a few island continents, mostly of granitic rocks – the source of the detrital zircons. The granitic outcrop in which the detrital zircons originated has been destroyed but left behind eroded sedimentary fingerprints; the oldest rock formations exposed on the surface of the Earth are exclusively Archean. The heavy bombardment by asteroids lasted about eight million years and finally began to diminish around 3.8 Ga.

Like Shiva's cosmic dance, asteroids can be both creative and destructive. So far, we have painted violent portraits of asteroids, not only during the Hadean but also in the beginning of Eoarchean. But asteroids had creative force too. Asteroid impacts, especially the carbonaceous chondrites, might have brought Earth's life-giving liquid water and the building blocks of life and created a network of hydrothermal crater lakes for prebiotic synthesis. Between destruction and renewal of Shiva's cosmic dance in the early history of our planet, the first life appeared during the beginning of the Eoarchean period or even earlier.

If liquid water and elevated granitic crust existed as early in Earth's history as the late Hadean eon, was life already present then? It is impossible to confirm whether life arrived before the Archean period because Hadean rocks are missing. The oldest known fossils come from rocks nearly four billion years old. There is no question that Earth was a very different place during the Eoarchean. There were few emergent drylands, except for several island continents or protocontinents, highly battered by meteorite impacts. These ancient protocontinents today survived as the cratons, the stable interior portion of a continent characteristically composed of ancient crystalline basement rocks of Eoarchean age. Nevertheless, we think that life began around this time, corroborated by biosignature or microfossil evidence in these cratons (Table 5.1).

## 5.4 AN URABLE EOARCHEAN WORLD

The concept of habitability is widely used to describe zones in the solar system in which planets with liquid water can sustain life. Because habitability does not explicitly incorporate the origin of life, Deamer et al. (2022) coined a new term, urabilty, which refers to the condition that allows life to begin. To understand the geological site of life's beginnings, we must reconstruct the Eoarchean Earth, when a permanent crust capable of sustaining first life began to form. We can start by building a picture of the Archean Earth and its geological, atmospheric, and biological conditions. Gases released by impacts and eruptions formed a thick atmosphere consisting mainly of carbon dioxide, with small amounts of nitrogen, methane, water vapor,

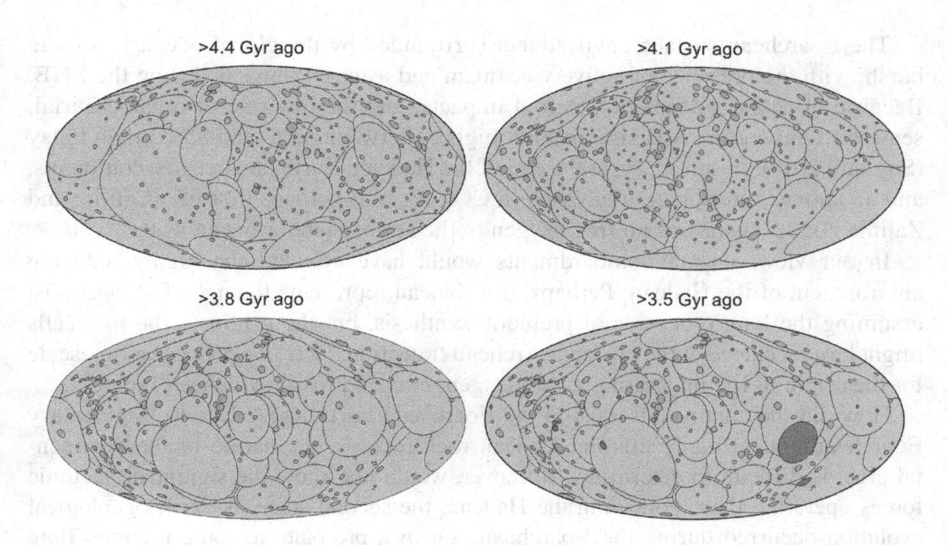


**FIGURE 5.2** Asteroid bombardment of the early Earth – Mollweide projections of the cumulative record of record at four different times during the Hadean and Eoarchean times. Each circle indicates the final crater size estimated from the transient cavity size by meteorite impacts. The maps do not show ejecta blankets and melt extrusion on the surface, which can greatly expand the effects of cratering. The surface of the Hadean and Eoarchean Earth would look like the surface of the Moon. Asteroid impacts were numerous during the first 600 million years of Earth's evolution. During this period, some 200 objects larger than 100 km in diameter collided with the Earth, and more than 10,000 asteroids larger than 10 km (approximate size of the Chicxulub bolide) smashed the Earth's fragile surface. The gray shade scale indicates the timing of the impacts. The smallest impactors represented had diameters of 15 km. The densely cratered surface of the Earth during the Eoarchean eon mimicked the existing surface of the Moon. The diameters of the circles represent the final sizes of the craters for objects smaller than 100 km in diameter. In contrast, the sizes of the regions buried by impact-generated melt are representative of the larger objects. Eventually, the Eoarchean craters on protocontinents would be filled in by rainwater to form hydrothermal crater lakes, the likely cradle for life synthesis (Marchi et al. 2014).

*Photo courtesy:* Simone Marchi

and sulfur gases – still, no oxygen to speak of (Kasting 2014). The greenhouse gas concentrations such as carbon dioxide and methane were sufficient to offset a fainter Sun by trapping more of its heat in the atmosphere. With a 100 times more methane than there is in the atmosphere today, the Archean sky would have had a faintly pinkish tinge.

The first life forms, reasonably considered hyperthermophiles, organisms that can tolerate temperatures up to 120°C, were suited to survive the Late Heavy Bombardment (LHB) by colonizing protected hydrothermal vent habitats (Stetter 2006). And as we gain more knowledge of what it takes to make a planet urable, we can also apply our understanding of the conditions on Earth when life first arose to our hunt for life in our solar system and beyond.

The Eoarchean surface environment surrounded by the global ocean was quite harsh, with the presence of active volcanism and asteroid impacts during the LHB. Because of active volcanism, asteroid impacts, and the low rate of organic burial, seawater temperature was significantly higher (between 55°C and 85°C) than today (Sugitani 2018). In summary, at the end of the Hadean, Earth had oceans, continents, and an anoxic atmosphere likely rich in $CO_2$, $N_2$, and perhaps first life (Catling and Zahnle 2020). There was no free oxygen in the atmosphere or ocean water.

In our view, impact bombardments would have affected the highly reducing environment of the Hadean. Perhaps, the Hadean represents the age of abiogenesis, assuming the long processes of prebiotic synthesis, but the origin of the first cells might have occurred at the Hadean–Archean transition. Here, we propose a timescale for the origin of life integrating geologic, genomic, and fossil evidence (Table 5.1).

How did the continental crust in the Eoarchean Earth arise? Due to fragmentary Eoarchean and virtually missing Hadean rock records, the nature of the continental crust is difficult to determine. Moreover, we do not know the significant tectonic forces operating then. Following the Hadean, the second stage in Earth's geological evolution occurred during the Eoarchean eon in a pre-plate tectonic regime (Tang et al. 2016). Compared to the present day, elevated mantle temperatures weakened the lithosphere through the emplacement of melts that inhibited subduction and plate

**TABLE 5.1**
**The Integrated Time Scale for the Origin of Life during the Hadean–Archean Eons Shows Major Geological, Paleontological, and Genomic Events**

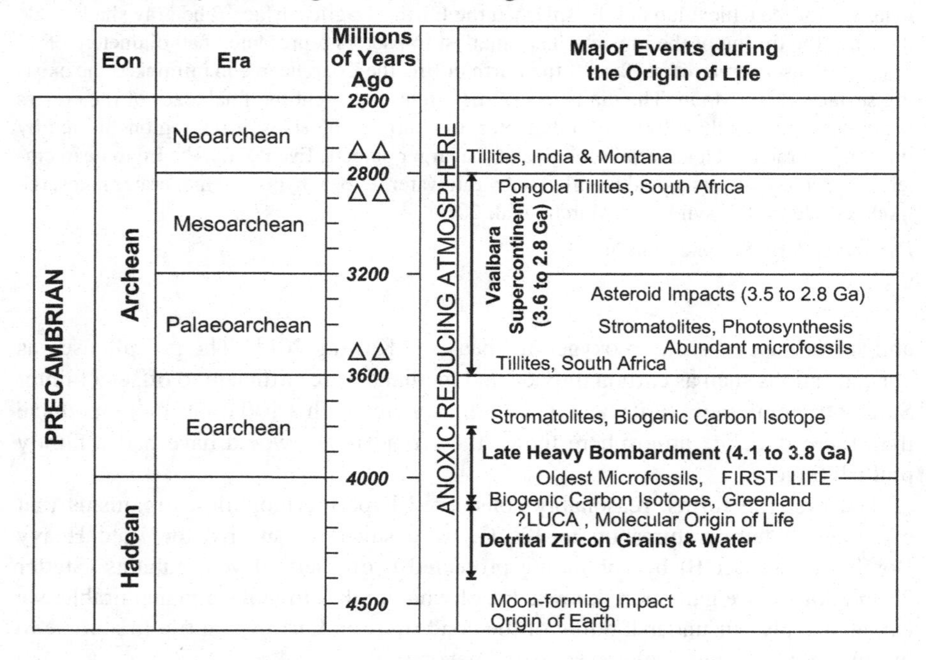

tectonics. Impact-induced mantle upwellings that percolated the lithosphere drove vigorous magmatism. The crust was at least 15–20 km thick (Hawkesworth et al. 2016). A high geothermal gradient in the early Archean led to crustal buoyancy, increased crustal ductility, and reduced crustal strength. Even while a relatively thin, ductile, and buoyant Archean lithosphere would have prevented subduction, a vertical mantle force unrelated to plate tectonics may have cycled material from the mantle to the surface.

### 5.4.1 Impact Origin of Eoarchean Cratons

The origin of the Eoarchean continental crust is one of the key objectives to understanding the quest for the cradle of life. A likely mechanism of crustal modification during the LHB was impact cratering, a major geodynamic force during the Archean times for creating tiny clusters of granitic protocontinents by mantle upwelling (Hansen 2015; Grieve et al. 2006). These protocontinents were small, exposed, protruding islands surrounded by vast oceans with basaltic floor. Impact cratering is now widely recognized as a ubiquitous geological process impacting all planetary bodies with solid surfaces (Kring 2000). The cratered surface of the Moon validates large-scale impacts in its crustal evolution.

Up to 90% of the preserved Eoarchean cratonic crust comprises tonalite–trondhjemite–granodiorite (TTG) series with enclaves of greenstone packages (basalt, komatiite, and hydrothermal sedimentary strata). The hydrothermal sedimentary strata of the greenstone belts have yielded the oldest microfossils, chemofossils, and stromatolites – the paleontological timeline for the emergence of first life. Eoarchean TTGs occur in several gneissic complexes worldwide and are mostly generated by partially melting hydrated rocks at different depths, variably thickened Hadean mafic crusts (O'Neil and Carlson 2017). TTG is the significant component of Earth's oldest remnant of continental crust and thereby holds the key to the origin of continental crust and the start of plate tectonics. These scarce rocks have been identified in only ten or so cratons around the world. The Greenstone sediments overlying the TTG series in Greenland, Canada, South Africa, Australia, and India hold the precious archives of the oldest fossils on Earth.

The Archean Kaapvaal Craton of South Africa and the Pilbara Craton of Australia are the only two sizeable areas in the world where granite–greenstone terrains ranging in age between 3.6 and 2.7 Ga have been preserved in a relatively pristine state, with rich records of the earliest microfossils. Moreover, these two cratons were once joined together into an ancient supercontinent. Vaalbara is the name given to the oldest Archean supercontinent, which consisted of the Pilbara and the Kaapvaal Cratons (Zegers et al. 1998). Remarkably similar lithostratigraphic and chronostratigraphic structural sequences in these two cratons have been noted for the period 3.5–2.7 Ga. Paleomagnetic data from two ultramafic complexes in the cratons showed that, at 3.87 Ga, these two cratons were part of the same supercontinent (Zegers and Ocampo 2003). Both cratons preserve four large meteorite impact events between 3.5 and 3.2 Ga, visible as layers formed when the high temperatures created by impact forces fused the target rocks into small glassy spherules (Beyerly et al. 2002). Occurring in both the Kaapvaal Craton and Pilbara Craton, these spherule layers are the oldest known terrestrial impact products. The spherules resemble the glassy chondrules in the carbonaceous chondrites found in carbon-rich organic compounds of meteorites. The simultaneous occurrence of impact layers in different regions of the Vaalbara supercontinent suggests large and recurrent impact-cratering events during the origin and early evolution of life. Just as they did on the Moon's surface, these impact events must have left numerous scars on the Eoarchean crust of the Vaalbara supercontinent, although evidence for this early cataclysm has long since been obliterated. During the Eoarchean and Paleoarchean periods, the surface of the Vaalbara supercontinent might have looked like the Moon's surface, with thousands of craters ranging in diameter from 1 km to several hundred kilometers (Marchi et al. 2014).

Other ancient shields (older than 3.6 Ga) include Isua/Akilia Craton of Greenland, Nuvvuagittuq Craton of Canada, and Singhbhum Craton of India, which have yielded the oldest biosignatures (Figure 5.4). These cratered surfaces of the Archean crust hold the key to the likely site of life's beginnings.

Chatterjee (2016) suggested that the structure and paleoenvironments of these Archean greenstone belts in Greenland, Canada, Australia, South Africa, and India suggest that these basins might be the relics of the primordial impact craters, initially filled with impact-triggered melts of the ultramafic to mafic composition and later deformed and metamorphosed. The Algoma-type Banded Iron Formation and volcanic chert suggest a hydrothermal vent environment (Furnes et al. 2004).

The architecture of the Greenstone basin mimics the morphology of a large complex crater. In cross-section, the central part of a Greenstone basin shows an anticlinal structure, which is flanked on either side by two synclinoriums, dominated by ultramafic, mafic, and andesitic rocks (Condie 1981). The Greenstone basin's bilateral symmetry resembles a complex crater's architecture. Perhaps the immense crater's central peak was shattered to initiate the hydrothermal systems. The annular basins, on either side of the central peak, contain typical Algoma-type Banded Iron Formations and the oldest microfossil-bearing chert, typical of hydrothermal vent environments. The basement of the crater contains the typical TTG complex. The impact scenario explains why the Eoarchean cratonic crust consists of TTG series with enclaves of greenstone packages (basalt, komatiite, and hydrothermal sedimentary strata). The Eoarchean crust was most likely pock-marked with innumerable craters like the surface of the Moon, which were largely erased by erosion and plate tectonics but retained the roots of these large craters in the TTG-greenstone belts in the oldest cratons. The Eoarchean impacts, instead of sterilizing the planet (Nisbet and Sleep 2001), made it urable by creating innumerable hydrothermal crater lakes for abiogenesis and sanctuary for early life during the tail end of LHB.

## 5.5 HYDROTHERMAL SYSTEMS AND THE PREBIOTIC ENVIRONMENT

Hydrothermal systems have persisted throughout geological history, and ancient hydrothermal deposits could provide clues to understanding Earth's earliest biosphere. Modern hydrothermal systems offer a dynamic, far-from-equilibrium reaction zone and a wealth of catalytic minerals allowing complex chemical reactions from simple organic compounds. They support a wide range of microorganisms and provide valuable comparisons for the paleontological interpretation of ancient hydrothermal systems. Continental and submarine hydrothermal systems are an integral part of Earth's thermal regime. Hydrothermal environments are dominated by the circulation of geothermally heated water and occur on both the ocean floor and the continents in areas with high heat fluxes. When water descending into the mantle meets underlying magma chambers, its temperature rises well above the boiling point, but it remains in a liquid state due to the high pressure from the overlying water. Superheated water rises to the top, whereas colder, heavier water sinks around it. This creates convection currents that allow the more buoyant hot water to rise to the surface, following the cracks and weak areas through the magma source. Hot water rich in dissolved minerals from underlying magma can penetrate cracks, fissures, and basin floors. We call this downward and upward circulation the natural 'plumbing' system of hydrothermal features.

Life started in hydrothermal environment. The habitats of hyperthermophiles (heat-loving microbes), which are the most primitive living organisms, support this view (Stetter 2006), whether at the bottom of the oceans or landlocked settings. Hydrothermal systems can develop anywhere on the crust where water coexists with a magmatic heat source. Such hydrothermal environments are likely to have encouraged chemical evolution and the subsequent origin of life. Among various hydrothermal settings, three locations stand out as possible incubators of life: (a) Submarine hydrothermal vents, such as Black Smokers and Lost City; (b) terrestrial hot springs

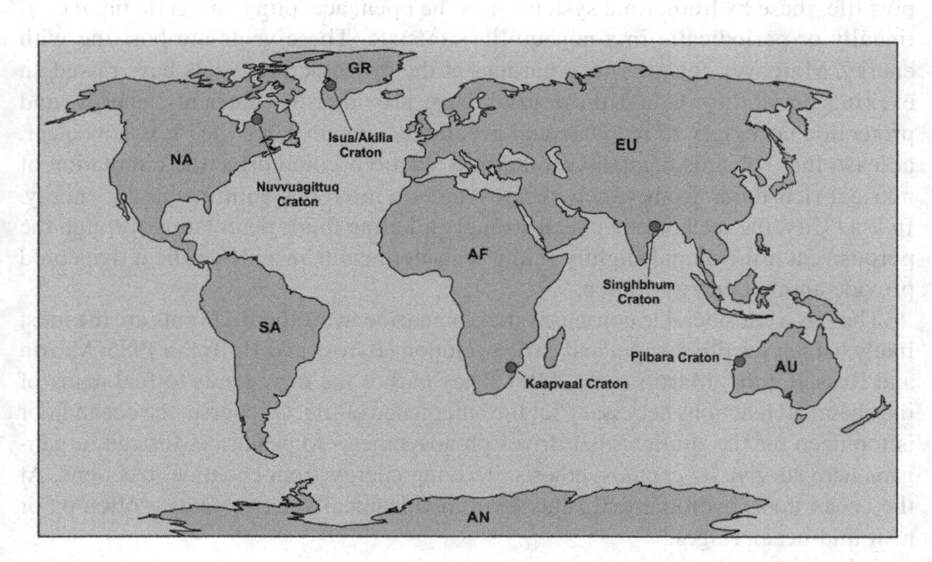


**FIGURE 5.3** The five locations in Archean greenstone belts where signs of the earliest life have been found: (1) Nuvvuagittuq Craton in Canada; (2) Isua/Akilia Craton in Greenland; (3) Kaapvaal Craton in South Africa; (4) Pilbara Craton in Australia; and (5) Singhbhum Craton in India. These greenstone belts are very likely relics of ancient impact scars on the protocontinents. *Abbreviations*: AF, Africa; AN, Antarctica; AU, Australia; EU, Europe; GR, Greenland; NA, North America; SA, South America.

and geothermal fields; and (c) terrestrial hydrothermal impact crater lakes. There are many controversies about the most likely site for life's beginnings, and each prospective environment has its pros and cons. Here, we evaluate these three possible hydrothermal sites, one at a time.

## 5.6 SUBMARINE HYDROTHERMAL VENTS

More than 99% of the liquid water on Earth is in the oceans, so intuitively it was assumed that life began in the salty sea. In 1977, scientists discovered unusual biological communities in the East Pacific mid-oceanic ridge, living around seafloor hydrothermal vents, far from sunlight, and thriving on a chemical soup rich in hydrogen, carbon dioxide, and sulfur, spewing geothermally heated water up to 400°C. Named Black Smokers, the vents emit black smoke when the hot water rich in chemicals comes in contact with the cold ocean water (Baross and Hoffman 1985). This was followed in 2000 by the discovery of a new type of alkaline submarine hydrothermal vent found a little off axis from the mid-ocean ridges. Since their discovery and their associated hyperthermophilic life forms, deep-sea hydrothermal vents such as the acidic, high-temperature Black Smokers and the alkaline, low-temperature (45–90°C) Lost City (Martin et al. 2008), associated with mid-ocean ridges, have been hailed as the likely setting for the origin of life. An abundance of living microbial and animal communities prospered in conditions without sunlight once considered too extreme to harbor much of any life. To support life, these hydrothermal systems must be open, accepting energetic input continually or periodically in a nonequilibrium state. These vents are bursting with energy. Moreover, the microbial gardens at the hydrothermal vents have raised an exciting possibility: Could these vents have provided the nutrients, energy, and protection necessary to incubate and eventually sustain early life? It is now clear, at least, that volcanic heat and exothermic reactions could drive the circulation of nutrient-rich fluids from which chemosynthetic microbes gain metabolic energy. In Lost City, the interface between strongly alkaline fluids percolating through the porous vent mineral and slightly acidic seawater results in a pH gradient that could provide another energy source.

There is a considerable consensus that submarine hydrothermal vents are the most likely habitat for life's origin and early evolution (Baross and Hoffman 1985; Martin and Russell 2007; Martin et al. 2008). They harbor the ingredients to fuel many of the chemical reactions necessary for the emergence of life. Of course, the ocean floor is too deep for the sunlight that drives photosynthesis to penetrate. Instead, organisms here survive by chemosynthesis, deriving energy from chemical reactions. At the ocean floor, hydrothermal vents form in volcanically active areas – often on or near mid-ocean ridges.

Submarine hydrothermal vents are generally considered as the most promising sites for life's beginnings. Black Smokers release acidic water rich in carbon dioxide; geothermally heated water up to 400°C; and packed with sulfur, iron, copper, and other elements that are essential to life. The Black Smoker vent supplies acidic fluids (pH 2–3) and dissolved reduced gases ($H_2S$, $H_2$, $CH_4$, $CO_2$, and $NH_3$) that facilitate the chemosynthetic reactions of microbial communities to use geothermal energy.

Currently, Lost City is favored over the Black Smoker environments as the likely cradle for prebiotic synthesis because it is highly alkaline (pH 9–11) and has a relatively cooler temperature (45–90°C). The Lost City vents are formed by a process known as serpentinization. When warm alkaline fluids are produced by serpentinization of seabed mafic rocks, such as olivine, which are mixed with seawater, they create white calcium carbonate chimneys, 30–60 m tall, like a white castle. The seawater expelled here is highly alkaline and lacks carbon dioxide but is rich in methane and offers more hospitable temperatures. Life came from harnessing the energy gradients when alkaline vent water mixes with more acidic seawater. This mechanism mirrors the way that cells harness energy by proton-motive force. Cells maintain a proton gradient by pumping protons across a membrane to create a charge differential from inside to outside (Martin and Russell 2007).

### 5.6.1 Criticisms against Submarine Hydrothermal Vents

Many researchers are not convinced that life started in the sea – many say the chemistry won't work and are looking for a landlocked birthplace. Support for the submarine vent hypothesis is waning in recent years (Deamer 2019; Damer and Deamer 2015; Hazen 2005). Some of the arguments against submarine hydrothermal vents are listed here.

First, Mulkidjanian et al. (2012) have discovered that the chemistry of modern cells provides essential clues to the original environment in which life evolved. They suggest that organisms have retained their chemical traits throughout the eons. According to the chemistry conservation principle, cells will include and evolve mechanisms to protect their fundamental biochemical architecture once established in any environment. It turns out that all cells contain a lot of phosphates, potassium, and other metals – but hardly any sodium. In contrast, seawater is rich in sodium but deficient in potassium and phosphates. Cells that contain ten times more potassium than sodium have their origin in seawater, which has 40 times more sodium than potassium. The composition of the living cell does not match that of the ocean water then or now. For example, the molecular backbone of RNA/DNA is made of phosphate; many ancient proteins require zinc, and the cytoplasm of cells needs potassium ions to link amino acids to create proteins. Seawater is deficient in these critical elements of life. The inorganic chemistry of cytoplasm rather mirrors freshwater pond and lake environments. These authors conclude that first life began on land, not in the sea. They argue that land-based geothermally active pools are the likely sites where magnesium, potassium, zinc, and phosphate are available to match the ionic content of cells. Freshwater hydrothermal vents are conducive to the lower salt and ion concentrations that allow fatty-acid membranes to form. In my view, the accumulation of organic compounds in terrestrial hydrothermal vents is more likely than in the ocean. Geothermally active areas provide numerous advantages, as expressed by the authors. Cellular life may have begun in anoxic freshwater hot springs rather than in seawater, and key chemical pathways to ribonucleotides could be facilitated by ultraviolet radiation in such a subaerial landscape.

Second, alternating dry and wet conditions are necessary for the significant polymerization reactions that must occur during the emergence of life. One of the biggest arguments against submarine hydrothermal vents is that so many macromolecules – DNA, RNA, proteins, and lipids – are all polymers; they form by the wet–dry cycles of condensation reactions to remove water between chemical groups of monomers and form links such as peptide bonds of proteins and ester bonds of nucleic acids (Damer and Deamer 2015; Deamer 2019; Hazen 2005). Such condensation reactions could not occur in deep-sea hydrothermal vents. The continuous water supply puts thermodynamic limits on the condensation reactions required for the polymerization of monomers. However, wet and dry cycling occurs daily on continental hydrothermal fields. This allows for the concentration of reactants and polymerization to form by condensation reactions. The major cell components, DNA, RNA, proteins, and lipids, are all polymers and all form by condensation reactions in a fluctuating environment, which is sometimes wet and sometimes dry on continental hydrothermal fields. In a deep-sea hydrothermal environment, wet–dry cycles are absent so that condensation reaction cannot occur. Moreover, forming bilayer membranes or vesicles was essential in abiogenesis, which can form in freshwater but not in salt water; when sodium chloride or ions of magnesium or calcium were added, the membranes fell apart, making it unlikely that viable protocells could have arisen there (Damer and Deamer 2015; Deamer 2019; Deamer et al. 2022).

In my previous paper, I discussed the shortcomings of both kinds of submarine hydrothermal vent theories (Chatterjee 2019). Several problems with either of the deep-sea hydrothermal vent hypotheses may be insurmountable. Black Smoker systems on the ridge axis are not only transient (on the order of decades) but also have violent flow rates and extreme temperatures. These are not ideal for abiogenesis. Submarine hydrothermal vent theories – Black Smokers and Lost City – suffer from the 'dilution problem' with organic compounds in the open sea, making it impossible to concentrate either ions or organic molecular components. Rather than being concentrated, the cosmic ingredients would be dispersed and diluted in the vastness of the Eoarchean global ocean before they can assemble into the complex molecules of life. Concentration and the 'crowding' of organic molecules are essential aspects of prebiotic chemistry. It is challenging to imagine how a sufficient concentration of reactants could occur in the open ocean. One crucial requisite for the origin of life is that comparatively simple biomolecules must have opportunities to form more complex molecules by the segregation and concentration of chemical compounds. However, it seems unlikely that cosmic and terrestrial chemicals could have been concentrated, mixed, selected, or organized in this way in the vast ocean. Second, in a one-plate, pre-tectonic Eoarchean Earth without spreading ridges, it is challenging to explain how submarine hydrothermal vents could have formed in the first place. Today's hydrothermal vents occur along or near the spreading ridges of oceanic plates (Figure 5.4), but plate tectonics did not begin before 3 Ga (Stern 2005; Tang et al. 2016). Life began around 4 Ga; then, we will have to seek alternative hypotheses if we wish to retain hydrothermal vent systems as likely crucibles of life's origin. We must look for these on land, not under the ocean floor.

The Lost City hypothesis faces several challenges, too. First, Lost City fluids do not contain an abundant supply of catalytic metals (iron, copper, manganese, zinc, nickel, etc.), which are needed to polymerizebiomolecules. Second, the oldest ecosystems on Earth, as preserved in Eoarchean greenstone belts in Canada, Greenland, Australia, and South Africa, do not show the evidence of the Lost City environments such as dominant carbonate deposits. Third, the first cells probably developed in zinc-rich environments; the cytoplasm in a cell is rich in potassium, zinc, manganese, and phosphate ions, which are not widespread in the Lost City environments (Mulkidjanian et al. 2012). Fourth, nucleic acids, such as RNAs, are unstable in highly alkaline pH, raising the possibility that RNA must have arisen in a neutral or mildly acidic environment. This observation weakens the hypothesis that RNA evolved in the vicinity of alkaline (pH 9–11) hydrothermal vents of the Lost City (Bernhardt and Tate 2012).

I argue that life on Earth began on land, as Darwin mused, not on the primordial ocean floor (Chatterjee 2019). Impact cratering was the primal geological force in the early history of our planet during the Late Heavy Bombardment period (4.1–3.8 million years ago) before the onset of plate tectonics. Asteroid impacts left young Earth's crust pockmarked with craters, like the surface of the Moon. These hydrothermal crater systems might have provided the habitats for the origin of life and were filled with water and the cosmic building block for life delivered by these meteorites. I suggest that if there was any life on Mars, the impact craters might provide the most likely setting for detecting early microbial life.

## 5.7 TERRESTRIAL HYDROTHERMAL FIELDS

The cherished assumption that life emerged in the oceans has been doubted recently. New research on geological, chemical, and experimental evidence increasingly supports the hypothesis that life originated on land rather than in a marine environment,

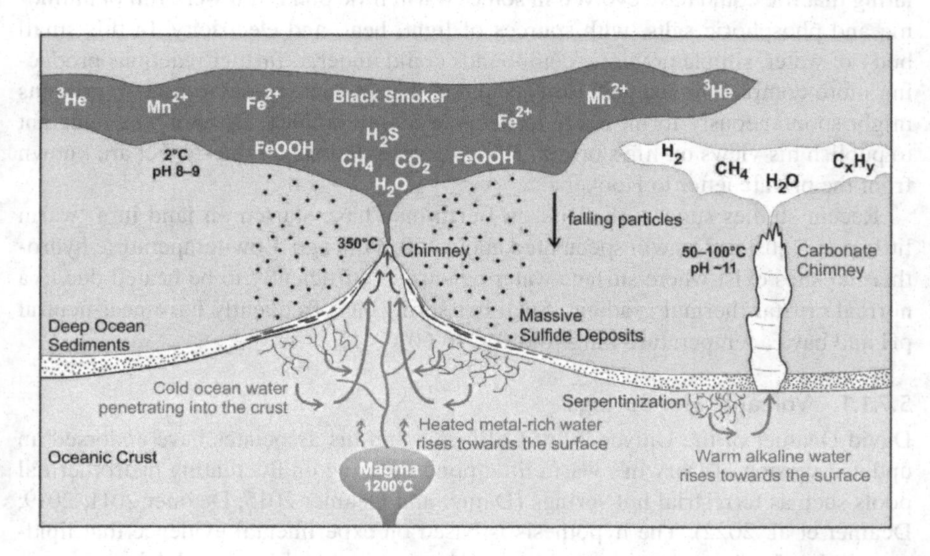


**FIGURE 5.4** The submarine hydrothermal vent environment shows discharges of the most abundant carbon compounds, toxic gases, metal, and energy sources. There are two kinds of submarine hydrothermal vents: Left side of the figure, the hot (~350°C) Black Smoker, the chemistry of which is driven by the magma chamber that resides below the ocean-floor spreading axis; right side of the figure, the cooler (~50–90°C) Lost City type located far from the ridge axis, where serpentinization drives the chemistry.

resembling a suggestion by Darwin that life could have sprung from 'warm little ponds' rich in nutrients (Chatterjee 2019; Deamer 2020; Mulkidjanian et al. 2012). This paradigm-shifting hypothesis that life began not in the sea but on land is corroborated by the earliest fossil record, suggesting that the first life forms inhabited terrestrial hydrothermal environments (Dodd et al. 2017; Furnes et al. 2004). Hydrothermal basins are surprisingly common and widely distributed across the terrestrial surfaces of planet Earth. The most relevant and numerous recent and active terrestrial hydrothermal environments are usually associated with volcanic islands, which allow magma to penetrate upward to a shallow depth where it can interact directly with groundwater. While a magma chamber's depth may be as shallow as 2–5 km below ground level, the thermal gradients may be as high as 150–200°C/km. Here, the groundwater may be superheated and forced back to the surface at high temperatures and under considerable pressure. Most of these hydrothermal systems are populated by diverse hyperthermophilic microbial life. Terrestrial hydrothermal environments are widely considered suitable analogs to conditions that may have given rise to early life on our planet (Stetter 2006). Terrestrial hydrothermal candidates for life's origin include (1) hot springs and tidal pools, like the central idea of Darwin's 'warm little pond', and (2) impact crater lakes. Here, we evaluate both terrestrial environments as possible locations for life synthesis.

### 5.7.1 Revisiting Darwin's Warm Little Pond

In February 1871, Charles Darwin wrote to his botanist friend Joseph Hooker speculating that life could have evolved in some 'warm little pond' if it were full of ammonia and phosphoric salts, with sources of light, heat, and electricity. In this small body of water, simple prebiotic compounds could undergo further reactions producing more complex molecules. He speculated that in such an environment, proteins might spontaneously form, ready to become more complex. Darwin was reluctant to publish his views on life's origin. His only speculations on the subject are known from the private letter to Hooker.

Recent studies suggest that life on Earth may have started on land in a 'warm little pond', just as Darwin speculated about 150 years ago. Low-temperature hydrothermal sites exist where surface water penetrates sufficiently to be heated due to a normal crustal thermal gradient. Such hot springs are frequently have near-neutral pH and have a temperature range of 40°C to 60°C.

#### 5.7.1.1 Volcanic Hot Springs

David Deamer of the University of California and his associates have endorsed an updated version of Darwin's warm little pond focusing on fluctuating hydrothermal pools such as terrestrial hot springs (Damer and Deamer 2015; Deamer 2011, 2019; Deamer et al. 2022). The hypothesis is based on experimental evidence that lipid-encapsulated polymers can be synthesized by cycles of hydration and dehydration in fluctuating volcanic hot springs pools on land to form protocells. The setting for this proposed model of a hydrothermal site is a volcanic island emerging from a global ocean four billion years ago. Storms dumped deluges of fresh water, which found its way underground and met a hot magma chamber, creating hydrothermal fields like

Yellowstone. Thousands of pools filled and chemically interacted with this enriched water. Thermal spring waters have diverse chemical compositions, and their outflows create thermal gradients and chemical precipitates that sustain diverse microbial communities and entomb their remnants. They provide abundant and diverse sources of energy and exploit thermodynamic disequilibrium in the environment. These pools were subjected to wet and dry cycling during seasonal changes. In many thermal hot springs, precipitation from rain and snow percolates through porous rock and, once deep underground, is turned into steam by magma. Although most subterranean steam condenses below ground, some reaches the surface through conduits to form fumaroles, hydrothermal fields, and hot springs. Primordial hot springs would have been able to concentrate organic molecules from meteoritic sources and interplanetary dust particles to initiate prebiotic synthesis. Because these hydrothermal ponds tap fresh water from rain and snow, their ionic concentrations are much lower and more favorable for prebiotic synthesis. The chemical composition of volcanic pools also resembles the cytoplasm composition more closely than the submarine vent environment. In addition, small ponds undergoing evaporation and replenishment by variable hot springs and precipitation enjoy fluctuating cycles of hydration and dehydration.

Subaerial hot springs offer several advantages over submarine hydrothermal vents as sites for abiogenesis. Three key features of these environments are conducive to prebiotic synthesis. First, hydrothermal fields provide sources of heat and chemical energy for prebiotic synthesis. Second, in small thermal pools, building blocks of life delivered by meteorites could be concentrated. Third, hydration–dehydration cycles drive condensation reactions producing long-chain organic polymers by a nonenzymatic reaction. These pools dry up periodically. Deamer and colleagues imagine volcanic islands with freshwater pools sprayed by hot springs or geysers in between drying out. These pools would have been much more apt locations than deep-sea vents to form the first fatty membranes or amphiphiles. In the hydration phase of a cycle, vesicles are produced when the water interacts with the dry multilamellar matrix, while during the dehydration phase, solutes and amphiphiles on mineral surfaces form highly concentrated films of multilamellar structures that capture concentrated monomers between layers. The amphiphiles and minerals provide environments for concentrating solutes and driving condensation reactions that link monomers to polymers. Fourth, the self-assembly of membranous compartments would encapsulate the polymers into populations of protocells leading to stepwise increments of chemical evolution toward the emergence of first cells. Moreover, the 'warm little pond' theory has supporting evidence from several experiments, rendering competing for submarine-vent hypotheses more problematic (Damer and Deamer 2015; Deamer 2019; Mulkidjanian et al. 2012).

Several modern analogs of hot springs in volcanic islands at the time of life's origin have been proposed by the Deamer team, including Bumpass Hell hydrothermal fields in Lassen Volcanic National Park in California, Mount Mutnovsky in the Kamchatka peninsula of Russia, as well as sites in Iceland and Hawaii. However, there is an issue with this volcanic hydrothermal field scenario from a geological perspective. The modern analogs provided by these authors were created by plate tectonic activity – the Bumpass Hell formed by the subduction of the Pacific (Juan de Fuca) plate under the North American plate along the Cascadia Trench and the Kamchatka Peninsula formed along the subduction of the Kuril–Kamchatka Trench at a triple plate collision between North American, the Pacific, and the Eurasian plates. Moreover, Iceland sits right on the Mid-Atlantic Ridge under the North Atlantic Ocean, where the Eurasian and North American plates slide apart. The volcanoes caused by subduction or spreading ridges are powered by the same magmatic heat that creates many modern hot springs. Since plate tectonics did not start until after the Eoarchean period, when life began (Tang et al. 2016; Stern 2005), these contemporary examples are irrelevant for abiogenesis.

Similarly, Hawaii's volcanic islands result from hotspots beneath the North Pacific Ocean. Without plate motion, volcanic islands would be solitary instead of a chain. Isolated volcanic islands would have eventually eroded, submerging their 'warm little ponds', releasing their diverse biomolecules and dispersing them into the ocean, thus halting further abiogenesis.

Other researchers have also elaborated the volcanic island model, suggesting as a modern analog for prebiotic synthesis a site such as Lake Waiau, about 3,970 m above sea level near the summit of Mauna Kea volcano on the island of Hawaii (Bada and Korenga 2018). However, this lake cannot retain water and sometimes completely dries up in the summer and freezes in the winter. The lake's seasonal temperature is too low (~0°C to 13°C) to support hyperthermophilic life, presumably the first life forms. From a geological perspective, then, the volcanic scenario is problematic. Most of the islands during the Eoarchean were continental, not volcanic.

Continental hot spring settings such as Yellowstone were formed by the interaction of a hotspot and a south-westerly drifting North American plate; it is another example of terrestrial hot springs. However, the duration of the hydrothermal activity in Yellowstone National Park ranges from a few hundred to a few thousand years, which is, apparently, too little for abiogenesis to take place (Lewis and Young 1982). Moreover, without plate tectonics, the plume would be short-lived.

Geothermal features are also observed in areas of active volcanism, other than plate tectonics, where subsurface magma heats groundwater, creating hot springs. Hot springs can also form along the active fault zones, where the groundwater is heated by circulation through faults to rock deep in the crust. Hot spring as a likely cradle of life is attractive, but it remains unclear how metabolism could have started in a 'warm little pond'.

#### 5.7.1.2 Comet Pond

Another variant of the warm little pond scenario is the comet pond model, which envisions the landing of a relatively pristine portion of a comet containing building blocks of life on a planetary surface during the late heavy bombardment by soft impact, creating a shallow pond as an incubator. In this pond of highly concentrated organic-rich matter, prebiotic synthesis would start up when ultraviolet radiation polymerized the cometary monomers, eventually leading to the origin of life on Earth (Clark and Kolb 2018). Although the comet pond theory is entirely novel and plausible, the authors acknowledge that, because of the rapid destruction and dispersal from entry through the atmosphere or a hypervelocity impact during landing, the delivery of pristine cometary material would be an improbable event. Moreover, it has been argued that the LHB was composed exclusively of asteroids rather than comets (Kring and Cohen 2002).

#### 5.7.1.3 Asteroid Pond

Pearce et al. (2017) suggested a hybrid model of Darwin's warm little pond when impacting meteorites splashed into warm little ponds and infused with building blocks of life, including nucleobases. The wet-and-dry cycles of the pond would promote condensation and rapid polymerization of nucleobases into RNA leading to the RNA world (before 4.17 Ga). These authors provided a quantitative estimate that RNA-like polymers could be synthesized in just a few wet-and-dry cycles rather than many million years. However, meteorites during the late heavy bombardments splashing into warm little ponds would likely vaporize the water instantly, halting any biogenesis. Moreover, the meteorites that crashed on Earth created large craters, several hundreds of kilometers in diameter, along with small craters very similar to the cratered surfaces of the Moon and Mercury (see Figure 5.2). The authors completely ignored the presence of large craters, which would be ubiquitous on terrestrial surface. This asteroid pond scenario is a modified version of the impact crater lake hypothesis with a pond-sized crucible allowing wet-and-dry cycles. Other than wet-and-dry cycles, mineral surfaces on the crater floor are equally effective for polymerization (Hazen 2005; Kloprogge and Hartman 2021). Statistically speaking, during the Heavy Bombardment in the Hadean crust, impact craters were more common than warm little ponds.

### 5.7.2 Impact-Generated Hydrothermal Crater Lakes

#### 5.7.2.1 Impact-cratering Process

An impact crater is a depression in the surface of a planet or a moon with a hard-crustal surface when a hypervelocity meteorite strikes the ground, releasing huge quantities of energy. It is inextricably linked with the propagation of shockwaves. All the bodies in our solar system have been heavily bombarded by meteorites throughout their early history. The solar system was once a much more violent place. The surfaces of the Moon, Mars, and Mercury have beautifully preserved cratering records of the late heavy bombardment because the surfaces of these relatively small planetary bodies have remained unchanged for billions of years without any plate tectonic activity. The Moon's surface is riddled with craters ranging in size and complexity; four billion ago, before life emerged, the Earth looked the same way. But Earth has several attributes the Moon lacks – an atmosphere, liquid water, and plate tectonics that had eroded and recycled much of the planet's early crust over millions of years and erased all the blemishes left by the early cosmic cataclysm. On Mars, impact craters are a key focus in the hunt for the planet's warmer, wetter past during the Noachian eon.

Analyses of images, data, and samples from the Moon and Mars provide critical information for understanding the environment in which early life evolved on Earth and potentially on Mars. Many significant implications for prebiotic history on Earth and Mars arise from the reasonably well-understood post-origin history of the Moon. The Moon and, in supporting roles, Mars and Mercury provide a record of the impact history of the inner solar system and, therefore, a record of that missing history on Earth during the late heavy bombardment (Schmitt 2015). Impact cratering is a fundamental geological process during the Late Haden–Eoarchean eon. The impact record on Earth remains invaluable for an understanding of impact processes, for it is the only direct source of data on the three-dimensional structural and lithological characters. When an asteroid strikes a planet's surface, it propels debris at 50 times the speed of a sound, vaporizing tons of target rock and carving a circular crater many kilometers across. Impact craters are the most recognizable geological landforms on almost all terrestrial planets and icy moons in our solar system. The impact of a kilometer-size asteroid on Earth releases more power than the explosion of the world's nuclear arsenal (Grieve 1990).

Impact cratering is the only geological process that produces shock metamorphic effects from the high shock pressures to the target rock and ejecta. Impact events generate shock pressures (several hundred gPa) and temperatures (up to approximately 15,000 K) that can heat or melt substantial volumes of the target material. Impacts produce distinctive shock-metamorphic effects, such as shatter cones, melted rocks, microscopic crystal deformations (such as shocked quartz and feldspar), and the formation of high-pressure minerals (such as diamond) that allow impact sites to be distinctly identified (Melosh 1989).

An energy-transfer mechanism mediates crater formation. The kinetic energy (proportional to the impactor's mass and the square of its velocity) is transferred into heat, fracturing, displacing, and excavating the target rocks, triggering powerful shock waves. Impactor velocity ranges from 10 to 70 km/s, with an average of 20 km/s on Earth. The meteorite is usually vaporized completely by the released energy. The crater's diameter relative to that of the meteorite depends on several factors, including the nature of the target rock. A high-speed impact process produces a crater that is approximately 12 times larger in diameter than the impacting object. For example, a 5-km wide asteroid could create a crater of 60 km in diameter. The excavated material, the ejecta blanket, surrounds the crater, causing its rim to be elevated above the surrounding terrain.

Impact craters leave very characteristic features. They are divided into two groups based on their morphology: Simple and complex craters (Grieve 1990). Simple craters are relatively small, no more than 3 km across their uplifted rim, surrounding a bowl-shaped depression partially filled by breccia. Their maximum depth is located at the center. A typical complex crater shows three distinct features: A central peak surrounded by an annual depression and an elevated crater rim. Complex craters and basins are generally  4 km in diameter; they have a central uplift peak surrounded by an annular trough and a slumped outer rim, with relatively low depth/diameter ratios.

Over 181 impact craters have been identified on our planet in a wide diversity of target rocks (Grieve et al. 2006). Impacts are instantaneous events. Even then, the formation of hypervelocity impact craters has been divided somewhat arbitrarily into three main stages (1) contact and compression, (2) excavation, and (3) modification (Gault et al. 1968). Because of the Late Heavy Bombardment, impact cratering was the most common landform on the Eoarchean crust. Impact craters are also highly interested in planetary exploration because they are considered possible sites for life, providing new habitats for microbial communities.

During the Hadean–Archean transition, the Earth's surface was heavily bombarded by asteroid impacts (Marchi et al. 2014; Kring 2000; Cockell 2006; Cockell and Lee, 2002; Osinski et al. 2013; Chatterjee 2019) (see Figure 5.2). If the pockmarked surface of the Moon and Mercury is any indication, early Earth was pelted by asteroid swarms for about 100 million years. However, because our planet's surface is constantly eroded by the hydrosphere and renewed by plate tectonics, the physical evidence of that early bombardment has been wholly erased. But since it roughly coincides with the time first life appeared on our planet, this early period of heightened asteroid activity would have had significant implications for life on Earth. A barrage of asteroids punched through the Earth's crust, creating thousands of impact craters and delivering enormous thermal pulses to local environments riddled with volcanically driven geothermal vents.

#### 5.7.2.2 Post-Impact Crater Lakes with Hydrothermal Systems

On Moon and Mercury, impact craters do not convert to hydrothermal crater lakes because of a lack of subsurface groundwater and precipitation. However, on Earth and Mars, groundwater and rain facilitated the formation of hydrothermal crater lakes from impact craters. The interaction of terrestrial melt rock with groundwater generated a hydrothermal system. Rainwater eventually filled the crater, creating a lake, and the heated floor of the crater sustained the hydrothermal system for an extended period. A unique combination of circumstances on early Earth enabled the entry of pristine building blocks of life aboard incoming asteroids and their landing onto an Eoarchean crust to create myriads of hydrothermal crater lakes. The most significant advantage of the cratering hypothesis for the origin of life is that it invokes a high-probability scenario. The resulting crater exhibits long-term hydrothermal activity when the impactor's kinetic energy is transferred onto a water-bearing crustal surface. When filled with water, these crater basins developed hydrothermal systems that could have helped sustain first life by providing sources of heat, energy, water, and nutrients. Out of 181 craters identified on Earth, at least 70 show evidence of impact-generated hydrothermal activity, from the ~1.8-km-diameter Lonar Crater in India to the 250-km-diameter Sudbury structure in Canada.

Akin to hot springs and geysers, hydrothermal vents have formed in impact-induced crater lakes. Impacts that excavated huge craters also shattered their central peaks to create volcanically driven geothermal vents in concert with groundwater. As rains filled the crater basins, underwater hydrothermal vents developed, and crater lakes formed. The reduced gases from hydrothermal vents mixed with cosmic biomolecules created ideal prebiotic mediums for biogenesis.

Terrestrial hydrothermal impact crater lakes have been proposed as the most likely sites for biogenesis (Chatterjee 2016, 2019; Cockell 2006; Osinski 2013) and, by analogy, on Mars (Farmer 2000). These crater lakes could have also provided a refuge for the earliest hyperthermophilic life during late, giant-impact events. Craters were presumably plentiful on the Eoarchean crust; this period also overlaps with the evidence of the earliest life on Earth. Moreover, modern subaerial hydrothermal lakes are widely colonized by hyperthermophilic, thermophilic, and mesophilic bacteria and archaea.

Due to the shift of kinetic energy from the impactor to the target, every impact onto water-bearing crustal surfaces generates long-term hydrothermal activity in the resulting craters. Impact events trigger shock pressures and temperatures that can melt substantial volumes of the target material. The three primary potential sources of heat for creating impact-generated hydrothermal systems are (Osinski et al. 2013):

1. Impact melt rocks and impact melt-bearing breccias.
2. High geothermal gradients occur in central uplifts.
3. Energy deposited in central uplifts due to the passage of the shock wave.

Complex craters with central peaks sustain hydrothermal systems longer than simple craters (Figure 5.5). In these larger craters, the thermal activity causes the convection of ground and meteoric water. There is an extensive zone of fractured rocks at the floor, favorable to the circulation of the prebiotic soup. Moreover, the shock deformation of mineral surfaces stimulates reactions between minerals and active fluids (Cockell 2006; Osinski et al. 2013). As we have noted, impact-generated silicate debris in the crater basins would produce clay minerals, and hydrothermal fluid would precipitate pyrites on the crater floor; these minerals could catalyze and polymerize the organic synthesis of prebiotic biopolymers such as nucleic acids and proteins (Chatterjee 2016; Hazen 2005; Kloprogge and Hartman 2021; Wachterhäuser 1993).

In a terrestrial setting, the interaction of melt rock with the groundwater generates a hydrothermal system (Figure 5.5). Eventually, the groundwater and rainwater would fill the crater to form a lake. The crater's floor around the central peak would have sustained the hydrothermal system for a long time. The crater rim rises considerably above the lake level and creates an ideal sequestered basin for the concentration of biomolecules to occur. Impacts on a water-rich planet like Earth or Mars can generate hydrothermal activity – that is, underwater areas boiling with heat (Farmer 2000). The high geothermal gradient of the impact-fractured central peak of a complex crater could spew chemicals like the activity at submarine hydrothermal vents. It might be an essential source of heat for creating a hydrothermal system. Small complex craters (~5 km diameter), such as the Gardnos Crater of Norway and the Gow crater of Canada, would be ideal for prolonged biosynthesis. Evaporative heating and drying of organic compounds near the crater's surface have been proposed as the mechanisms for concentrating prebiotic molecules (Cockell 2006; Damer and Deamer 2015; Deamer 2011, 2019).

There is no constraint on the length of time needed for the origin of life. The duration of impact-generated hydrothermal systems is poorly known. In general, the larger the crater, the longer the hydrothermal activity lasts. For example, the hydrothermal system in the 4-km-diameter Kärdla Crater in Estonia lasted for several thousand years. In contrast, the hydrothermal system in the 24-km-diameter Haughton Crater in Canada was maintained for more than 10,000 years. The large Sudbury crater (~250 km) retained hydrothermal activity for ~2 myr (Abramov and Kring 2004; Osinski et al. 2013). However, during the Eoarchean time, because the crust was relatively thin, the heat flow through it was greater than on the present Earth, so these crater lakes possibly retained hydrothermal activity for more extended periods. Moreover, given the greater abundance of heat-producing isotopes in the early part of Earth's history, the continental crust at that time produced more heat than it does today. Gradually cooling them in complex craters would have been advantageous for biogenesis, creating both simple and complex organic compounds at different thermal gradients.

The crater rim rose considerably above the lake surface and formed an ideally isolated basin for the concentration of biomolecules to occur. Impacts on a water-rich planet, such as the Earth or Mars at that time, could have generated hydrothermal activity where underwater areas boil with heat. The central peak of a complex crater (>3 km) might have been fractured by such an impact, providing a high geothermal gradient and spewing forth chemicals, as occurs along the axes of submarine hydrothermal vents. The central peak may have been an essential source of heat in the creation of a hydrothermal system. We think of cooling impact craters as natural laboratories where convection currents circulate hot hydrothermal fluids. Impact-induced hydrothermal systems are very diverse, with a wide range of pH and temperature gradients. There is a temperature gradient inside the crater basin, where the surface water is cold. Still, the region adjacent to the central peak at the floor is hot, creating a steady convection current that helps mix the cosmic ingredients. Depending on the impactor's size, they would cool down at different rates, generating hydrothermal systems of various sizes and temperatures and creating a variety of incubators with complex compounds undergoing different chemical reactions. The larger the crater, the longer its hydrothermal activity can persist. During the lifetime of a crater, the chemical reactions that may have facilitated prebiotic synthesis can vary. During an early hot stage, simple cosmic molecules could be concentrated; in later stages, lower temperatures would be conducive to more complex molecules in different regions of the impact structure.

While the impact-generated silicate debris in the crater basins also produced clay minerals, the hydrothermal fluids precipitated pyrites, components likely to have played critical catalytic roles in the organic synthesis of prebiotic and early biotic compounds and structures (Hazen 2005). Carbonaceous chondrites during the LHB impact event delivered not only water and building blocks of life but also a large amount of clay minerals (Kloprogge and Hartman 2021). The interaction of this impact-melted material with near-surface groundwater can generate and sustain hydrothermal systems. During the collision of an asteroid, an enormous thermal pulse is delivered into the local target rock, which can generate circulating hydrothermal systems.

Thus, the underwater havens of the impact crater lakes, especially the impact-shocked, porous rocks at the basin floor, could have provided a crucial sanctuary for life's origin, protecting the first microorganisms not just from the Sun's harsh ultraviolet rays when the planet still lacked the ozone shield, but also from the continued violence of the Heavy Bombardment. We think these crater lakes were the perfect locations to concentrate and cook these cosmic chemicals to create the first simple microbes (Chatterjee 2016, 2019; Cockell and Lee 2002; Osinski et al. 2013).

#### 5.7.2.3 The Paleoenvironment of Hydrothermal Crater Basins

Impact-induced hydrothermal systems are very diverse, with the widest range of pH and temperature gradient possible (Cockell 2006; Kring 2000; Nisbet and Sleep 2001; Osinski et al. 2013). Hydrothermal crater lakes hosted many microenvironments, offering many possible niches for prebiotic synthesis. The environmental complexity in a single setting was a prerequisite for the origin of life from cosmic biomolecules. A hydrothermal crater-lake creates different habitats, such as the central uplift, impact ejecta deposits, the annular basin, the crater rim region, post-impact water and sediments, and mineral substrates at the floor of the lake, where many chemical processes can occur simultaneously. The anoxic crater lakes in a pre-plate tectonic Earth were unique in geologic history. They were very different from submarine Black Smokers or Lost City, which we encounter today along the spreading ridge. The cold freshwater of the crater lake is heated by hot magma from the central peak. It reemerges to form the vents that may reach high temperatures like the condition in Yellowstone Geysers, driving the convection of lake water with fluctuating temperatures at the water column, creating different niches. The convection current inside the crater vent mixed hot, concentrated prebiotic soup thoroughly and caused simple chemicals to grow into larger, more complex ones by combinatorial chemistry, with a chaotic mix of energy sources and organic compounds released from the vents.

Abundant in the Eoarchean crust, hydrothermal systems in numerous crater lakes are the likely cradles for the origin of life. Derived from meteorite impacts, the building blocks of life began to concentrate here. In addition, a continuous infall of micro-meteorites containing iron, manganese, and silicates formed a blanket around the lifeless surface of newly formed crater lakes. It started to interact with biomolecules in the basins. Hydrothermal vents inside the crater basins of these sequestered hydrothermal crater lakes provided a continuous stream of energy to mix chemical compounds with cosmic ingredients.

The prebiotic chemistries inside the hydrothermal crater basins were powered by solar, tidal, and chemical energies, including ATP (Martin et al. 2014). Convection currents formed a sticky, brownish, primordial prebiotic soup in which concentrated cosmic and terrestrial chemicals rich in organic compounds could grow into larger and more complex molecules (Chatterjee 2016, 2019). Hydrothermal impact crater lakes provided for the selection, concentration, and organization of specific organic molecules into successively more information-rich biopolymers and finally into the first cells (Chatterjee and Yadav 2019). Moreover, if interconnected by underground fractures, neighboring impact craters of different sizes, temperature regimes, and concentrations could create a feedback loop among networks for abiogenesis (Figure 5.4).

The hydrothermal crater lakes possess all the previously proposed favorable characteristics of deep-sea hydrothermal vents for the synthesis of life, such as prolonged circulation of heated water for mixing and concentration of cosmic building blocks of life; various chemicals, energy, and ATP for chemosynthesis; abundant catalytic surfaces of mineral substrates with nanopores and pockets for polymerization; and nutrients for primitive life. Moreover, each crater basin provides additional geochemical and environmental advantages over the submarine vents in a single location.

The crater lakes also have several advantages in common with terrestrial hot springs that allow the assembly of the lipid membrane, a wet–dry cycle for polymerization. This freshwater environment of crater basins mirrors the chemistry of cytoplasm and allows the concentration of organic molecules to form. A key strength in the hydrothermal crater basins for favorable constraints for life synthesis includes a suite of conditions in single locations, making them prime geological sites for abiogenesis (Chatterjee 2019):

1. Microhabitats: Impact events are generated several niches inside crater basins, which are conducive to prebiotic synthesis. The continental excavation by impacts, forming the crater lakes, exposed a variety of rock types, including sialic, mafic, and ultramafic, which provided a dynamic fluid mixed with variable temperatures, magmas, and pH conditions (Figure 5.5). Crater-fill impact melt rocks and breccias create pores and crevices for abiogenesis. This combination allowed an active fluid that favored the prebiotic synthesis in reducing and reactive habitats. But in a purely volcanic and tectonic venue, such mixtures would likely have been less available. Impact-induced fracturing at the floor of the crater basin increased the surface area for the concentration of polynucleotides and polypeptides to occur (Cockell 2006). Mineral surface bonding and pore spaces are functional precursors of the cellular enclosure and bonding enzyme surfaces.
2. Concentration of biomolecules: In the global ocean, most of the organic compounds from cosmic and volcanic sources would have been diluted and dispersed, inhibiting prebiotic synthesis. On the other hand, sequestered crater basins with high crater rims would have been ideal for concentrating the complex reactions of cosmic molecules. Hot water circulation from the central peak to the cold surface creates a convection current that promotes the mixing and concentration of monomers to occur. Additionally, on the surface of the crater lake, evaporative heating and drying of minerals allow the concentration of biomolecules to occur.
3. Nourishment: In modern hydrothermal crater lakes, the chemical content of vent water originates in a complex set of reactions between lake water and hot, newly minted impact melt rock, nourishing a chain of living chemosynthetic microbial communities. It was likely to have done the same during abiogenesis. All hydrothermal crater lakes are abundant in $H_2S$, a potential source of nourishment for emerging hyperthermophiles. Even when hydrothermal activity ceased, hydrothermally altered and precipitated rocks would have provided enduring sources of nutrients and habitats (Osinski et al. 2013).
4. Chemistry of freshwater: Terrestrial crater basins contained freshwater with low total salts and $K^+/Na^+$ ratios that match the cytoplasm of all cells of three domains of life and encouraged prebiotic synthesis (Mulkidjanian et al. 2012).
5. Formation of bilayer fatty acid vesicles: Amphiphilic compounds are present in carbonaceous compounds and can assemble into membranous vesicles in freshwater (Deamer 2011).
6. Diverse assemblages of minerals. Multiple mineral assortments, such as metal sulfides, clays, pyrites, and zeolites, were available on the floor of the crater basins that provide critical prebiotic reactions. The clay came from impact melt, and the pyrite came from a hydrothermal vent (Hazen 2005). Another source of clay was the impactor itself – carbonaceous chondrites (Kloprogge and Hartman 2021). Hydrothermal reactions were the primary source of the metal-rich sediments and nodules that carpet the crater floor. (Chatterjee 2016; Cockell 2006; Hazen 2005). The widespread formation

of clays and pyrites at the basin floor encourages the polymerization of monomers. Phyllosilicates, formed by hydrous alteration of silicate mineral debris, may have played a critical catalytic role in creating prebiotic and possibly early biotic organic compounds and structures (Schmitt 2015). Diverse minerals facilitated the following prebiotic reactions.

a. Protometabolism: The water of the vent environment was rich in ferrous iron and transition-state metals such as ions of magnesium, copper, and zinc, compounding the catalytic capabilities of the iron-rich clays of the crater. The mineral surfaces of the crater could absorb and concentrate thin films of organic solutes from aqueous solutions. The 'Iron-Sulfur World' inside the hydrothermal vents promoted various chemical reactions and acted as an early form of metabolism before nucleic acids appeared (Wachterhäuser 1993). Clays, such as smectites derived from carbonaceous chondrites, also drove protometabolism. De Duve (2005) suggested a high-energy thioester-based protometabolism in the vent environment, which follows pathways not dissimilar to modern metabolism, in which thioesters also play a crucial role. Along with thioester, ATP (adenosine triphosphate), the currency of life, was also available in the vent environment (Martin et al. 2014).
b. Chiral selection: One of the biosignatures of life is the chirality of monomers, left-handed amino acids, and right-handed sugars. Crystal faces of enantiomorph pairs of crystals in the craters of a granitic craton such as quartz, feldspar, diopside, and calcite molecules separated chiral molecules from the racemic mixture. For example, the left-handed faces of calcite may have concentrated left-handed amino acids and *vice versa* (Hazen 2001).
c. Polymerization of monomers: Nonenzymatic polymerization of monomers such as L-amino acids and D-ribose nucleotides, when incubated by tiny particles such as clay or pyrite, could have polymerized into protein and RNA-like molecules (Hazen 2005). Another possible mechanism lies in the wet–dry cycles of the fluctuating volcanic pools (Deamer 2019) or small crater basins (Chatterjee 2019).
d. Encapsulation of polymers: Mineral surfaces would have been able to concentrate and polymerize monomers and thus produce RNA and polypeptides. The hollow lipid membranes would stick to the mineral substrate like tiny blisters by convection current, providing access to a wide range of polymers and other biomolecules (Figure 5.5). The vesicle crowded at the crater floor on the mineral surface could trap RNA and polypeptide from the adsorbed surface of the clay and encapsulate, bringing these two components together to generate a protocell-like structure. As these protocells were released from the mineral surface, their polymers became encapsulated, ready to participate in further chemical reactions (Chatterjee 2016).

7. Wet and dry cycles: Favored the concentration of biomolecules in the small crater basin and condensation for polymerization at the exposed crater's edge (Chatterjee 2019).

8. Temperature and pH ranges: High temperate ranges (~60°C–90°C) and pH values near neutrality (pH 5–8) are ideal for stabilizing the membrane vesicles of emerging hyperthermophilic microbes (Deamer 2011, 2019; Damer and Deamer 2015).
9. Sanctuary: Impact-shocked rocks and impact glasses offered endolithic habitats that would offer shelter against harmful UV radiation when Earth still lacked an ozone layer (Osinski et al. 2013).
10. Abundant energy sources: There were plenty of energy sources in the hydrothermal crater vent environment that could have driven chemical evolution leading to complex organic molecules and first life. But the earliest hyperthermophilic life powered itself through reactions involving hydrogen gas, sulfur, or ammonia, available in the hydrothermal vent environments. Deamer and Weber (2010) identified four levels of energy resources that might be available on the prebiotic Earth to drive primitive metabolism and synthesis of polymers. The first level includes photochemical energy available in ultraviolet light, atmospheric electric discharge, and geochemical energy. These are relatively high energies, modeled in the laboratory by electrical discharge and ultraviolet light. In a terrestrial environment, solar energy, ultraviolet radiation, electric discharges, lightning, and impact shock waves were available for prebiotic synthesis. Because of the huge amount of energy from impacts, impacted melt rocks could keep the vent environments hot and reactive for the long durations likely necessary for biosynthesis (Osinski et al. 2013). The second energy level would come from concentrating dilute solutions of cosmic ingredients in the vent environment and concentrating monomers by evaporation. The third level uses a series of low-energy reactions that incorporate condensation processes by which monomers are assembled into random polymers on the mineral surfaces. The final level is the energy flow in protometabolic networks. Protocells would have captured the local environment's energy and transferred it to energy carriers like ATP (adenosine triphosphate). The metabolic and catabolic processes in emerging life discovered how to synthesize ATP from materials available in the vent environment and break ATP into ADP (adenosine diphosphate) through organic phosphate catabolic reactions (Martin et al. 2014).
11. Crater basin networks: Underground networks of crater basins (Figure 5.6[d]), from small to large size allow unique opportunities for prebiotic synthesis. When interconnected by subterranean fractures or surface run-off, impact crater lakes of variable sizes, each with a distinct set of features (such as pH, chemistry, temperature, and ionic concentration), could have mixed components and shared information during prebiotic synthesis.
12. Longevity of hydrothermal systems: An important difference between a submarine and impact-generated hydrothermal systems is the comparatively temporary nature of the latter. In general, the larger the crater, the longer the duration of hydrothermal activity. For example, the 24-km-diameter Haughton impact structure in Canada can last 10,000 years, whereas the hydrothermal system in Sudbury, Canada, lasted for 2

million years (Abramov and Kring 2004). Impact melt-bearing impactites within the crater interior represent a major source of hydrothermal activity (Osinski et al. 2001). The relatively short duration of impact-hydrothermal vents is favorable for abiogenesis because complex molecules can form as lower temperatures favor their stability (Shock and Schulte 1998). Moreover, crater networks of small to large sizes can optimize the thermal gradients for abiogenesis.

13. Paleoecology of the earliest microfossils and stromatolites: The 3.5-Ga Dresser Formation of the Pilbara Craton was deposited within a crater basin affected by voluminous hydrothermal circulation and has yielded one of the oldest microfossils (Djokic et al. 2017). Similarly, thermophilic, anoxygenic, and photosynthetic bacteria built the desiccation-cracked mats of the Paleoarchean (~3.3 Ga), the Josefsdal Chert of the Barberton Greenstone Belt, South Africa, located in a nearshore hydrothermal setting that was periodically subaerial (Westall et al. 2015). Perhaps the Josefsdal Chert was deposited in a hydrothermal crater near a coastal region that was periodically flooded (Chatterjee 2019) (see Figure 5.6[b]). As discussed in the next section, both the Pilbara Craton of Australia and the Kaapval Craton of South Africa were once united in the supercontinent Vaalbara (3.6–2.8 Ga). They showed correlative impact spherule layers indicating hydrothermal impact crater lake paleoenvironment.
14. Microbial colonization: Hydrothermal crater lakes today harbor rich ecosystems and variable energy sources, which stem mainly from vents and impact melt rocks, and are favorable habitats for microbial colonization (Parnell et al. 2004; Reysenbach and Cady 2001). Many of these microbes are relics or living fossils from the Archean ecosystems and have retained their characteristics. The modern hydrothermal crater basins may represent the oldest surviving ecosystems prevalent on young Earth.

These 14 geochemical, environmental, and ecological characteristics favor hydrothermal crater lakes as the likely cradle for life's origin rather than the submarine hydrothermal vents or subaerial hot springs.

Hydrothermal crater lakes encompass a multiplicity of physical, chemical, and mineralogical gradients shaping the structure of the microbial communities inhabiting these systems. The ecosystem of a hydrothermal crater vent includes a gradient of temperature, nutrient abundance, chemical environment, and pH, even within a single thermal regime. Because a wide range of environmental conditions were required to produce a living cell from organic precursors, the environmental complexity of the crater basin was necessary for the origin of biological complexity. The anoxic crater lakes of a pre-plate tectonic Earth are unique in geologic history and very different from the submarine Black Smokers or the Lost City. The cold water of the crater lake is heated by hot magma from the central peak and reemerges from the vents, reaching a relatively moderate temperature (40°C–90°C). The fluctuating temperature gradient created a convection current within the lake water, mixing the assorted cosmic and terrestrial chemicals with reduced gasses such as $H_2S$, $H_2$, $CH_4$, $CO_2$, and $NH_3$. These mixed hydrothermal fluids formed a complex solution of

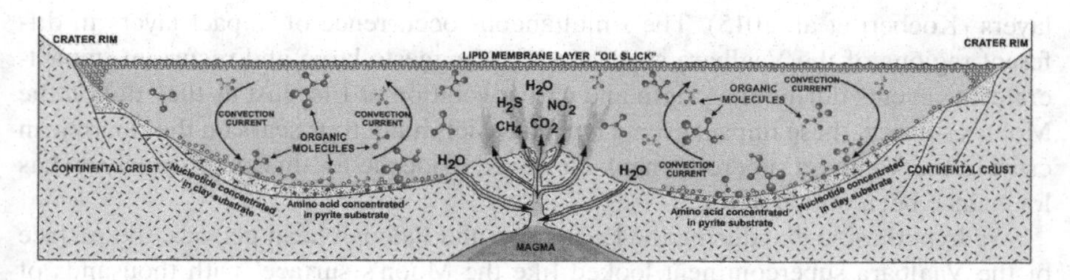


**FIGURE 5.5** Cradle of life: Hydrothermal crater lakes in the Early Archean offered a protective haven for prebiotic synthesis. The boiling water was rich with building blocks of life. On the surface crater basin, lipid vesicles and hydrocarbons were buoyant like tars. The mineral substrates on the basin floor acted as catalytic surfaces for the concentration and polymerization of monomers. Convection currents thoroughly mixed the bubbling biotic soup. Some lipid vesicles, by convective current, went down to the crater floor and stuck to the mineral substrate, encapsulating biopolymers such as RNA and peptides. Hydrothermal vents provide heat, gases, and chemical energy, including ATP molecules.

*Source:* Modified from Chatterjee (2019)

thick, prebiotic soup. Alternating wet/dry cycles of the lake surface facilitated the concentration of organic molecules to occur. They drove condensation reactions at the elevated crater edge to polymerize monomers such as nucleic acids and amino acids. Mineral surfaces of the crater floor, such as clay and pyrite, could catalyze the polymerization of nucleic acid and polypeptides and concentrate phosphate and polymers in pores and cavities for prebiotic reactions.

## 5.8 RECONSTRUCTION OF HYDROTHERMAL IMPACT CRATER LAKES IN THE VAALBARA SUPERCONTINENT

The Archean Kaapvaal Craton of South Africa and the Pilbara Craton of Australia are the only two sizeable areas in the world where granite–greenstone terrains ranging in age between 3.6 and 2.7 Ga have been preserved in a relatively pristine state, with rich records of the earliest microfossils. Moreover, these two cratons were once joined together into an ancient supercontinent. Vaalbara is the name given to the oldest Archean supercontinent, which consisted of the Pilbara and the Kaapvaal Cratons (Zegers et al. 1998). Remarkably similar lithostratigraphic and chronostratigraphic structural sequences in these two cratons have been noted for the period 3.5–2.7 Ga. Paleomagnetic data from two ultramafic complexes in the cratons showed that at 3.87 Ga, these two cratons were part of the same supercontinent (Zegers and Ocampo 2003). Both cratons preserve four large meteorite impact events between 3.5 and 3.2 Ga, visible as layers formed when the high temperatures created by impact forces fused the target rocks into small glassy spherules (Beyerly et al. 2002). Occurring in both the Kaapvaal Craton and Pilbara Craton, these spherule layers are the oldest known terrestrial impact products. The spherules resemble the glassy chondrules in the carbonaceous chondrites found in carbon-rich organic compounds of meteorites (Lowe et al. 2003). Drill core samples of meteoritic components from the Barberton Greenstone Belt of the Kaapvaal Craton confirm the impact origins of these spherule layers (Koeberl et al. 2015). The simultaneous occurrence of impact layers in different regions of the Vaalbara supercontinent suggests large and recurrent impact-cratering events during the origin and early evolution of life. Just as they did on the Moon's surface, these impact events must have left numerous scars on the Eoarchean crust of the Vaalbara supercontinent. However, evidence for this early cataclysm has long since been obliterated.

We hypothesize that during the Eoarchean and Paleoarchean periods, the surface of the Vaalbara supercontinent looked like the Moon's surface, with thousands of craters ranging in diameter from 1 km to several hundred kilometers (Marchi et al. 2014). Unlike the Moon, however, the cratered landscape of Vaalbara was filled with water and impact-generated hydrothermal systems, forming ideal incubators for life (Chatterjee 2019). These hydrothermal lake sediments contain a rich record of the earliest microbial communities, preserving crucial evidence of the origin of life (Djokic et al. 2017; Wacey et al. 2011; Westall et al. 2001). Thus, the hydrothermal crater-lake environments of the Vaalbara continent hold clues to the locations and likely niches of life's crucibles.

I propose that the Eoarchean and Paleoarchean crust of the Vaalbara supercontinent was heavily pockmarked by meteorite impacts like those on the Moon (Figure 5.6). As suggested by four distinct impact spherule layers, during a period of light bombardment around 3.5–2.5 Ga, large asteroids struck the Vaalbara continent and blasted out vast crater basins (Koeberl et al. 2015; Westall et al. 2015). The second line of evidence of bolide impact in these greenstone belts is an iridium anomaly of meteoritic origin (Glikson 2010). Iridium is rare in the crust but is abundant in certain carbonaceous asteroids. Chromium isotopes and iridium anomalies confirm the origins of these Archean spherule beds in the multiple impacts of carbonaceous chondrite (Kyte et al. 2003). Therefore, the large impacts did not stop at ~3.8 Ga but continued throughout the Eoarchean and Paleoarchean eras (~3.8–3.2 Ga). Great floods of molten rock gushed up from the mantle to the granitic crust, filling the crater basins repeatedly with impact-triggered melts and creating ultramafic and basaltic maria or plains associated with hydrothermal lake sediments. Impact-induced ultramafic and mafic volcanic rocks with thin interflows of hydrothermal chert horizons are visible on the Onverwacht Group of the Kaapvaal Craton and the Warrawoona Group of the Pilbara Craton of the ancient Vaalbara continent (Koeberl et al. 2015). The hydrothermal cherty sedimentary layers are rich in microfossils and have yielded critical information about the origin and early evolution of life (Djokic et al. 2017; Westall et al. 2001).

Environmental conditions at the Eoarchean surface of the Pilbara Craton and the Kaapvaal Craton were conducive to the origin and diversification of microbial life. These greenstone belts with hydrothermal crater lakes were the ideal location for biosynthesis. Interstellar particles, micro-meteorites, small comets, and chondrites were suitable carriers for safely delivering cosmic biomolecules to these crater regions. Earth's young atmosphere slowed down these carriers of life's first building blocks, lightly settling as fine dust upon the crater-lake surface and enriching them with cosmic ingredients to be mixed by the convection currents of the hydrothermal systems into a rich, prebiotic broth.

While raised rims on the surface separated and isolated these crater lakes, underground cracks and crevices interlinked them, like the Mono Lake in California, where a series of lakes from higher to lower elevations have a linked flow of groundwater. Instead of a single crucible, then, the closely spaced crater basins on the Vaalbara supercontinent, ranging from 5 to 500 km in diameter, were interconnected through extensive underground networks. Such networks of connected craters had a higher probability than a single crater of forming the ideal crucible systems for the origin of life. Cosmic ingredients and temperature gradients could move from one crater to another through these elaborate underground networks, thus increasing the chances for just the right crucibles in which life took form (Figure 5.6[c]). The network of crater lakes became increasingly suited to bootstrap prebiotic synthesis and enhanced conditions for biosynthesis. On the surface, numerous stream channels during the rainy season connected many craters, forming networks for exchange of chemicals and pH.

## 5.9 MICROBIAL COLONIZATION IN IMPACT CRATER LAKES

The inhabitants of current hydrothermal crater lakes represent relict microbial communities that have remained distinct from other surrounding terrestrial organisms for millions of years, retaining their stamp of extreme antiquity. Today, hydrothermal crater lakes harbor rich ecosystems with variable energy sources, mainly from vents and impact melt rocks. They are favorable habitats for microbial colonization, many being holdovers from Archean ecosystems that have retained their ancestral characteristics (Parnell et al. 2004; Reysenbach and Cady 2001). Thus, the hydrothermal crater lakes of today are a refuge for novel derivatives of the ancient forms of young Earth. The high-temperature vents of modern crater basins are perhaps the oldest ecosystems reminiscent of early Earth. Inside the crater, basins are reactive gases, dissolved elements, and chemical and thermal gradients on the water surface ranging from hyperthermophilic to thermophilic to mesophilic conditions. The vent microorganisms are self-sufficient. Vent chemicals sustain hyperthermophiles, while photoautotrophs requiring solar power enjoy a planktonic lifestyle near the surface. While the fractured rocks of impact craters have been expected to host deep microbial communities on Earth and potentially other terrestrial planets, direct evidence remains elusive. Still, a few examples of microbial activity in some of these impact craters do suggest that crater lakes are favorable microbial habitats on Earth and perhaps beyond.

### 5.9.1 Siljan Crater, Sweden

Formed during the Late Devonian period (~378 Ma) with a diameter of 52 km, the Siljan Crater in Sweden is the largest known impact structure in western Europe. Its drill core samples give evidence of deep and ancient life. The structure had preserved long-term deep microbial activity that formed 80–22 million years ago, when conditions, such as temperature, became more favorable than at the time of the impact event. Calcium carbonate and sulfide crystals in the crater's fracture zones have isotopic signatures revealing both microbial methanogenesis and anaerobic methane oxidation (Sapers et al. 2015).

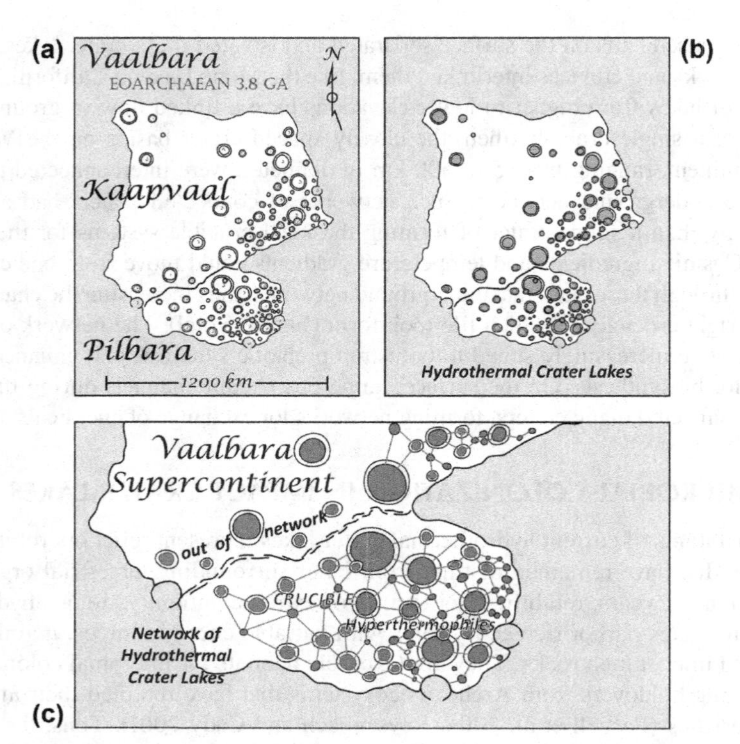


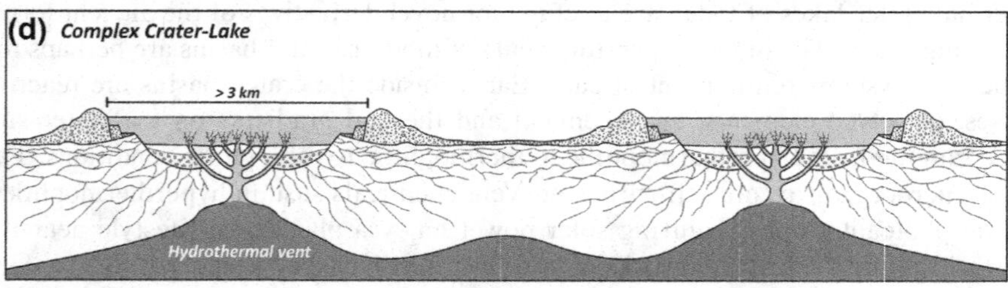


**FIGURE 5.6** Hydrothermal impact crater lakes become the prime location for the origin of life. (a) Reconstruction of the highly cratered surface of Earth's first supercontinent, Vaalbara, where the granitic crust is pockmarked with innumerable craters like the Moon. (b) Crater lakes with hydrothermal systems, created by rain and aquifers, rich in cosmic ingredients. (c) Crater lake networks formed within the bedrock fractures, interconnecting the various, closely linked crater lakes and exchanging heat and life-building chemical ingredients. These interconnected hydrothermal crater lakes become the ideal cradle for the origin of life. (d) A cross-section of two small complex craters showing the underground fissure network that connects the closely spaced crater basins.

### 5.9.2 Ries Crater, Germany

Formed about 14 Ma, the asteroid impact that created the Ries Crater (~24 km in diameter) excavated 500–650 m of Triassic–Jurassic and Tertiary sedimentary rocks and more than 2 km of the underlying granitic basement. The Ries Crater is widely recognized as an analog for Martian craters and specifically ejecta fluidization, post-impact hydrothermal activity, and aqueous sedimentation. The sediments deposited from impact-induced melt rocks such as breccias and glassy melt fragments had a substantial effect on lake water chemistry. In particular, aqueous alteration of the impact glass led to abundant zeolite minerals. Rapid weathering ash and melt fragments initially created a highly alkaline crater lake, but the weathering of the felsic crust shifts the pH toward neutral. This observation is like the condition of the Gale Crater of Mars, which hosted an ancient, potentially habitable lake. Moreover, in both cases, the target rock is granitic (Stüeken et al. 2020). Microbial trace fossils occur in the Ries Crater as tubular features in impact glasses. The tubules have complex forms – consistent with the tunneling behavior of microbes – and contain organic molecules associated with biological activity. The Ries Crater may have generated hydrothermal activity for as long as 10,000 years, giving microbes ample time to build colonies (Sapers et al. 2015).

### 5.9.3 Haughton Crater, Canada

The Miocene Haughton impact structure (~20 km in diameter) on Devon Island in the High Canadian Arctic contains fossils of microbial communities (Parnell et al. 2004). The site includes massive gypsum-bearing carbonate rocks of the Ordovician age. Microbial communities found in the highly porous and shocked gypsum crystals show two species of cyanobacteria: *Gloeocapsa alpine* and *Nostoc commune*. Microorganisms can better colonize porous minerals and more efficiently extract nutrients from them. NASA's ongoing search for life initiated the Haughton–Mars Project, as the freezing environment of the Haughton Crater offers a potential Martian analog.

### 5.9.4 Lonar Crater, India

Formed entirely within the Deccan Traps around 50,000 years ago, the Lonar Crater of India represents one of the youngest and best-preserved impact structures on Earth. The 1.8-km-diameter simple crater presents an essential analog for small craters on the Moon and Mars' basaltic surfaces. The impact event that generated the Lonar Crater probably tapped groundwater supplies in the underlying basaltic aquifer, producing a fluidized ejecta blanket. A hydrothermal system developed as water from this lake began to interact with the hot, porous impact melt deposits. Core samples from the Lonar Crater flow show some clay minerals that formed in scorching, hyperthermophilic environments between 130°C and 200°C (Hagerty and Newsom 2003). Environmental constraints in this highly saline and alkaline soda lake with a microbial community have favored a distinctive and diverse mesophilic microbial community. Lonar Crater's microbial assemblage includes methylotrophs, anoxygenic purple sulfur, non-sulfur photosynthetic bacteria, and oxygenic cyanobacteria. *Actinobacteria* (24%), *Proteobacteria* (30%), *Firmicutes* (11%), and *Cyanobacteria* (5%) predominate, but other microbes such as *Nitrospirae* (0.41%), *Bacteroidetes* (1.12%), *BD1–5* (0.5%), and *Verrucomicrobia* (0.28%) also occupy the ecosystem (Paul et al. 2016).

## 5.10 THE PLAUSIBLE CRADLE FOR LIFE'S BEGINNINGS

The large quantities of extraterrestrial material delivered to young terrestrial planetary surfaces in the early history of our solar system may have provided the organic material necessary for the emergence of life about four billion years ago (Chyba and Sagan 1992). Microscopic and hyperthermophilic, these more complex combinations could grow and reproduce. But at the time of their emergence, Earth was completely devoid of what we would recognize today as a suitable environment for living things. Instead, the hostile and extreme environment of the young planet sustained widespread impact and volcanic activity. Where on Earth could life-forming processes find enough shelter to proceed?

In the quest for the cradle of life, the essential first step is to determine the precise environment where the cosmic ingredients and other building blocks of life could be concentrated and synthesized into more complex molecules and where energy for protometabolism would be available. Moreover, that environment should be geologically compatible with a planet lacking plate-tectonic activity. The lifestyle of living hyperthermophiles, the most primitive organisms, provides an important clue to the cradle's paleoenvironment: An anoxic hydrothermal vent environment that supplies rich minerals and energy (Stetter 2006; Lane and Martin 2012). Indeed, hydrothermal conditions are conducive environments for nascent life. Moreover, genetic analyses lend a certain amount of corroboration for the emergence of life in hydrothermal systems, for these studies suggest that LUCA lived in a hydrothermal environment. LUCA's closest living relatives are clostridium bacteria and methanogenic archaea, gaining nutrients from $H_2S$, $H_2$, $CO_2$, transition metals, and sulfur (Akanuma et al. 2013; Weiss et al. 2016).

The top four hydrothermal locations for abiogenesis are submarine vents such as Black Smokers and Lost City and terrestrial vents such as 'Darwin's warm little ponds' and impact crater lakes. While each environment has pros and cons, in recent times, the long-held view that life originated in the marine environment has been contested. Many now consider that oceanic sites will dilute rather than concentrate organic molecules for biosynthesis. Moreover, the lack of plate tectonics during the early Archean has cast doubt on the very presence of hydrothermal vents in the young Earth. Also, primitive cellular membranes assemble more easily in freshwater than in saltwater. Moreover, the cytoplasm of the living cells more resembles the chemical nature of terrestrial vents. Finally, the earliest biosignatures of microbial life are not found in the deep sea but in the terrestrial environment.

Regarding the two terrestrial alternatives, both hot springs and impact craters are attractive models. However, hot springs are relatively short-lived and were probably less frequent on the granitic crust. Both surface sites do have access to light energy, chemical energy, and concentrating mechanisms. But in the Eoarchean, the only land exposed above the sea level was the granitic crust, where impact craters were likely to be more available than the volcanic hot springs during the Heavy Bombardment period. The floors of impact crater lakes also had pyrite and clay surfaces for polymerization, dissolved gases like hydrogen and methane, solutes such as ferrous iron, energy like ATP for protometabolism, and long residence times for hydrothermal activity.

Impact-generated hydrothermal systems may have provided the most favorable environments for the prebiotic synthesis of cosmic ingredients and life's origin. The impact-cratering record on Earth during the Eoarchean suggests that hydrothermal crater lakes were ubiquitous after asteroid impacts on water-rich crustal surfaces. Making water, heat, dissolved chemicals, and nutrients available for extended periods, these hydrothermal systems are prime locations for abiogenesis and colonization by hyperthermophilic microorganisms. They could also have provided a sanctuary for hyperthermophiles during the late heavy bombardment. Mineral surfaces at the crater floor could have helped to catalyze complex chemical reactions, polymerization, and available energy for primitive metabolism. In our view, the interconnected impact crater lakes with a wide range of temperatures, pH values, and concentrations of cosmic ingredients had the right conditions for synthesizing membranes, peptides, RNAs, viruses, DNAs, and the first cells. The oldest fossils from the Vaalbara supercontinent are consistent with the hydrothermal crater environment for early life (Beyerly et al. 2002; Lowe et al. 2003). Impact-cratering hydrothermal systems appear to be common in young Earth and its rocky neighbor in the Solar System. Especially in planetary bodies such as Mars that otherwise are geologically dead, the fact that such systems could have provided rare havens for life has considerable astrobiological implications (Osinski et al. 2013). The crater lake is most likely to be the long-sought cradle of life. Life probably got started on the surface of the Earth, such as inside impact craters, where it found sequestered hydrothermal basins offering refuge during a bombardment. And when the cratering activity was diminished, life could then spread to the tranquil ocean surfaces, eventually harnessing the Sun's energy and distributing globally (Chatterjee 2019).

## 5.11 THE SEARCH FOR EVIDENCE OF ANCIENT LIFE ON MARS

Mars is the fourth planet from the Sun and the second-smallest planet in the solar system, being larger than only Mercury. It is commonly referred to as the Red Planet. The rocks, soil, and sky have a red or pink hue. Mars is also a dynamic planet with seasons, polar ice caps, impact craters, extinct volcanoes, and evidence that it was even more active in the past. Mars is one of the most explored bodies in our solar system, where NASA has sent rovers to study the alien landscape for detecting life.

The search for evidence of life on Mars is one of the outstanding scientific challenges of our time. There is now abundant evidence that during the Noachian and early Hesperian eras, free-standing and free-flowing water was widespread on Mars (Carr 1996). Several examples of ancient lake sites have also been described. Most lakes occupied crater basins in Noachian highland regions. Lacustrine and associated fluviatile geomorphologies are clearly recognizable. Terraces, layered sedimentary rocks, and some inflow and outflow channels may also be discerned

(Carr and Head 2010; McKay and Stoker 1989). The presence of liquid water is central to the existence of life on Earth and the possibility of past or present life on Mars. Once thought barren, Mars now holds the promise of life beyond Earth as scientists confirm the presence of water on the Red Planet. Mars shows physical similarities with Earth. Early in their respective histories, each planet shows evidence of stable liquid water on the surface. A large sea in Mars' southern hemisphere once held ten times as much water as all of North America's Great Lakes combined. This sea existed roughly 3.7 Ga and was located in the Eridania Basin (Michalski et al. 2017).

Nevertheless, few places inspire the imagination when imagining where extraterrestrial life could potentially dwell, like Mars. Mars has long looked the most promising due to its proximity and similarities to Earth, its relatively pristine conditions, and its location at the outer edge of the habitable zone in our solar system. A quick start to life on Earth may indicate that life could emerge on other worlds in the solar system. And Mars may well have harbored life in the past.

Finding life beyond Earth is one of the primary goals of NASA's exploratory programs. The evidence of life on Earth dates back approximately four billion years to the age of the earliest sedimentary rocks and when Earth and Mars were likely very similar. There is a chance that life also emerged in the Noachian environment of Mars. We speculate such life, if found on Mars, would be hyperthermophiles, using Earth as a reference point. The discovery of ancient microbial life on Mars would have significant implications for science and society.

Today, Mars is a frigid, windy desert like the Dry Valley of Antarctica. But dried-up deltas, riverbanks, deltas, lake basins, and inland seas reveal that water once flowed over the planet's surface. The question, then, is where did all that water go? Today, aside from a possible series of briny, underground lakes and aquifers, most of Mars's water is locked up in the polar ice caps or ice buried below the surface. About four billion years ago, Mars had a denser atmosphere and higher surface temperatures, allowing a vast amount of liquid water over the surface when microbial life might have emerged concurrently with Earth. NASA scientists now believe that ancient hydrothermal crater vents found on Mars could have been a cradle for life. Based on data gathered during China's Tianwen-mission, the Zhurong rover spotted hydrated minerals in a few hundred million years-old terrains. The presence of hydrated minerals implied that liquid water once persisted on Mars. From this evidence, Chinese scientists have concluded that a large asteroid impact basin on Mars hosted liquid water during the most recent epoch of Mars's geologic history, Amazonian (see Table 5.2). The latest discovery adds to a growing body of evidence that suggests that liquid water activity on Mars may have existed for far longer than we previously thought (Liu et al. 2022).

For astrobiologists looking for possible life on Mars, water is essential – but only in its liquid state. Plenty of water, in the form of ice, is locked up in Mar's polar caps – about the same amount as in Greenland ice sheets. In the south Martian pole, the ice-sheet-covered region known as Ultimis Scopilli, a lake, may be buried beneath the ice. It could measure as much as 30 km across and is kept warm by geothermal heating, similar to the kind generated on earth by subsurface magma (Arnold et al. 2022).

Was the appearance of life on Earth likely or was it an anomaly? Are the universal features of life on Earth, like the 20 amino acids to build proteins and the DNA and RNA used for genetic information, or chiral monomers such as D-ribose and L-amino acids, and the universal genetic code truly universal, or are they just one sample in the vast universe? In fact, of all the worlds in our solar system and Milky Way, Mars may be the best one for seeking answers to these questions.

Early Mars was like early Earth. Both were warm and wet rocky planets with oceans and protocontinents. Both had an atmosphere dominated by carbon dioxide. Meteorite impacts on both world had delivered the building blocks of life and water and created innumerable hydrothermal crater lakes as the likely cradles for the origin of life. Most likely, microbial life might have emerged on Mars about four billion years ago, but we do not have any proof yet. Mars exploration program is meant to 'seek signs of life'.

Despite previous attempts to find it, we still don't know if Mars has life or ever had life. The Viking landers of the 1970s conducted biological experiments to detect life on the Martian surface but failed to find any organic molecules. The scientific results did not demonstrate conclusive biosignatures at the two landing sites. To date, then, no proof has been found of past or present life on Mars. Nevertheless, Mars was not always a desolate wasteland. Cumulative evidence shows that Mars' surface environment had liquid water during the Noachian eon (see Table 5.2) and may have been habitable for microorganisms. Dry riverbeds and minerals that form only with liquid water indicate that in deep time, Mars had a thick atmosphere that retained enough heat for liquid water – a necessary ingredient for life – to flow on the surface. Like the young Earth, it had vast oceans with clouds floating through the sky. Over the eons, the water was lost into space or locked underground. However, early conditions on a wetter planet could have been suitable for life to emerge. Mars remained warm and moist for millions or even billions of years, so microbial life might have had enough time to arise and proliferate. When conditions on the surface of Mars turned frozen and dry, life may then have become extinct there while leaving fossils behind. It's even possible, judging from some microbes on Earth that thrive miles underground, that extremophile forms of early Mars life could have survived on Mars below the surface.

If Mars ever harbored life during the Noachian eon, when liquid water flowed freely on the Martian surface, some microbes stowed away in the nooks and crannies of the rock could have been ejected from Mars by impact and, like the ALH84001 Martian asteroid found in Antarctica, traveled to Earth for free. We know that some microbes are incredibly hardy and may be able to survive high doses of solar radiation for long periods during an interplanetary journey after being blasted off their home by an asteroid impact. Some think that Earth life originated on Mars and was brought to this planet in this way, aboard a meteorite. Orbital dynamics show that it's much easier for rocks to travel from Mars to Earth than the other way around. If Mars was ever a living planet, it died early and lost its atmosphere. However, to date, no indigenous Martian organisms have been discovered to support this exciting possibility of interplanetary travel of early life.

Nevertheless, if microbial life did exist in the Noachian sediments, it may have been extremophile that could withstand solar and cosmic radiation. Is it possible that biosynthesis began separately but simultaneously on both nearby planets? If so, what genetic code did it use? Was Martian life genetically coded with nucleic acids? Did it use proteins for enzymatic catalysis? Did it have left-handed amino acids and right-handed sugars? Whatever the answers to these questions, should we discover life on Mars, we will have the first opportunity to compare the biology of two different planets. Or was life transported from one world to another by asteroid impacts, circumventing the need for independent origins? The optimistic conviction that we may have all these answers within a few decades lends an enormous impetus to the search for signs of life, past or present, on the Red Planet.

### 5.11.1 Environment and Stratigraphy of Mars

Like Earth, our sister planet Mars is a rocky planet, and like Earth's seafloor, much of the Martian surface is made of basalt. Mars is a cold, inhospitable desert today. Its average temperature ranges from −10 °C to −76°C with an average surface temperature of −65°C, fluctuating temperatures from as high as 25°C to as low as −123°C. The current Martian atmosphere is much different from that on Earth, with the composition 95.3% $CO_2$, 2.7% $N_2$, and 0.1% $O_2$ as compared with Earth's atmospheric composition of 78.1% $N_2$, 21% $O_2$, and 0.04% $CO_2$. Its radius is about half that of Earth, with a mass of about one-tenth that of Earth (Figure 5.7[a]). Consequently, Mars' gravity is less than half that of Earth. Today, Mars is a bone-dry frozen desert, like the Dry Valley in Antarctica. Part of its thin atmosphere freezes every winter to cover the poles with a white cap, but it provides little protection from heat loss to the Martian surface. An average temperature of about −60°C is too cold to support life on its surface. Mars lost nearly all its original atmosphere for a billion years, transforming its climate from one that might have supported life into a desiccated and frozen environment.

Based on their intersection relationships and the numbers of superimposed impact craters, the Martian geological time scale has been divided into four eons: The pre-Noachian (4.5–4.1 Ga), the Noachian (4.1 to about 3.7 Ga), Hesperian (3.7–3.0 Ga), and Amazonian (3.7 Ga to Recent) (Carr and Head 2010). At the end of the Noachian, most geological activities, such as rates of impact, valley formation, weathering, and erosion, all slowed. Still, volcanism continued at a relatively high average rate throughout the Hesperian (Table 5.2). Evidence that liquid water used to be present suggests that Mars was once more habitable than today. During the Noachian Period, about four billion years ago, when life was emerging on Earth, Mars enjoyed a warmer climate, with abundant liquid water on its surface, river valley networks, large crater lakes, and oceans. Many large impact craters scar the Noachian-aged surfaces. Dried riverbeds, impact crater lake deposits, polar ice caps, and minerals that form in the water suggest that Mars had an environment microorganisms could inhabit. Moreover, its subsurface may remain amenable as a habitat.

We don't know for sure how Mars lost its atmosphere and oceans. Mars was certainly habitable in the Noachian when both Mars and Earth were covered with two protective shields – a relatively thick atmosphere and a strong magnetic field. Earth has retained both, but Mars lost them billions of years ago and so became more vulnerable to cosmic assaults. Mars' loss of its magnetic field strongly affected surface environments through increased radiation, significantly degrading surface habitability. Because of its small size, lower mass, and lesser overall gravity, Mars' internal heat was also lost quickly, ceasing widespread volcanism. Without volcanism, Mars

**TABLE 5.2**
**Earth and Mars Are Compared, Showing the Significant Factors that Affect Their Current Habitability**

| | Earth | Mars |
|---|---|---|
| **1. Average distance to Sun** | 1 AU* | 1.5 AU* |
| **2. Average surface temperature (ºC)** | 15 | −56 |
| **3. Greenhouse effect (ºC)** | 33 | 7 |
| **4. Atmospheric pressure (bar)** | 1 | 0.006 |
| **5. Liquid water** | Abundant | Frozen |
| **6. Main gases in the atmosphere** | 78% $N_2$, 21% $O_2$ | 95.3% $CO_2$, 2.7% $N_2$ |
| **7. Diameter (equatorial)** | 12,756 km | 6,794 km |
| **8. Mass** | $5.974 \times 10^{24}$ kg | $6.418 \times 10^{23}$ kg |
| **9. Escape speed** | 11.2 km/s | 5 km/s |

**10. Geological time scale (Ga)**

**Earth**: Hadean (4.5 Ga – 4.0 Ga); Archean (4.0 Ga – 2.5 Ga); Proterozoic (2.5 Ga – 0.54 Ga); Phanerozoic (0.54 Ga – Present)

**Mars**: Pre-Noachian (4.5 Ga – 4.1 Ga); Noachian (4.1 Ga – 3.7 Ga); Hesperian (3.7 Ga – 3.0 Ga); Amazonian (3.0 Ga – Present)

*Astronomical Unit (AU) = Earth–Sun distance, about 150 million km

could not recycle atmospheric carbon dioxide. Moreover, part of its early atmosphere was lost during the early bombardment period. Because of the planet's low escape velocity, the Martian atmosphere was particularly prone to impact erosion. With its weak atmospheric pressure, Mars could not keep its water in liquid form.

## 5.11.2 Post-Impact Ancient Crater Lakes with Hydrothermal Systems

During the Late Heavy Bombardment period, inner rocky planets and the Moon were heavily cratered by asteroids. Impact crater lakes are of high interest in planetary exploration because they are viewed as cradles for the origin of life and its subsequent survival. Impact craters on Mars larger than 1 km diameter exist by the hundreds of thousands, and many of them were hydrothermal crater lakes in its early history. Because subsurface fluids and crustal heat surface could have coexisted on Mars, hydrothermal crater lakes are essential targets in NASA's planetary exploration and their ongoing search for extraterrestrial life. Hydrothermal impact crater lakes on early Mars could have hosted early life.

Due to the essential role of liquid water in the biochemical reactions that sustain living organisms, planetary bodies with global oceans or impact crater lakes are prime targets in searching for life beyond Earth. In addition to water, life would require energy and a source of essential building blocks. Hydrothermal systems formed from impact events possess all the critical requirements for life: Liquid water, a supply of cosmic building blocks of life, and potential energy (Abramov and Kring 2005). Because it fulfills these conditions by once having running water and a more robust atmosphere, Mars was once capable of hosting hydrothermal ecosystems – and it might still be an incubator for microbial life today.

NASA is currently exploring ancient post-impact hydrothermal crater lakes to find biosignatures. Curiosity is a car-sized rover designed to explore the Gale Crater on Mars. It landed on Mars in August 2012. Three missions set out in the summer of 2020 on a journey to Mars, carrying large instruments to explore the Red Planet. The Hope Orbiter, launched to Mars by the United Arab Emirates, was followed to Mars by the Chinese Tianwen-1, a combination of the orbiter, lander, and rover. NASA's latest Mars mission includes Perseverance, a 2,200-pound rover, and Ingenuity, an experimental Mars helicopter. The Ingenuity helicopter was the first to attempt powered flight on another planet. The Perseverance rover's design is based on Curiosity, a successful NASA mission to gather critical data.

#### 5.11.2.1 Gale Crater

Evidence obtained by NASA's Curiosity rover from Aeolis Palus in the Gale Crater on Mars strongly suggests an ancient hydrothermal freshwater lake, which could have been an environment hospitable for microbial life (Grotzinger et al. 2014). The Gale Crater was created by an asteroid impact in the Martian Noachian–Hesperian transition, around 3.8–3.6 Ga, roughly contemporaneous with the Eoarchean Earth (Figure 5.7[b]). Therefore, Curiosity's exploration of the Gale Crater for signs of life is so exciting. Curiosity has detected mudstones deposited from water containing the six biogenic elements for life synthesis – carbon, hydrogen, oxygen, nitrogen, phosphorous, and sulfur (CHONPS). It has also found some structures akin to stromatolites that could have been microbially generated.

Additionally, drill samples from the Gale Crater data provide new evidence for organic molecules – theophanic, aromatic, and aliphatic compounds – in 3.5-billion-year-old sedimentary rocks. These organic molecules are important because they tell us about the chemical pathways of their formation and preservation. Curiosity data indicate that several billion years ago, a hydrothermal lake inside the Gale Crater might have held all the ingredients necessary for life, including chemical building blocks, energy sources, and liquid water. These recent discoveries of possible biotic activity are tantalizing.

Curiosity has shown that Gale Crater was habitable around 3.5 billion years ago during Hesperian, with conditions comparable to those on early Earth, where microbial life evolved (Figure 5.8[b]). The hope of discovering contemporary life on Mars increased in December 2014 when the Curiosity rover recorded intriguing bursts of methane. Specifically, Curiosity found a tenfold spike in atmospheric methane and detected other organic molecules in a powdered rock sample collected with the robotic laboratory's drill (Grotzinger et al. 2014). Around 90% of the methane on Earth comes from the metabolism of organisms, so it seems plausible that life on Mars is also emitting this gas. Because methane is an unstable gas, its enduring presence indicates an active source on the planet to maintain such levels in the atmosphere. Moreover, background levels of methane in Mars' atmosphere show substantial seasonal variations. The methane presence may indicate the existence of methanogens, such as archaea, which produce methane as a metabolic by-product under anoxic conditions. Localized methane sources released from surface or subsurface reservoirs may indicate hyperthermophilic activity (Grotzinger et al. 2014). Detecting organic molecules and methane on Mars has far-ranging implications regarding potential past life on Mars. However, no firm conclusions can be concluded until such Martian rocks can be brought back to Earth for laboratory confirmation. Therefore, the Perseverance mission of sampling the rocks of the Jezero Crater and bringing them back to Earth is so exciting. Perseverance can play a critical role in understanding our place in the Universe.

One major outstanding question crucial for biosynthesis concerns the longevity of liquid water on Mars. During its exploration of the Gale Crater, the Curiosity rover also discovered the sediments of an ancient lakebed, suggesting that water was present for at least a few million years. In the upper part of the ancient lakebed sequence, there is a high concentration of sulfate salts from the late Hesperian eon starting around 3.7 Ga, a time when Mars had active volcanoes and plenty of water in shallow seas and lakes. The high concentration of salts in the upper sequence points to a period of rapid evaporation at the end of the Hesperian (~ 3.4 Ga) when Mars began the transition to the cold desert we know today. The saline impact crater lakes on Mars endured a time of atmospheric loss and may have been some of the last surface waters on the Red Planet. Porous rocks on the Martian surface are typical of salty lake deposits, such as magnesium sulfates, bromides, and chlorides. Some scientists speculate that iron oxide balls strewn on the Martian surface resemble those formed by underground bacteria and found on Utah's sandstone hills. Perhaps these iron balls were also formed by ancient Martian microbes.

#### 5.11.2.2 Jezero Crater

The exploration of the Jezero Crater for past microbial life is of particularly high interest to NASA scientists. The car-sized Perseverance rover landed on Mars on 18 February 2021 on the floor of Jezero Crater with three scientific objectives: To explore the geologic setting of the crater, to identify ancient habitable environments and assess the probability of Martian life, and to collect samples for potential transport to Earth for analysis in the laboratory. The rover snapped a gorgeous HD panorama of the Mars landing site. The Jezero Crater was formed by an impact around 3.5 billion

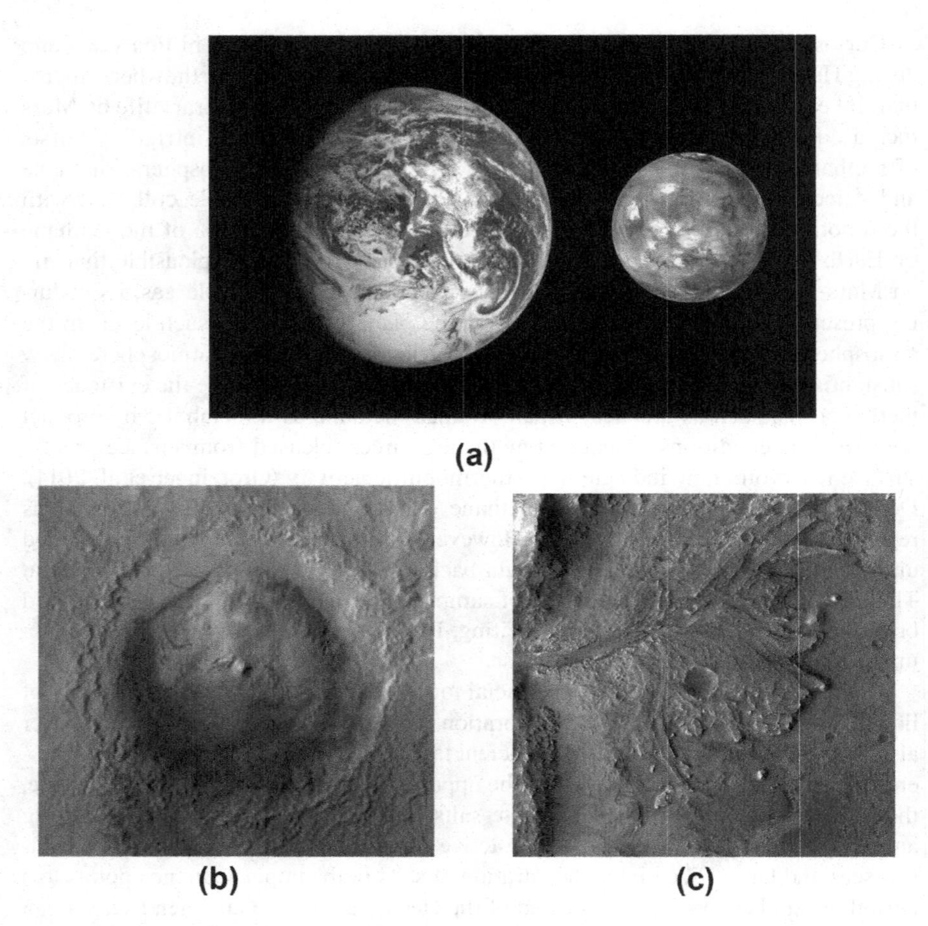


**FIGURE 5.7** (a) Size differences between Mars (right) compared with Earth. Mars' diameter is approximately half that of Earth, and its mass is about one-tenth that of Earth. (b) Gale Crater (~154 km diameter) on Mars, a complex crater with a central peak, ~3.5–3.8 Ga old (Noachian), where Curiosity Rover is currently exploring for life. (c) At the center of the photograph, the circular depression is the Jezero Crater (~50 km diameter) on Mars, ~3.5 Ga old (Hesperian), where Perseverance rover began to collect samples of rock and regolith for detection of microbial life.

*Photo credit:* NASA

years ago on the western flank of the giant Isidis impact basin of the Noachian age (~4 Ga), when microbial life was present on Earth. The Jezero Crater later hosted an open-basis lake, fed by an extensive system of river valleys that were active in the late Noachian (~3.6–3.8 Ga) to early Hesperian (>3.2 Ga). Perseverance has been busy collecting samples from the Jezero Crater and bringing them back to Earth to be studied for signs that life may have developed in this long-lived, ~45-km diameter impact crater lake. A lake and the river delta that emptied into it, point bars, and clay minerals with the organic matter have been detected around the crater. The fan-shaped geological feature, once present where a river converged with a lake, preserves layers of Martian history in sedimentary rock, the potential site for microbial fossils. Jezero Crater is likely to have been habitable in the distant past (Figure 5.7[c]). The rover investigated the crater floor and found evidence of igneous rocks. Preliminary analysis of the samples from the floor of the Jezero Crater suggests aqueously altered igneous rocks below the crater's sedimentary delta. Core samples of these rocks have been stored aboard Perseverance for potential return to Earth (Farley et al. 2022). During its second campaign to study the overlying rich sedimentary rock layers, Perseverance collected more samples full of organic molecules. The rover is able to drive up along various exposed layers of the sedimentary rocks deposited in the crater lake. Of particular interest are two rocks that scientists named Skinner Ridge and Wildcat Ridge. Once these samples from the lake deposits are in labs on Earth, they could reveal insights about Mars's ancient climate, environment, and potential microbial life.

#### 5.11.2.3 Eridania Basin

The Eridania region in the southern highlands of Mars once contained a vast inland sea or lake with a volume of water greater than all other Martian lakes combined. Michalski et al. (2017), based on data from NASA's Mars Reconnaissance Orbiter (MRO), detected vast mineral deposits at the bottom of the basin, which could be seen as evidence of ancient hot springs (Figure 5.8[a]). In cross-section, the Eridania Basin looks like a giant impact crater with a truncated central peak, annular basin, and faulted and raised rim (Figure 5.8[b]). Moreover, innumerable craters surrounded the basin. Since this type of hydrothermal activity is believed to be responsible for the origin of life on Earth, these findings could indicate that this basin once hosted life.

## 5.12 WHAT HAPPENS IF NASA SCIENTISTS FIND LIFE ON MARS?

Finding evidence for life on Mars has been a decade-long ambition for NASA, which has spent billions of dollars to send rovers to probe and sample the Red Planet. But if the ancient microbial fossils are found, how are those findings verified? Most likely, detailed comparisons with the Archean microfossils or stromatolites on Earth may provide some answers. If one finds compelling morphological evidence of Martian life, chemical and isotopic analyses should be done to see if the results are consistent with morphology and other biosignatures. Verifying the data in as many ways as possible is necessary to avoid the ALH84001 fossil controversy.

Suppose, we are fortunate enough to find even one sample containing microbe, living or fossil, on Mars. In that case, we will have identified the wonderful circumstance of two sister planets supporting life in the same early epoch. In that case, did life arise independently on these two sister planets, only to be wiped out on Mars when the climate irrevocably altered? Or might there be subsurface refugees on Mars, some still lingering life forms? It is reasonable to think that life also appeared on Mars under similar conditions in hydrothermal crater lakes, with similar cosmic ingredients, and at much the same time. Perhaps life started simultaneously on Earth and Mars around 4 billion years ago, when Mars had a much thicker atmosphere, warmer climate, and surface water. The evidence that Mars was once a wet planet in the Noachian is incontrovertible in the sedimentary structures left by lakes, flowing rivers, and large oceans.

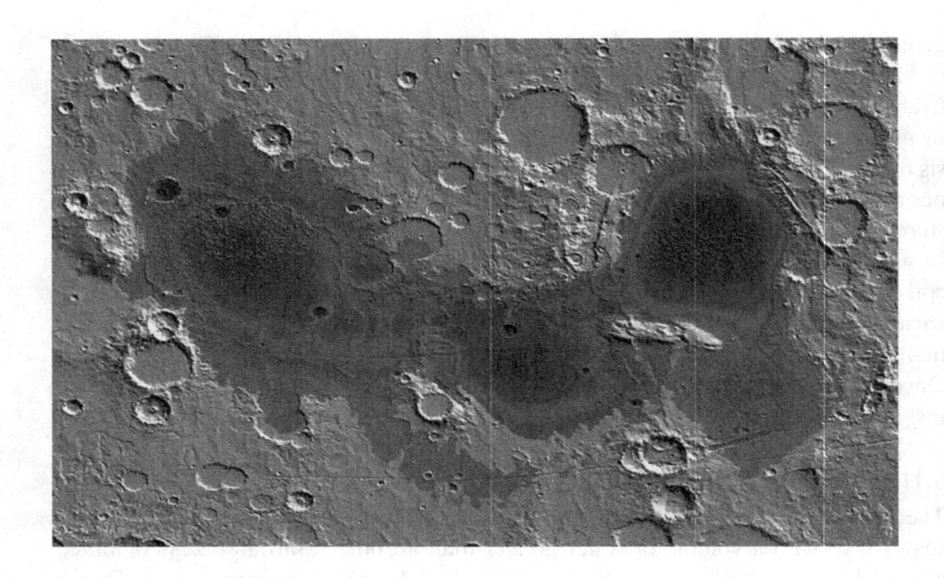

(a)

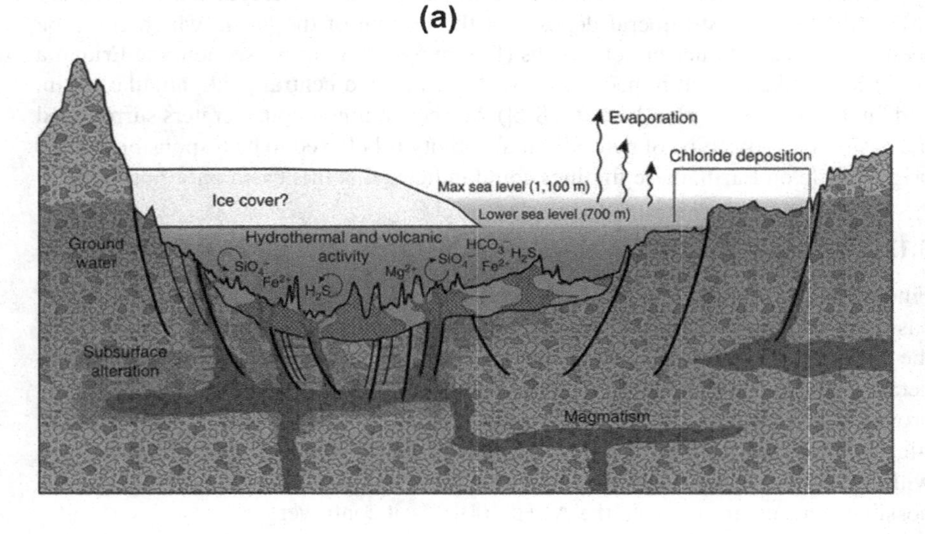


(b)

**FIGURE 5.8** The Eridania Basin of southern Mars. (a) The Eridania Basin, surrounded by innumerable craters, is believed to have held a hydrothermal crater basin 3.7 billion years ago, with crater fl oor deposits likely resulting from underwater hydrothermal activity. (b) A crosssection of the Eridania Basin shows the morphology of a complex crater with raised rims.

*Photo credit:* NASA.

Perhaps, as the widespread cratered surface and volcanic rocks testify, early Mars was warmed in the past by volcanic emissions and asteroid impacts. However, Mars did not develop plate tectonics, unlike Earth, so its surface has faithfully retained its early history. It gives us a rare glimpse of how an early Earth might have looked.

No proof of past or present life has been detected on Mars. Cumulative evidence suggests that during the ancient Noachian eon, the hydrothermal crater environment of Mars might have been urable or habitable for microbial life. But habitable conditions do not necessarily indicate the presence of life on Mars. In a recent article, Sauterey et al. (2022) suggested that primitive methanogenic microbes did once live on Mars and might have initially flourished but then ended up causing their extinction. These authors suggested that the Martian atmosphere was warm about four billion years ago and had enough hydrogen to support methanogens to thrive. Those organisms may have been similar to methanogenic microbes on Earth, which release methane as a waste product. Methanogenesis would have triggered a global cooling event, ending potentially warm conditions of Noachian Mars, and forced these microbes to live underground for habitable temperatures. By removing hydrogen from the atmosphere, microbes would have dramatically cooled down the planet's climates, making it uninhabitable. Ultimately, the whole biosphere may have been wiped out and driven to extinction.

Mars is of particular interest for the study of the origin of life because of its similarity to the Eoarchean Earth, when it lacked plate tectonics like today's Mars with a highly cratered surface. Moreover, the surface of Mars is now dry and cold, like Antarctica, with the latter supporting organisms in icy lakes and endolithic microorganisms. Could Mars host life now? A recent discovery of three buried lakes of salty water under Mars' icy surface sustains hope for the survival of subsurface life. If there is any life on Mars now, it needs liquid water. Most of its liquid water is locked away in underground reservoirs. Like frozen lakes in Antarctica, life may be hidden away in those underground lakes. Furthermore, because a diversity of extremophiles is known to survive in the most inhospitable habitats on Earth, it is likely that Martian extremophiles could still be hanging on.

## 5.13 SUMMARY

Recent astrobiological evidence suggests that carbonaceous asteroids and interplanetary dust particles delivered the prebiotic building blocks and water to young Earth during the Late Heavy Bombardment era (4.1–3.8 Ga). But life itself was an endogenous production synthesized in our planet's womb. The oldest known microfossils come from rocks nearly four billion years old from the Eoarchean hydrothermal sedimentary formations. The genetic antiquity of hyperthermophiles (extreme heat-loving microbes) that can tolerate temperatures up to 120°C in a hydrothermal vent environment may reveal the paleoecology and ancestry of life's beginnings. The first life forms, reasonably thought to be hyperthermophiles, were suited to survive the Late Heavy Bombardment by colonizing protected hydrothermal vent environments. Hydrothermal systems in both submarine and terrestrial settings have been considered habitats for the origin of early life on Earth and possibly other planets such as Mars. Life could have started in a range of hydrothermal environments, from the bottom of the oceans to hot springs on volcanic islands to hydrothermal crater lakes on protocontinents.

Although submarine hydrothermal vents as a venue for abiogenesis are popular in literature, recent evidence suggests that life on Earth began on land, not at sea. Submarine hydrothermal vent habitats suffer from the 'dilution problem' with organic compounds in the open sea, making it impossible to concentrate either ions or organic molecular components. Rather than being focused, the cosmic ingredients would be dispersed and diluted in the vastness of the Eoarchean global ocean before they can assemble into the complex molecules of life. Concentration and the 'crowding' of organic molecules are essential aspects of prebiotic chemistry. It is challenging to imagine how a sufficient concentration of reactants could occur in the open ocean. One crucial requisite for the origin of life is that comparatively simple biomolecules must have opportunities to form more complex molecules by the segregation and concentration of chemical compounds. However, it seems unlikely that cosmic and terrestrial chemicals could have been concentrated, mixed, selected, or organized in this way in the vast ocean. Two processes required for life's origin are condensation reactions that produce essential biopolymers by a nonenzymatic response and self-assembly of membranous compartments that encapsulate polymers. Both processes are inhibited in the marine environments. Thermodynamic barriers to the polymerization of key molecular blocks and difficulty in stable membranous compartments in seawater suggest that life on Earth began on land, as Darwin mused, not on the primordial ocean floor. Moreover, Eoarchean tectonic evidence contradicts the submarine hydrothermal vent environment. Today's hydrothermal vents occur along or near the spreading ridges of oceanic plates, but plate tectonics did not operate before 3 Ga. Life began around 4 Ga. The chemical nature of terrestrial hydrothermal systems, with their neutral pH and high $K^+/Na^+$ ratio, resembles the living cell's cytoplasm more closely than that of the submarine hydrothermal vents.

We explore alternative hypotheses of terrestrial hydrothermal systems as likely crucibles of life's origin. The hot spring hypothesis on land, similar to Darwin's musing that life may have originated in a 'warm little pond', is an attractive alternative in which fluctuating environment allows a wet/dry cycle for the condensation reaction of polymers and their encapsulation into protocells. But most of the long-lived hot springs needed for abiogenesis are related to plate boundaries, which were not operating in the Eoarchean eon.

Impact cratering was the primal geological force in the early history of our planet during the late heavy bombardment period before the onset of plate tectonics. Asteroid impacts left young Earth's crust pockmarked with craters, like the surface of the Moon. These hydrothermal crater systems might have provided the habitats for the origin of life. Impact formation that results in the formation of complex impact craters (>3-km diameters) exhibits several microenvironments within and around craters for prebiotic synthesis. Moreover, these crater lakes were interconnected by surface channels and underground fracture systems to form a network for exchanging heat and chemical ingredients; the surface channels allowed condensation reactions. The paleoecology of Earth's earliest microfossil records suggests impact-generated hydrothermal crater lakes were the likely cradle for the origin of life.

Scientists have long been fascinated by the possibility of life on Mars. Mars' current surface conditions are extremely challenging for life as we know it. But early Mars was very similar to early Earth, with liquid water on the surface, and was habitable. Mars is of particular interest for the study of the origin of life because of its similarity with the Eoarchean Earth. Both were warm, wet, rocky planets with oceans, rivers, and lakes. Both had an atmosphere dominated by carbon dioxide. Meteorite impacts on both worlds had delivered the building blocks of life and water and created innumerable hydrothermal crater lakes as the likely cradles for the origin of life. Like the Earth and Moon, Mars also experienced an intense asteroid impact-cratering process early in its history and exhibited more than 43,000 impact craters with diameters greater than 5 kilometers. These craters were filled with water in the Noachian eon, forming hydrothermal lakes. These sedimentary lake deposits are promising sites for the preservation of early life. This is why NASA rovers are probing impact crater lakes on Mars, such as the Gale Crater, the Jezero Crater, and the Eridania Basin, as the most likely setting for detecting biosignatures. If life could originate on one planet, it might have done so on the other. The search for life is a high priority for NASA's Mars exploration.

## ACKNOWLEDGMENTS

I thank Vera Kolb for inviting me to contribute this chapter to her edited volume *Guidebook for Systems Application in Astrobiology* and for critically editing the manuscript. I thank Volkan Sarigul for illustrations, Simon Marchi for providing Figure 5.2 showing the impacting history of early Earth, and NASA for images on the craters of Mars. P. W. Horn Distinguished Professor grant from Texas Tech University supported the work.

## REFERENCES

Abramov, O., and Kring, D. A. 2004. Numerical modeling of an impact-induced hydrothermal system at the Sudbury crater. *J. Geophys. Res.* 109. http://doi.org/10/1029/2003JE002213.

Abramov, O., and Kring, D. A. 2005. Impact-induced hydrothermal activity on early Mars. *J. Geophys. Res. Planets.* 110: 1–19.

Abramov, O., and Mojzsis, S. J. 2009. Microbial habitability of the Hadean Earth during the late heavy bombardment. *Nature.* 449: 419–422.

Akanuma, S., Nakajima, Y., Yokobori, S., Kimura, M., Nemoto, N., Mase, T., Miyazono, K., Tanokura, M., and Yamagishi, A. 2013. Experimental evidence for the thermophilicity of ancestral life. *Proc. Nat. Acad. Sci. USA.* 110: 11067–11072.

Alexander, C. M. O'D., Bowden, R., Fogel, M., Howard, K. T., Herd, C. D. K., and Nittler, L. R. 2012. The provenances of asteroids and their contributions to the volatile inventories of the terrestrial planets. *Science.* 337: 721–723.

Arndt, N. T., and Nisbet, E. G. 2012. Processes on the young earth and habitats of early life. *Ann. Rev. Earth Planet. Sci.* 40: 521–549.

Arnold, N. S., Butcher, F. E. G., Conway, S. J., Gallagher, C., and Balme, M. R. 2022. Surface topographic impact of subglacial water beneath the south polar icecaps of Mars. *Nature Astro.* 6: 1256–1262.

Bada, J. L., and Korenga, J. 2018. Exposed areas above sea level on Earth >3.5 Gyr ago: Implications for prebiotic and primitive biotic chemistry. *Life.* 8: 55. https://doi.org/10.3390/life8040055.

Baross, J. A., and Hoffman, S. E. 1985. Submarine hydrothermal vents and associated gradient environment as sites for the origin and evolution of life. *Orig. Life Evol. Bio.* 15: 327–345.

Bernhardt, H. S., and Tate, W. P. 2012. Primordial soup or vinaigrette: Did the RNA world evolve at acidic pH? *Biol. Dir.* www.biology-direct.com/content/7/1/4.

Bernstein, M. P., Sandford, S. A., and Allamonda, L. J. 1999. Life's far flung raw material. *Scient. Amer.* 263: 42–49.

Betts, H. C., Puttick, M. N., Clark, J. W., Donoghue, P. C. J., and Pisani, D. 2018 Integrated genomic and fossil evidence illuminates life's early evolution and Eukaryote origin. *Nature Ecol. Evol.* 2: 1556–1562.

Beyerly, G. R., Lowe, D. L., Wooden, J. L., and Xie, X. 2002. An Archean impact layer from the Pilbara and Kaapvaal cratons. *Science.* 297: 1325–1327.

Boss, A. P., and Keiser, S. A. 2010. Who pulled the trigger: A supernova or an asymmetric giant branch star? *Astro. J. Lett.* 717: L1–L5.

Brownlee, D. E., Tsou, P., Anderson, J. D., Hanner, M. S., Newburn, R. L., Sekannia, Z., Clark, B. E., Horz, F., Zolensky, M. E., Kissel, J., McDonnell, J. A. M., Sandford, S. A., and Tuzzolino, A. J. 2003. Stardust: Comet and interstellar dust sample return mission. *J. Geophys. Res.* 108: E10, 8111. https://doi.org/10.1029/2003/E0002087.

Carr, M. H. 1996. *Water on Mars.* New York: Oxford University Press.

Carr, M. H., and Head, J. W. H. 2010. Geologic history of Mars. *Earth Planet. Sci. Lett.* 293: 185–203.

Catling, D. C., and Zahnle, K. J. 2020. The Archean atmosphere. *Sci. Adv.* 6(9), https://doi.org/10.1126/sciadv.aax1420.

Chatterjee, S. 2016. A symbiotic view of the origin of life at the hydrothermal impact crater-lakes. *Phys. Chem. Chem. Phys.* 18: 20033–20046.

Chatterjee, S. 2019. The hydrothermal impact crater lakes: The crucibles of life's origin. In: V. Kolb (ed.). *Handbook of Astrobiology.* Boca Raton, FL: CRC Press, Taylor & Francis Group, pp. 265–269.

Chatterjee, S., and Yadav, S. 2019. The origin of prebiotic information system in the peptide/RNA world: A simulation model of the evolution of translation and the genetic code. *Life.* 9: 25. https://doi.org/10.3390/life9010025.

Chyba, C., and Sagan, C. 1992. Endogenous production, exogenous delivery and impact-shock synthesis of organic molecules: An inventory for the origin of life. *Nature.* 355: 125–132.

Clark, B. C., and Kolb, V. M. 2018. Comet Pond II: Synergistic intersection of concentrated extraterrestrial materials and planetary environments to form procreative Darwinian ponds. *Life.* 8: 12, https://doi.org/10.3390/life80200012.

Cockell, C.S. 2006. The origin and emergence of life under impact bombardment. *Phil. Trans. R. Soc.* B361: 1845–1856.

Cockell, C., and Lee, P. 2002. The biology of impact craters – A review. *Biol. Rev.* 77: 279–310.

Condie, K. C. 1981. *Archean Greenstone Belts.* Amsterdam, the Netherlands: Elsevier.

Damer, B., and Deamer, D. W. 2015. Coupled phases and combinatorial selection in fluctuating hydrothermal pools: A scenario to guide experimental approaches to the origin of cellular life. *Life.* 5: 872–887.

Deamer, D. W. 2011. *First Life.* Berkeley: University of California Press.

Deamer, D. W. 2019. *Assembling Life: How Can Life Begin on Earth and other Habitable Planets*? New York: Oxford University Press.

Deamer, D. W. 2020. *Origin of Life: What Everyone Needs to Know.* New York: Oxford University Press.

Deamer, D. W., Cary, F., and Damer, B. 2022. Urability: A property of planetary bodies that can support an origin of life. *Astrobiol.* 22. https://doi.org/10.1089/ast.2021.0173.

Deamer, D. W., and Weber, A. L. 2010. Bioenergetics and life's origins. In: D. W. Deamer and J. Szostak (eds.). *On the Origin of Life, Cold Spring Harb. Perspect. Biol.* 2: a004923

De Duve, C. 2005. *Singularities: Landmarks on the Pathways of Life.* New York: Cambridge University Press.

Delsemme, A. H. 1988. *Our Cosmic Origins.* Cambridge: Cambridge University Press.

Djokic, T., Van Kranendonk, M. J., Campbell, K. A., Walter, M. R., and Ward, C. R. 2017. Earliest signs of life on land preserved in ca, 3.5 Ga hot spring deposits. *Nat. Comm.* https://doi.org/10.1038/ncomms15263.

Dodd, M. S., Papineau, D., Grenne, T., Slack, J. F., Rittner, M., Prajno, F., O'Neil, J. O., and Little, C. T. S. 2017. Evidence for early life in Earth's oldest hydrothermal vent precipitates. *Nature.* 543: 60–65.

Farley, K. A. et al. 2022. Aqueously altered igneous rocks sampled on the floor of Jezero crater, Mars. *Science.* 277: eabo2196.

Farmer, J. D. 2000. Hydrothermal systems: Doorways to early biosphere evolution. *GSA Today.* 10: 1–9.

Furnes, H., Banerjee, N. R., Muehlenbachs, K., Staudigel, H., and de Wit, M. 2004. Early life recorded in Archean pillow lavas. *Science.* 304: 578–581.

Gault, D. E., Quaide, W. I., and Oberbeck, V. R. 1968. Impact cratering mechanics and structures. In: B. M. French and N. M. Short (eds.). *Shock Metamorphism of Natural Minerals.* Baltimore, MD: Mono Book Corp, pp. 87–99.

Glikson, A. Y. 2010. Archean asteroid impacts, banded iron formations and MIF-S anomalies. *Icarus.* 207: 39–44.

Grieve, R. A. F. 1990. Impact cratering on earth. *Scient. Amer.* 261: 66-73.

Grieve, R. A. F., Cintala, M. J., and Therriault, A. M. 2006. Large-scale impacts and the evolution of Earth's crust: The early years. In: W. U. Reimold and R. L. Gibson (eds.). Processes of the early earth. *Geol. Soc. Am. Spec. Pap.* 405: 23–31.

Grotzinger, J. P., Sumner, D. Y., Kah, L. C., et al. 2014. A habitable fluvio-lacustrine environment at Yellowknife Bay, Gale Crater, Mars. *Science.* 343. https://doi.org/10.1126/science1242777.

Hagerty, J. J., and Newsom, H. E. 2003. Hydrothermal alteration at the Lonar Lake impact structure, India: implications for impact cratering on Mars. *Meteor. Planet. Sci.* 38: 365-381.

Hansen, V. L. 2015. Impact origins of Archean Cratons. *Lithosphere.* 7. https://doi.org/10.1130/L371.1.

Harrison, T. M., Schmitt, A. K., McCulloch, M. T., and Lovera, O. M. 2008. Early (>4.5 Ga) formation of terrestrial crust: Lu-Hf, $\partial^{18}O$, and Ti thermometry results for Hadean zircons. *Earth Planet. Sci. Lett.* 268: 476–486.

Hawkesworth, C. J., Cawood, P. A., and Dhuime, B. 2016. Tectonics and crustal evolution. *GSA Today.* 26: 4–11.

Hazen, R. M. 2001. Life's rocky start. *Scient. Amer.* 284(4): 62–71.

Hazen, R. M. 2005. *Genesis: The Scientific Quest for Life.* Washington, DC: Joseph Henry Press.

Kasting, J. F. 2014. Atmospheric composition of Hadean-early Archean Earth: The importance of CO. *Geol. Soc. Sp. Paper.* 504: 19–28.

Kloprogge, J. T., and Hartman, H. 2021. Clays and the origin of life: The experiments. *Life.* 12: 259. doi.org/10.3390/life120202559

Koeberl, C., Schulz, T., and Reimold, W. U. 2015. Remnants of early Archean impact deposits on Earth: Search for a meteorite component in the BARBS and CT3 drill cores (Barberton Greenstone Belt, South Africa. *Proc. Eng.* 103: 310–317.

Kring, D. A. 2000. Impact events and their effect on the origin, evolution, and distribution of life. *GSA Today.* 10(8): 1–7.

Kring, D. A., and Cohen, B. A. 2002. Cataclysmic bombardment throughout the inner Solar System 3.9–4.0 Ga. *Jour. Geophys. Res. Planets*. 107: E2. https://doi.org/10.1029/2001JE001529.

Kyte, F. T., Shukolyukov, A., Lugmair, G. W., Lowe, D. R., and Byerly, G. R. 2003. Early Archean spherule beds: Chromium isotopes confirm origin through multiple impacts of projectiles of carbonaceous chondrite type. *Geology*. 31: 283–286.

Lane, N., and Martin, W. F. 2012. The origin of membrane bioenergetics. *Cell*. 151: 1406–1416.

Lewis, R. E., and Young, H. W. 1982. Geothermal resources in the Banbury Hot Springs area, Twin Falls County, Idaho. *US Geol. Surv. Water-Supply Paper*. 2186: 1–27.

Liu, Y., Wu, X., Zhao, Y. U. S. et al. 2022. Zhurong reveals recent aqueous activities in Utopia Planitia, Mars. *Sci. Adv*. 8(9). https://doi.org/10.1126/sciadv.abn855.

Lowe, D. R., Byerly, G. R., Kyte, F. T., Shukolyukov, A., Asaro, F., and Krull, A. 2003. Geological and geochemical record of 3400-million-year-old terrestrial meteorite impacts. *Science*. 245: 959–962.

Marchi, S., Bottke, W. F., Elkins-Tanton, L. T., Bierhaus, M., Wuennemann, K., Morbidelli, A., and Kring, D. A. 2014. Widespread mixing and burial of Earth's Hadean crust by asteroid impacts. *Nature*. 511: 578–582.

Martin, W. F., Baross, J., Kelley, D., and Russell, M. J. 2008. Hydrothermal vents and the origin of life. *Nat. Rev. Microbiol*. 6: 805–814.

Martin, W. F., and Russell, M. J. 2007. On the origin of biochemistry at an alkaline hydrothermal vent. *Phil. Trans. R. Soc*. B362: 1887–1926.

Martin, W. F., Sousa, F. L., and Lane, N. 2014. Energy at life's origin. *Science*. 344: 1092–1093.

McKay, C. P., and Stoker, C. R. 1989. The early environment and its evolution on Mars: Implications for life. *Rev. Geophys*. 27: 189–214.

Melosh, H. J. 1989. *Impact Cratering: A Geologic Process*. New York: Oxford University Press.

Michalski, J. R., Noe Dobrea, E. Z, Niles, P. B., and Cuadros, J. 2017. Ancient hydrothermal seafloor deposits in Eridania basin on Mars. *Nat. Comm*. 8. https://doi.org/10.1038/ncomms15978.

Miller, S. R., Mueller, P. A., Meert, J. G., Kamenov, G. D., Pivarunas, A. F., Sinha, A. K., and Pandit, M. K. 2018. Detrital zircons reveal evidence of Hadean crust in the Singhbhum Craton, India. *J. Geol*. 126: 541–552.

Mojzsis, S. J., Harrison, T. M., and Pidgeon, T. T. 2001. Oxygen-isotope evidence from ancient zircons for liquid water at the Earth's surface 4,300 Myr ago. *Nature*. 409: 178–181.

Mulkidjanian, A. Y., Bychkov, A. B., Diprova, D. V., Galperin, M. Y., and Koonin, E. B. 2012. Origin of first cells at terrestrial, anoxic geothermal fields. *Proc. Nat. Acad. Sci. USA*. 109: E821–E830.

Nisbet, E. G., and Sleep, N. H. 2001. The habitat and nature of early life. *Nature*. 409: 1083–1091.

O'Neil, J., and Carlson, R. W. 2017. Building Archean cratons from Hadean mafic crust. *Science*. 355: 1199–1202.

Osinski, G. R., Spray, J., and Lee, P. 2001. Impact-induced hydrothermal activity within the Haughton impact structure: Generation of warm, wet oasis. *Meteorit. Plant. Sci*. 36: 731–745.

Osinski, G. R., Tornabene, L. L., Banerjee, N. R., Cockell, C. S., Flemming, R., Izawa, M. R. M., McCutcheon, J., Parnell, J., Preston, L. J., Pickersgill, A. E., Pontefract, A., Sapers, H. M., and Southam, G. 2013. Impact-generated hydrothermal systems on Earth and Mars. *Icarus*. 224: 347–363.

Parnell, J., Lee, P., Cockell, C. S., and Osinski, G. R. 2004. Microbial colonization in impact-generated hydrothermal sulphate deposits, Haughton impact structure, and implications for sulfate on Mars. *Int. J. Astrobiol*. 3: 247–256.

Paul, D., Kumbhare, S. V., Mhatre, S. S., Chowdhury, S. P., Shett, S. A., Marathe, N. P., Bhute, S., and Shouche, Y. S. 2016. Exploration of microbial diversity and community structure of Lonar lake: The only hypersaline meteorite crater-lake within basalt rock. *Front. Microbiol.* https://doi.org/10.3389/fmicb.2015.01553.

Pearce, B. K. D., Pudritz, R. E., Semenov, D. A., and Henning, T. K. 2017. Origin of the RNA world: The fate of nucleobases in warm little ponds. *Proc. Nat. Acad, Sci.* 114: 11327–11332.

Reysenbach, A. L., and Cady, S. L. 2001. Microbiology of ancient and modern hydrothermal systems. *Trends Microbiol.* 9: 79–86.

Ryder, G. 2003. Bombardment of the Hadean Earth: Wholesome or deleterious. *Astrobiol.* 3: 3–6.

Sapers, H. M., Osinski, G. R., Banerjee, N. R., and Preston, L. J. 2015. Enigmatic tubular features in impact glass. *Geology.* 43: 635–638.

Sauterey, B., Charnay, B., Affholder, A., Mazevet, S., and Ferriere, R. (2022). Early Mars habitability and global cooling by H2-based methanogenesis. *Nat. Astro.* 6: 1263–1271.

Schmitt, H. H. 2015. Potential catalytic roles in prebiotic organic synthesis. In: G. R. Osiniski and D. A. Kring (eds.). *Large Meteorite Impacts and Planetary Evolution V.* Boulder, CO: Geological Society of America Special Papers, pp. 1–16.

Shock, E. L., and Schulte, M. D. 1998. Organic synthesis during fluid mixing in hydrothermal systems. *J. Geophys. Res.* 103: 28513–28527.

Stern, R. J. 2005. Evidence from ophiolites, blueschists, and ultrahigh-pressure metamorphic terranes that the modern episode of subduction tectonics began in the Neoproterozoic time. *Geology.* 33: 5557–5560.

Stetter, K. O. 2006. Hyperthermophiles in the history of life. *Phil. Trans. R. Soc.* 361: 1837–1843.

Stüekan, E. E., Tino, C., Arp, G., Jung, D., and Lyons, T. W. 2020. Nitrogen isotope ratios trace high pH conditions in a terrestrial Mars analog site. *Sci. Adv.* 6: eaay3440.

Sugitani, K. 2018. Fossils of ancient microorganisms. In: V. M. Kolb (ed.). *Handbook of Astrobiology.* Boca Raton, FL: CRC Press; Taylor & Francis Group, pp. 567–596.

Tang, M., Chen, K., and Rudnick, R. L. 2016. Archean upper crust transition from mafic to felsic marks the onset of plate tectonics. *Science.* 351: 373–375.

Wacey, D., Kilburn, M. S., Saunders, M., Cliff, J., and Brasier, M. D. 2011 Microfossils of sulfur-metabolizing cells in 3.4-billion-year-old rocks of Western Australia. *Nat. Geosc.* 4: 698–702.

Wachterhäuser, G. 1993. The cradle of chemistry of life: on the origin of natural products in a pyrite-pulled chemoautotrophic origin of life. *Pure Appl. Chem.* 65: 1343-1348.

Weiss, M. C., Sousa, F. L., Mrnjavac, N., Neukirchen, S., Roettger, M., Nelson-Sathi, S., and Martin, W. F. 2016. The physiology and habitat of the last common ancestor. *Nature Microbiol.* 1:161116, doi: 10.1038/nmicrobiol.2016.116.

Westall, F., Campbell, K. A., Breheret, J. G., Foucher, F., Gautret, P., Hubert, A., Sorfieul, S., Grassineau, N., and Guido, D. M. 2015. Archean (3.33 Ga) microbe-sediment systems were diverse and flourished in a hydrothermal context. *Geology* 43:615-618

Westall, F., de Wit, M. J., Dann, J., van der Gaast, S., Ronde, C. E. J., and Gerneke, D. 2001. Early Archean fossil bacteria and biofilms in hydrothermally-influenced sediments from the Barberton Greenstone belt, South Africa. *Precamb. Res.* 106: 93–116.

Woese, C. R. 1994. The universal ancestor. *Proc. Nat. Acad. Sci. USA.* 99: 6854–6859.

Zegers, T. E., De Wit, M. J., Dann, J., and White, S. H. 1998. Vaalbara, Earth's oldest assembled continent? A combined structural, geochronological, and palaeomagnetic test. *Terra Nova.* 10: 250–259.

Zegers, T. E., and Ocampo, A. 2003. Vaalbara and tectonic effects of a mega impact in the early Archean 3470 Ma. Abstract, *Third International Conference on Large Meteorite Impacts.* Nordington, Germany.

# 6 Systems Geochemistry in an Astrobiological Context

*Maheen Gull and Matthew A. Pasek*

## 6.1 GEOLOGIC CYCLES AS SYSTEMS

The Earth is characterized by multiple systems acting on its surface and interior. These systems create habitable environments and are influenced by biological changes, which result in a complex interplay between the surface of the planet and the organisms that inhabit it. Although these interactions are not limited to chemical interactions, the science of geochemistry is especially well suited to think of these interactions in systems chemistry terms. Systems chemistry is characterized by understanding the networks of reactions that result in the production of higher ordered phenomena and systems that don't only go thermodynamically downhill but instead are maintained through the addition of energy. Like a systems chemistry experiment, geological environments also function through out-of-equilibrium processes, with a simple example being the day–night cycle driving changes in temperature.

Presuming life arose on Earth and then understanding geochemistry as a system in turn provide an understanding of the origin of living cells from nonliving matter. The biochemical complexity of living cells suggests that the formation of cells was a consequence of a long evolutionary process (Kee and Monnard, 2017). Over time, many different theories have been proposed for the origin of life; yet, we are still far from understanding the processes that involved the transition from nonliving matter to life (Ruiz-Mirazo et al., 2014). Some of the major problems in understanding the origin of life include understanding the syntheses of simple yet essential biomolecules such as amino acids, sugars, and nitrogenous bases; to how these formed complex assortments of biomolecules such as lipids, peptides, nucleosides, and nucleotides; to how these engaged to form proto-biochemical machinery; and then how polymerization and concentration formed cellular compartments that could function independently leading to the capabilities such as storage, safe replication, and transfer of genetic information (Ruiz-Mirazo et al., 2014; Pasek et al., 2017).

Many of these questions are fundamentally approached through biochemical investigation to determine the development of life. Instead, geochemistry may be better able to address how the availability of key nutrients (e.g., C, N, O, P, and S) that form these biomolecules arose and became available as part of planetary habitability. In this chapter, we shall first draw parallels between systems chemistry


DOI: 10.1201/9781003294276-6

and geologic processes; then we will highlight a few major chemical systems in the geochemical context, involving major biogenic elements such as phosphorus, carbon, and nitrogen, which would have played a major role in the origin of biopolymers forming life and – after the origin of life – would have been substantially altered by life itself.

The student of geology is introduced early on to the concept of geologic cycles. For example, the water cycle is the process that involves evaporation, precipitation, and surface water flow. The water cycle is ultimately powered by the sun and causes the evolution of the Earth's surface through weathering of rocks and the deposition of sediment. An active water cycle is critical for the development of life as we know it, and the Earth is largely habitable because of it.

Multiple other systems dominate geological processes. This is because Earth is an active planet. Unlike other worlds, many of which seem to be geologically inactive as indicated by their ancient surfaces, Earth is constantly renewing its surface through volcanism, altering it through the breakdown of rocks, and recycling it through plate tectonics.

### 6.1.1 The Rock Cycle

Students are introduced to the three main rock types in an introductory geology class: sedimentary, igneous, and metamorphic rocks. Sedimentary rocks are those that are formed from sediments and often contain fossils, igneous rocks are those formed from molten rock (either lava on Earth's surface or magma in its interior), and metamorphic rocks are those rocks that have been heated and/or pressurized to the point that individual grains and rock textures are changed and often bear new minerals.

Igneous rocks are the first rock type to form and are the most common rock type on other planets. When these rocks weather, for example, by break apart by the action of water or wind, pieces of these rocks change physically and chemically and migrate away from the parent rock. These smaller rock chunks are called sediment, and when they are transported by a fluid of some sort (e.g., water, wind, ice) and finally settle out, the sediment can lithify, forming a sedimentary rock. Lithification happens in this case through the cementing of grains along grain boundaries (e.g., by carbonate minerals gluing together individual mineral grains). After a rock forms, it can be buried in the Earth where, as it descends through burial by other rock, it is heated and subjected to high pressure. This heating under pressure causes rocks to metamorphose into new rocks, often bearing new minerals and textures compared to their precursor. Any rock can be metamorphosed, from sedimentary to igneous to even metamorphic rocks. Notably, if a rock is heated too far, it may melt, which when it recrystallizes would now be considered an igneous rock.

The rock cycle is driven by the geologic process of plate tectonics. In plate tectonics, plates move on the surface of the Earth, forming at 'spreading centers' and returning to the subsurface at 'subduction zones'. Plate tectonics is itself driven by huge convection currents within the subsurface, which generate regions where magma flows up and down. Igneous rocks form at the places where this magma flows up (at the spreading centers and at other volcanic regions) and are also made at subduction zones, mainly because volatiles trapped in rocks at the regions boil off and up the subducting rock.

The rock cycle caused by plate tectonics has caused some of the more unique geology on the surface of the Earth. Most notably, the cycling of rock through this process has enriched the silica ($SiO_2$) content of the crust, changing the dominant rock type into silica-rich rocks such as granite. Granite and other silica-rich rocks (relative to the mantle or bulk silicate earth) are less dense and float on the denser rock as the crust. Indeed, silica-rich igneous rocks are quite rare in the solar system but are prevalent across the Earth. These silica-rich rocks have also assembled together to make the 'cratons', which are large blocks of old, silica-rich crust that make up the continents. As a result, Earth is separated into two main types of crust: The continental crust makes up the land masses and the oceanic crust is where the oceans settle.

In addition to causing a separation of the Earth's crust, the rock cycle and plate tectonics both result in a diversification of minerals. Due to chemical separation occurring as part of the rock cycle, elements that are usually finely dispersed or rare have a chance to accumulate in larger masses, ultimately giving rise to new minerals otherwise not found in any significant quantity elsewhere in the solar system. This includes the ores of valuable materials that modern society mines for industry (e.g., Cu, Co, Au).

The process of plate tectonics also recycles elements that have been locked up in rocks. For example, carbon is mostly found as carbonate minerals such as calcite in sedimentary rocks. However, when these sedimentary rocks are subducted, the heat causes the carbonates to decompose, releasing $CO_2$ that then enters the atmosphere. Similarly, water and sulfur can also be recycled through plate tectonics. Sometimes, this recycling also couples to a change in redox state, including the formation of methane from $CO_2$ and sulfides from sulfate.

Comparing this to systems in chemistry, the rock cycle is a process that is fed from a thermodynamic driver (plate tectonics) that diversifies rocks, forming a greater variety of rocks and minerals than those planets without such a driver. 'Self-assembly' is also evident in the formation of continents, which may have been critical to the formation of life, given that life may have required fresh water to originate (best found on continents!). While drawing analogies between the rock cycle and systems chemistry involves some stretching of definitions, it should be apparent that the Earth's unique geology is due in no small part to the presence of active cycling and recycling of its rocks (Figure 6.1).

The rock cycle operates on geologic timescales. Geologic timescales are usually periods of millions of years, in contrast to chemical reaction timescales studied by the typical bench chemist, which usually take seconds to days to react. The rock cycle is very slow comparatively: as an example, the oldest oceanic crust is about 200 million years old, which suggests that its formation at mid-ocean ridges and eventual subduction at tectonic boundaries takes about this long to proceed. This is one of the fastest rock-cycling processes on Earth's surface.

### 6.1.2 The Water Cycle

In searching for a more laboratory-relevant geologic process, the water cycle is a much closer match to the lifetime of a benchtop experiment. In contrast to the rock

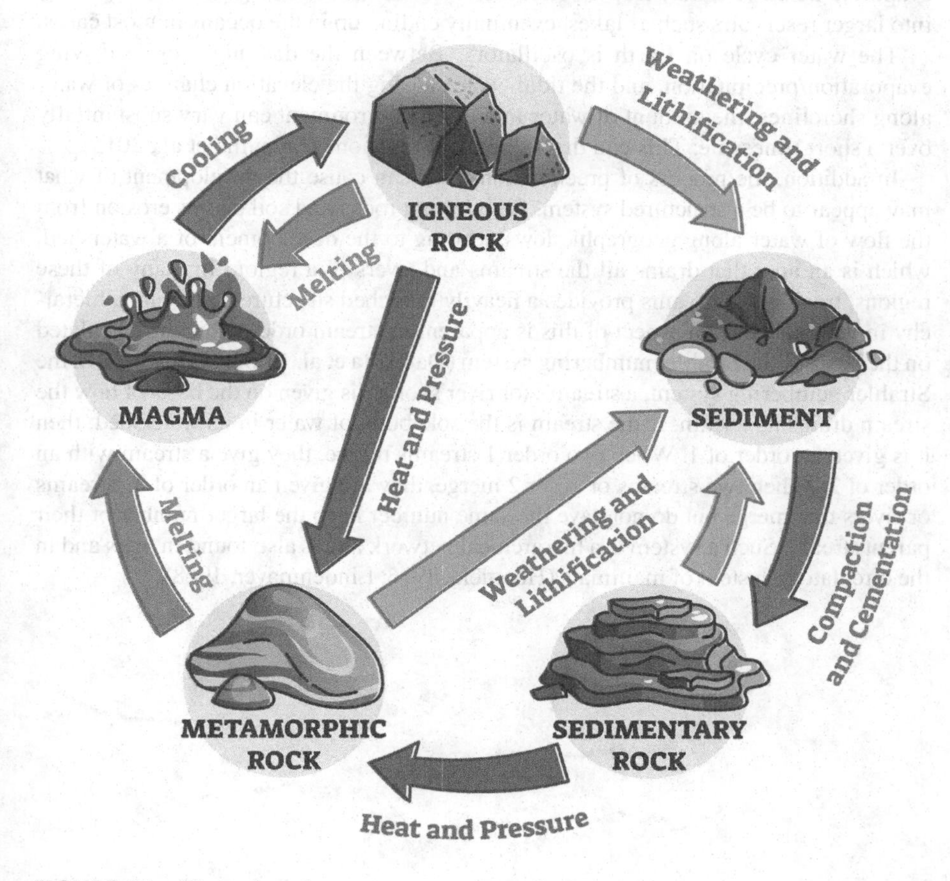


**FIGURE 6.1** The rock cycle.

cycle, the water cycle is ultimately powered by the sun, which results in cycling on the order of a day. Indeed, the water cycle operates on timescales as short as hours (think storms) to timescales as long as months (think water flowing through rivers), to even longer timescales (water's residence time in the ocean).

In the water cycle, water evaporates from surface bodies (or sublimes from ice), whenever the carrying capacity of the surrounding air is below the vapor pressure of water at a given temperature. This is usually given as a percentage, termed relative humidity. For example, 100% humidity implies that the partial pressure of water in air is equivalent to its vapor pressure, and thus no water is capable of evaporating at that point (it often feels quite hot then, as sweating no longer cools you!). Water vapor then reaches a region where the temperature drops, which usually happens when the air is lifted, which cools it. When the air temperature drops, the humidity reaches

100%, and thus the water condenses and forms precipitation. Precipitation plummets to Earth's surface, where it gathers at topographic lows. At these lows, the water may evaporate again or it may flow as streams and rivers (weathering and altering rock) into larger reservoirs such as lakes, eventually ending up in the oceans in most cases.

The water cycle on Earth is oscillatory. Between the day–night cycle driving evaporation/precipitation, and the tidal cycle causing the elevation changes of water along shorelines, the amount of water in a given environment can vary substantially over a short timescale. This can drive chemical reactions (Forsythe et al., 2015).

In addition, the process of precipitation itself can cause the development of what may appear to be a structured system. Rain hitting rocks and soil causes erosion from the flow of water along geographic lows, leading to the development of a watershed, which is an area that drains all the streams and rivers of a region. In many of these regions, tracing the streams provides a heavily branched structure that bears a hierarchy in its branches. An aspect of this is apparent in stream order, which is calculated on the basis of the Strahler numbering system (Da Costa et al., 2002, Figure 6.2). In the Strahler numbering system, a stream's (or river's) order is given on the basis of how the stream drains its terrain. If the stream is the sole body of water in its watershed, then it is given an order of 1. When two order 1 streams merge, they give a stream with an order of 2. When two streams of order 2 merge, they are given an order of 3. Streams or rivers that merge but do not have the same number keep the larger number of their parent stream. Such a system is a hierarchical network and is also found in trees and in the circulatory system of mammals (Horsfield, 1976; Lindenmayer, 1968).

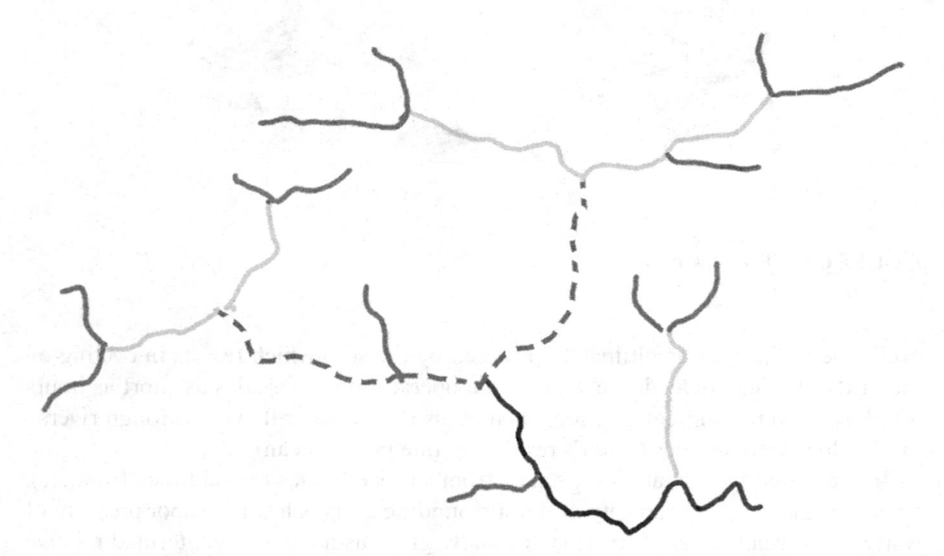

**FIGURE 6.2** In the Strahler stream numbering system, first-order streams (dark gray) are the primary streams that drain a single watershed. When two of these streams meet, they form a second-order stream (light gray). When two second-order streams meet, they form a third-order stream (dotted gray), and finally when two third-order streams meet, they form a fourth-order stream (solid black). The fourth-order stream will bear the highest flow of water of these.

Although the formation of a hierarchical network from rain (Figure 6.3) may seem to be only loosely connected to systems chemistry and astrobiology, the presence of such a hierarchical drainage pattern is in fact used to identify a planet's potential habitability. For example, the stream systems on Mars have been observed to be substantially different than terrestrial systems (Galofre et al., 2020). This may indicate less habitable conditions than Earth (which is certainly the case today!). In contrast, Saturn's moon Titan bears somewhat more structured systems, which may indicate a more active cycling on this moon (Burr et al., 2013) (Figure 6.3).

### 6.1.3 Biogeochemical Cycles

Astrobiology is by its nature focused on life and its interplay with its environment, whether that environment is – Earth or another world. One of the main ways life interacts with its environment is through the establishment of an active recycling of vital materials from the environment. Many key materials are slowly lost from an ecosystem, for example, by washing away to the ocean, vaporizing into the atmosphere, or crystallizing out as minerals. Biology mitigates these effects by extracting nutrients from minerals, fixing elements from the atmosphere, and the altering of younger rock to reclaim nutrients. Biogeochemical cycles are by their nature superimposed on abiotic cycles such as the rock cycle and water cycle, but the activity of life usually alters these and other cycles in such a way to make a planet more habitable (e.g., the rise of oxygen enabled modern biology on Earth, though its first appearance may have been highly detrimental). To this end, when considering biogeochemical cycles, there are several aspects of the chemistry of an element to consider.

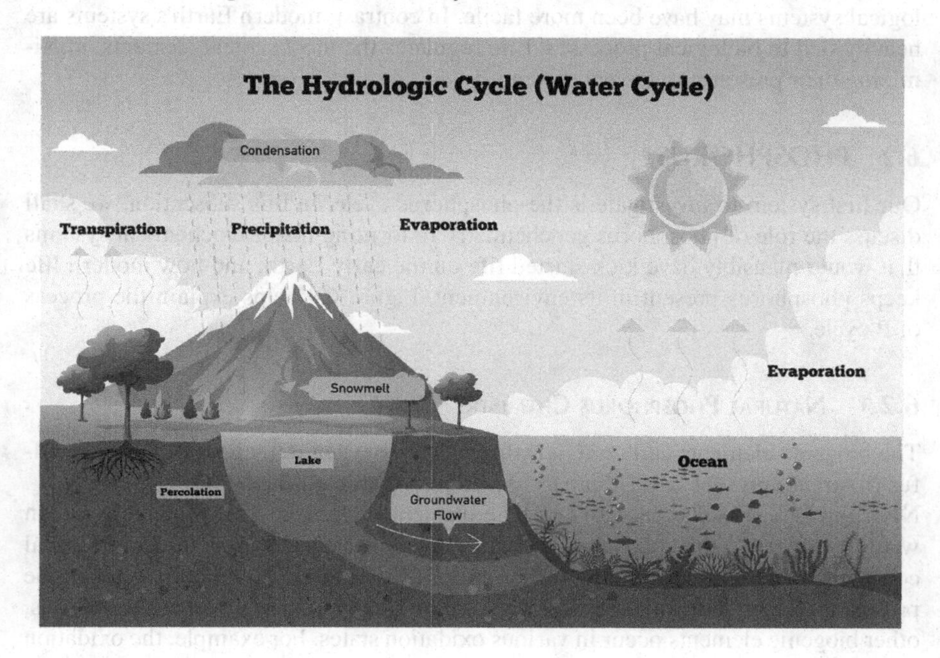


**FIGURE 6.3** The water cycle.

First of all, in the astrobiological context, consider how life uses a given element. What sorts of biomolecules incorporate that element, and what role does it play as part of the chemical structure of that biomolecule? Some elements get used as well as inorganic feedstock molecules for energy. Oxygen (as $O_2$) is a driver of chemistry, where organics such as sugars are broken down then joined with oxygen to make $CO_2$. In such a route, life uses oxygen as an inorganic reagent (both its reactant – $O_2$ – and product – $CO_2$ – are inorganic) to release the energy of organic compounds.

Most biologic elements are redox-sensitive: Certain redox states are favored under certain conditions. Some of these redox changes are induced by biology such as oxygen, where $O_2$ (oxidation state is 0) is a waste product of photosynthesis, and its reduction to $O^{2-}$ drives metabolism. Others are influenced by physical processes, such as UV light and reactions with rocks.

In many cases, there is a large reservoir of a given biologic element that serves as its source for biochemistry. For many elements, this is the atmosphere (e.g., $CO_2$ in photosynthesis, $N_2$ in nitrogen fixation): Many elements are present as volatiles. Other elements are extracted from minerals or water with dissolved ions (e.g., P and S). In some cases, minerals are actively broken down in life to acquire a key element. This may occur through the solubilization of minerals, which is often done enzymatically. Alternatively, when an ecosystem dies off, the element may crystallize out and rejoin the rock cycle.

Part of our understanding of how life arose on Earth also considers how the chemistry of the biologic elements may have been different on early Earth. For example, some elements may have been more soluble, and more reducing redox states may have been more abundant. As a result, the incorporation of these elements into biological systems may have been more facile. In contrast, modern Earth's systems are heavily tied to biological processes. Life regulates the most critical elements, maximizing their presence in the ecosystem.

## 6.2 PHOSPHORUS

Our first system to investigate is the phosphorus cycle. In this subsection, we shall discuss the role of phosphorus geochemistry in forming major biochemical systems that would plausibly have kick-started life on the early Earth, and how modern life keeps phosphorus present in its environment. Figure 6.4 helps explain the process of P-cycle.

### 6.2.1 Natural Phosphorus Chemistry

The biogenic elements carbon (C), hydrogen (H), oxygen (O), nitrogen (N), and sulfur (S) are all present at least in part as various volatile phases – $CO_2/CH_4$, $H_2O$, $N_2$, $NH_3$, and $H_2S$ (Pasek et al., 2017). These elements are also usually soluble in ocean water. Contrary to this, phosphorus (P) has no major volatile phase under terrestrial conditions, and its solvated ion – orthophosphate – is poorly soluble in water in the presence of divalent cations such as $Ca^{2+}$. Additionally, under terrestrial conditions, other biogenic elements occur in various oxidation states. For example, the oxidation state of carbon ranges from −4 (methane) to +4 ($CO_2$), nitrogen from −3 ($NH_3$) to 0

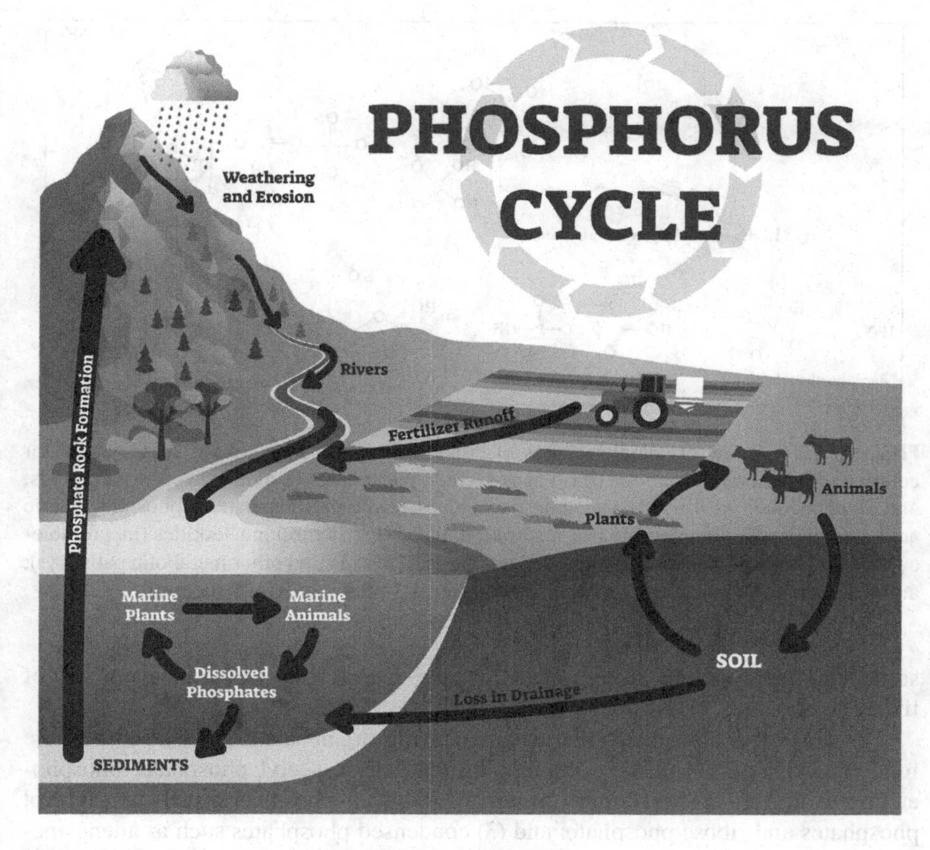


**FIGURE 6.4** Natural phosphorus (P) cycle on the Earth.

($N_2$) to +5 ($NO_3^-$), oxygen from 0 ($O_2$) to −2 (most everything else), hydrogen from +1 (most cases) to 0 ($H_2$), and sulfur from −2 ($H_2S$) to +6 ($SO_4^{2-}$) [3]. Phosphorus, however, under many terrestrial conditions exists in single oxidation state of +5 (e.g., $HPO_4^{2-}$) (Pasek et al., 2017). To this end, P is generally found as a constituent of phosphate minerals.

Phosphorus mainly occurs in nature in the tetrahedral state having five valence electrons. The exception includes elemental forms of phosphorus and the phosphide minerals (Pasek et al., 2017). Phosphine, the only naturally occurring gaseous form of P, also has a lone pair of electrons, and like all other, naturally occurring, long-term stable P compounds, it has a tetrahedral coordination geometry (Pasek et al., 2017).

Phosphorus comprises about 1% of the dry weight of cells (Pasek et al., 2017; Pasek and Kee, 2011; Gull, 2014). It serves in all life-sustaining processes including (1) cellular metabolism and cellular respiration that involves various sugar phosphates and various P-containing enzymes and coenzymes, (2) cellular structure as phospholipids, and (3) as an essential constituent of information-storing molecules RNA and DNA (Gull, 2014; Gull and Pasek, 2021). Around 44% of all metabolic compounds bear P in some form (Srinivasan and Morowitz, 2009). Figure 6.5 shows

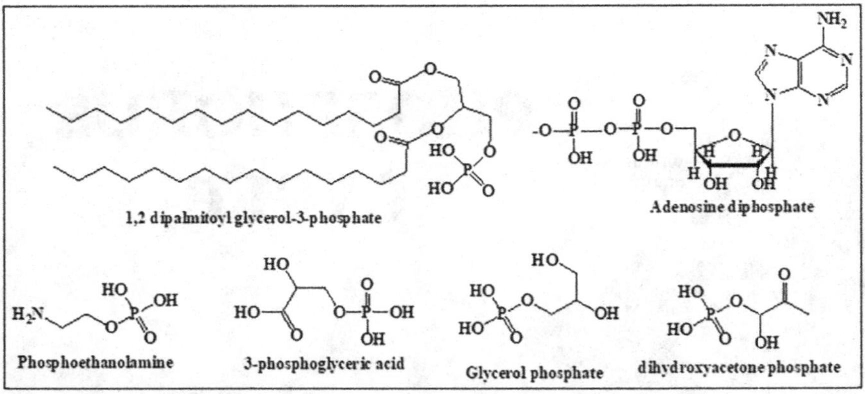

**FIGURE 6.5** A few organophosphates of biological significance, which are essential for cellular activities and hence for life. These include phosphates of aldehydes, ketones, acids, and alcohols such as dihydroxyacetone phosphates, glycerol phosphate, 3-phosphoglyceric acid, and phosphoethanolamine (for respiration and cell structure); nucleotides (macromolecules) such as adenosine diphosphate (for storing information and other metabolic pathways); and (phospholipids) such as 1,2-dipalmitoyl glycerol-3-phosphate (for structure).

some significant biological phosphates playing central roles in the biochemistry of living organisms.

The biological phosphates that occur in living organisms are classified into the following classes: (1) reactive organophosphates e.g., acetyl phosphate, phosphoenolpyruvate, phosphocreatine; (2) stable biological phosphates such as glycerol phosphates and ribose phosphate; and (3) condensed phosphates such as adenosine-di-phosphates (ADP) and adenosine-tri-phosphates (ATP) (Ruiz-Mirazo et al., 2014). Biological phosphates containing C–O–P linkages are essential for life (Pasek, 2008). The prebiotic relevance and significance of P in the context of origin of life are old concepts dating to Darwin's time. Darwin's ideology pertaining to the significance of P in the origin of life can be studied from his letter written to his close friend Joseph Dalton Hooker in which he wrote:

> [I]t is often said that all the conditions for the first production of a living being are now present, which could ever have been present. But if (and oh what a big if) we could conceive in some warm little pond with all sort of ammonia and phosphoric salts, light, heat, electricity present, that a protein compound was chemically formed, ready to undergo still more complex changes.
>
> (Pereto et al., 2009)

### 6.2.2 Phosphorus Mineral Reservoirs

Phosphate minerals are the dominant source of P on Earth. As such, the P biogeochemical cycle is intimately associated with the rock cycle. Phosphorus at present is supplied primarily through rock weathering to life and then actively recycled by organisms. Prior to the development of life on Earth, P would have originated primarily from mineral dissolution and weathering.

Supplementing this endogenous source would have been exogenous sources. Meteoritic impacts are considered to be directly linked with the origin of life on early Earth (Cooper et al., 1992; Cockell, 2006; Osinski et al., 2020). Besides various organic compounds, meteorites are also known to have provided a non-negligible portion of the P to early Earth, possibly through heavy bombardment events (Macià et al., 1997; Pasek and Lauretta, 2005). The meteoritic mineral schreibersite [$(Fe,Ni)_3P$]) is an iron–nickel containing phosphide and is ubiquitous in iron meteorites as well as many other types of meteorites (Pirim et al., 2014; Bryant and Kee, 2006). This phosphide mineral is reported to be among the first inorganic P compounds to condense from the solar nebula as part of homogeneous accretion model and, thus, is considered to be one of the most ancient P minerals within our solar system (Pasek, 2008).

Approximately 5–10% of all crustal P was at some stage delivered as meteoritic phosphide minerals (Pasek and Lauretta, 2005; Bryant and Kee, 2006). Schreibersite (and other phosphides) corrode in water by oxidation to release several inorganic P species in aqueous solutions with a concomitant release of $H_2$ (Pasek et al., 2017 and 2007; Pasek and Lauretta, 2008). Pasek and Lauretta have suggested about $10^8$ kg/year, while Ritson and colleagues (Ritson et al., 2020) have proposed that around $3 \times 10^7$ kg/year of meteoritic P could have been delivered by meteorites to early Earth.

### 6.2.3 Geochemical Routes to Providing Reactive P on the Early Earth

The ultimate source of P on early Earth would have been a mineral source. Various prebiotically relevant P minerals would have been present on early Earth (Hazen, 2013; Hazen et al., 2008), as a result of being primordial, or in other words sourced from the basic building blocks of early Earth (Pasek, 2020). With these geochemical constraints on early Earth, these prebiotic mineral sources could have served as a P source either by releasing P through dissolution or by using reactive reagents that would support endergonic processes such as phosphorylation.

Prebiotic phosphorylation could take place either as a nucleophilic or electrophilic reaction (Pasek, 2020). In contemporary biology, the process of phosphorylation usually proceeds through an electrophilic reaction on the P atom as shown by $^{18}O$ isotopic studies (Cohn, 1953; Molden et al., 2014). However, under limited water activity, nucleophilic reactions are possible. In any case, a generic phosphorylation reaction for organophosphate formation is given as:

$$\mathbf{HO\text{-}PO_3H_n{}^{n-2} + R\text{-}OH \rightarrow RO\text{-}PO_3H_n{}^{n-2} + H_2O} \quad (1)\ \textbf{(Pasek, 2020)}$$

where R represents a generic organic compound. In this case, water is the leaving group. To favor this reaction, the organic and the phosphate must have a high activity (as reactants), and there must be a low activity for water. Therefore, such reactions only happen under dry conditions with minimum water activity and relatively at higher temperatures (Gajewski et al., 1986). Heat is helpful for phosphorylation because the enthalpy for such reactions is positive (endothermic).

One of the major issues of phosphorylation is the low availability of phosphate. The major phosphate mineral on Earth today is apatite (actually a mineral group), which is minimally soluble (a good thing, since it makes up your teeth!). Apatite therefore can be considered to be implausible as a source of phosphate to reach in the molar concentrations utilized by prebiotic chemists in model phosphorylation studies (Pasek et al., 2017; Ponnamperuma and Chang, 1971; Powner et al., 2009; Fernández-García et al., 2017). However, early Earth geological conditions could have favored high-phosphate concentrations, and perhaps the clues to answering such issues might be found by considering the geochemistry of the primordial Earth coupled to other key geological factors.

Phosphate's poor solubility is observed in the soil sciences. In basic soils, the calcic minerals can trap phosphate in the form of calcium phosphate minerals, thus limiting phosphate's solubility. In acidic soils, aluminum and ferric iron capture phosphate. Thus, phosphate is trapped under nearly all pH conditions (Pasek et al., 2017). This principle is also directly applicable to the fertilizer industry. Phosphate-bearing fertilizers are made by employing acid calcium phosphates (e.g., $Ca(H_2PO4)_2$), as well as basic phosphates such as ammonium phosphates ($(NH_4)_2HPO_4$). Both of these phosphate salts are much more soluble than apatite or other common, natural phosphate minerals.

Being guided by the fertilizer industry, the prebiotic synthesis of organophosphates often uses phosphate salts such as $NaH_2PO_4$ and $Na_2HPO_4$ as phosphorylating agents (Tarelli and Wheeler, 1993; Reimann and Zubay, 1999; Hargreaves et al., 1977; Lohrmann and Orgel, 1971). None of salts are common as minerals in rocks today. An alternative route for phosphate solubilization is chelation. Chelation occurs when cations associate with organic complexes. As an example of chelation, the organic complex EDTA (ethylenediaminetetraacetic acid) binds to various cations thus making anions such as phosphate more soluble in solution. From the prebiotic context, Schwartz suggested that chelation with acetate could enhance phosphate solubility (Schwartz, 1972).

A final alternative route to phosphate mineral dissolution is via mineral transformation. Under a high $CO_2$ pressure, presumed to be present on the primordial Earth (see Section 4), apatite could have carbonated to form calcite, thereby liberating phosphate into aqueous solutions (Kakegawa et al., 2002). The suggested chemical reaction is given in Eq. (2):

$$\mathbf{Ca_5(PO_4)_3OH + 4H_2O + 5CO_2\,(g) \rightarrow 5CaCO_3 + 3H_2PO_4^- + 3H^+} \quad (2)\ \textbf{(Pasek et al., 2017)}$$

Similar to this pathway is a proposed pathway in alkaline lakes. On early Earth, carbonate-rich lakes could have concentrated phosphate to >1 molar levels by locking up calcium in carbonate minerals, which could plausibly have prevented phosphate removal by apatite formation (Toner and Catling, 2020).

Pasek suggested P redox as a route to altering phosphorus geochemistry on early Earth (Pasek, 2008; Pasek et al., 2013). This proposal was built on one by Gulick, who suggested the significance of reduced-oxidation-state P (reduced P, hereafter) compounds in the origin of life owing to their increased solubility in water. Reduced P compounds may be $10^3$–$10^6$ times more soluble than orthophosphate in water in the presence of divalent cations (Pasek et al., 2017; Gull, 2014; Gulick, 1955). Various sources of reduced P compounds on Earth include meteorites bearing phosphides (Pasek, 2008; Pasek and Lauretta, 2005 and 2008; Pirim et al., 2014; Bryant and Kee, 2006; Pasek et al., 2007), lightning and other high-energy events (Pasek and Block, 2009; Bindi et al., 2023; Graaf and Schwartz, 2000), by iron-induced redox (Herschy et al., 2018), and by reduction when rocks react with water (Pasek et al., 2022). These reduced P compounds can oxidize to produce condensed P compound such as pyrophosphate and polyphosphates (Pasek et al., 2007, 2008; Kee et al., 2013; Gull et al., 2023b), which can then phosphorylate organics (Gull et al., 2023a). Moreover, these reduced P compounds (such as phosphite) are stable over the course of time (Gull et al., 2022).

Condensed phosphate compounds such as trimetaphosphate can react with ammonia in solution form to form energetic N–P type compound called amidotriphosphate (Krishnamurthy et al., 1999). This and other amidophosphates can phosphorylate the 2-carbon sugar glycolaldehyde as well as nucleosides (Quimby et al., 1958). Just like certain condensed P compounds such as $P_4O_{10}$ and trimetaphosphate, the prebiotic provenance of amidophosphates has been questioned; however, recently, a viable prebiotic route of formation of N–P compounds has been shown (Gibard et al., 2019). This reaction has shown that $Fe_3P$ (an analogue of the mineral schreibersite) reacts in 25% $NH_4OH$ aqueous solution to form N–P bonds at a concentration equivalent to the N to O (nitrogen-to-oxygen) ratio found within the solution (Gibard et al., 2019). Figure 6.6 shows the structures of various inorganic P compounds discussed earlier.

### 6.2.4 Earth's Phosphorus Cycle

The availability of P is considered to be the determining factor for the amount of primary productivity that can be maintained in the oceans over geologic time scales (Tyrrell, 1999). Estimating P concentrations in the oceans over certain timescales

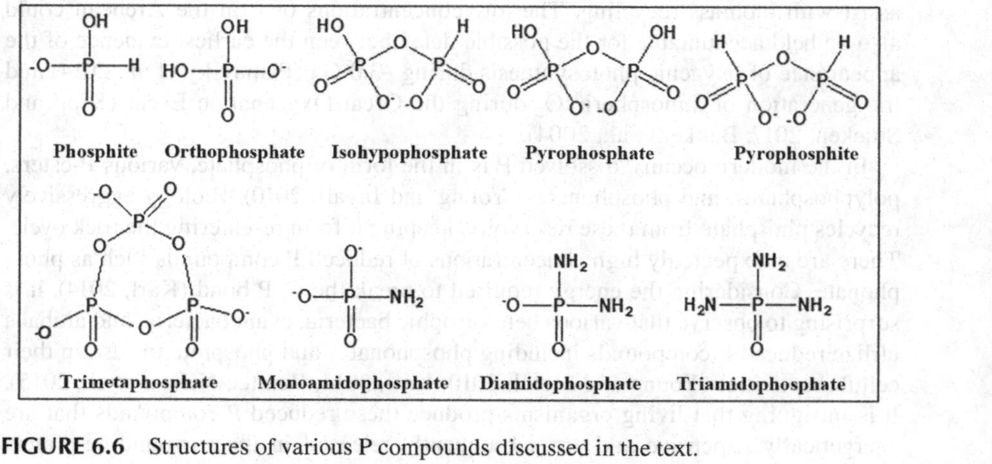

Phosphite Orthophosphate Isohypophosphate Pyrophosphate Pyrophosphite
Trimetaphosphate Monoamidophosphate Diamidophosphate Triamidophosphate

**FIGURE 6.6** Structures of various P compounds discussed in the text.

in Earth's history is therefore significant to understanding the growth and development of the biosphere, as well as the development of biogeochemical cycles such as the rise of oxygenated atmosphere (Kipp and Stüeken, 2017).

Several proxies have been constructed to determine P levels in the ocean over geologic time, including the P concentrations in iron (Fe) oxide-rich sedimentary rocks (Bjerrum and Canfield, 2002; Planavsky et al., 2010; Jones et al., 2015) and marginal marine siliciclastic sedimentary rocks (Reinhard et al., 2017). Many studies have agreed on low P concentrations (<20% modern concentrations where modern concentrations are ~2 μM (Tyrrell, 1999; Kipp and Stüeken, 2017) in the Archean and probably also in Proterozoic eons (Kipp and Stüeken, 2017; Jones et al., 2015; Reinhard et al., 2017). The most likely mechanism of P depletion in ancient oceans is attributed to the scavenging of P from the water column by incorporation into various phases of ferrous ($Fe^{2+}$) minerals or possibly by adsorption onto iron oxides (Bjerrum and Canfield, 2002; Jones et al., 2015; Reinhard et al., 2017).

Recently, however, it has been shown that, in the Archean, it may have been the oxidant-limited recycling of P rather than scavenging by iron minerals that drastically impacted P availability. Secondary phosphate minerals cannot form if P remains linked up with the organic matter (Kipp and Stüeken, 2017). Furthermore, there plausibly was a remarkable increase in the capacity for recycling P as early Earth's surface gradually evolved from being reducing to an oxidizing one, particularly around the Great Oxidation Event (GOE) ~2.4 billion years ago (Ga), when there was also a steep rise in the marine sulfate levels (Bekker et al., 2004). It has also been suggested that in the Archean, the reduction of iron and sulfate by microbial activity played a significant role in sustaining early life and biological productivity by carrying out the major P recycling in the early Archean oceans (Kipp and Stüeken, 2017). Furthermore, during the Archean eon, perhaps the chief source of sulfate supply ($SO_4^{2-}$) to the early oceans would have been the photolysis of volcanic $SO_2$ (Farquhar et al., 2000). These events would have been more likely before the major oxidative weathering of rocks and early Earth systems (Kipp and Stüeken, 2017; Stüeken et al., 2012). These results suggest the significance of volcanism in enabling the early biosphere on the anoxic Earth to thrive by supplying sulfates to assist with biomass recycling. The low concentrations of P in the Archean could also be held accountable for the possible delay between the earliest evidence of the appearance of oxygenic photosynthesis during ~3.0 Ga (Planavsky et al., 2014) and the generation of atmospheric $O_2$ during the Great Oxygenation Event (Kipp and Stüeken, 2017; Bekker et al., 2004).

In the modern oceans, dissolved P is in the form of phosphate, various P-esters, polyphosphates, and phosphonates (Young and Ingall, 2010). Biology aggressively recycles phosphate from these reservoirs, keeping it from re-entering the rock cycle. There are unexpectedly high concentrations of reduced P compounds such as phosphonate. Considering the energy required to break the C–P bond (Karl, 2014), it is surprising to observe that various heterotrophic bacteria, cyanobacteria, and archaea utilize reduced P compounds including phosphonates and phosphite to sustain their cellular activities (Young and Ingall, 2010; Karl, 2014; Benitez-Nelson et al., 2015). It is intriguing that living organisms produce these reduced P compounds that are energetically expensive and are subsequently released in the surrounding waters (Benitez-Nelson et al., 2015). Van Mooy et al. also discussed the significance of reduced P compounds in the marine P cycle (Van Mooy et al., 2015). The authors reported the discovery of a large amount of reduced P compounds that can quickly recycle and therefore play a significant role in the ocean P biogeochemistry (Van Mooy et al., 2015). When natural samples of plankton were incubated in the presence of $^{33}$P-phosphate, various $^{33}$P-labeled reduced P (III) compounds were observed. This suggested uptake and, consequently, chemical reduction of $^{33}$P-phosphate to form $^{33}$P-labeled reduced P (III) compounds (Van Mooy et al., 2015). Nevertheless, different types of P compounds (phosphates and reduced P compounds) with different chemical properties are most likely to impart major effects on the P cycle at any stage in the history of Earth.

## 6.3 NITROGEN

Our second system to investigate is the nitrogen cycle. In this subsection, we shall discuss the role of nitrogen geochemistry in forming major biochemical systems that would plausibly have kick-started life on early Earth, and how modern life keeps nitrogen present in its environment.

Nitrogen (N) is the element right above phosphorus on the periodic table; yet, its chemical behavior is significantly varied from P. Nitrogen is a critical part of biochemical systems, and its chemistry varies significantly more in the terrestrial environment than does P. There are more redox variations for N in life, and N is also significantly more volatile. The N biogeochemical cycle is also much more likely to have been rapidly cycled, as opposed to the P cycle which is tied to the rock cycle.

### 6.3.1 Nitrogen in Biology: What Molecules Are Present?

Nitrogen is a key component of many biomolecules that play critical roles in biochemical systems. Molecules such as the amino acids and nucleobases both contain N as an important part of their chemical structures. Broadly, nitrogenous compounds in biological systems are divided into two main groups: The amines and the heterocycles.

Amines and amino acids occur when nitrogen is present as an -$NH_2$ group in a molecule. Amino acids are carboxylic acids at the alpha carbon, and the $NH_2$ group is on the beta carbon. The formation of proteins occurs as these two moieties condense, forming a C–C–N structure. Proteins are critical to modern biology, as they serve as both building materials for cells and because several are enzymes that catalyze biochemical reactions.

Nitrogen in heterocycles includes the nucleobases and bears N that is part of the ring structure of a molecule, for instance, the purines and the pyrimidines. Nitrogen in both cases is either present as N with a lone pair exposed or N–H with a proton capable of H-bonding. For this reason, nitrogen allows for the formation of the double helix in DNA as the nucleobases link to each other and stabilize DNA. In these molecules, N is found primarily in its reduced form, with an oxidation state that is close to –3. Oxidized nitrogen is uncommon as a biomolecular constituent. Figure 6.7 shows the structures of significant N-containing biomolecules central to the biochemistry.

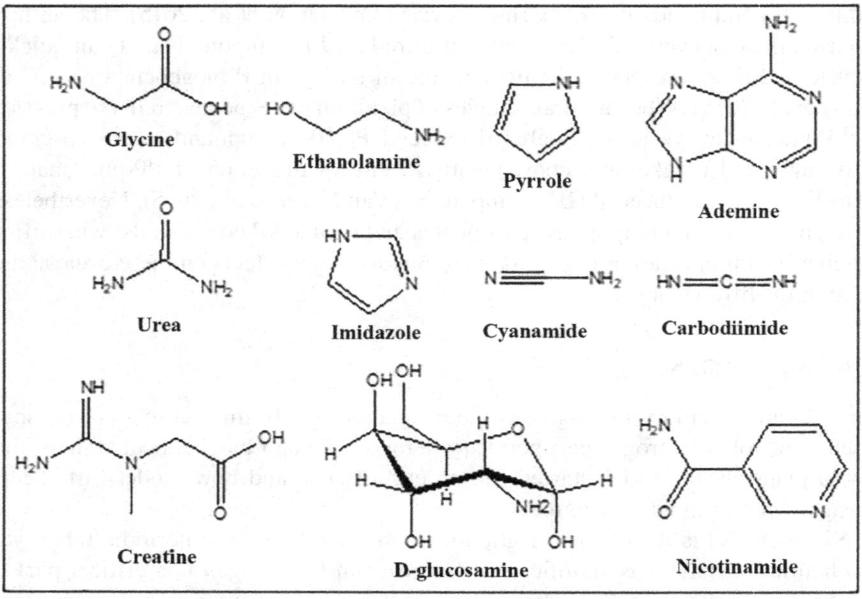

FIGURE 6.7 Various significant N-containing biomolecules in living organisms.

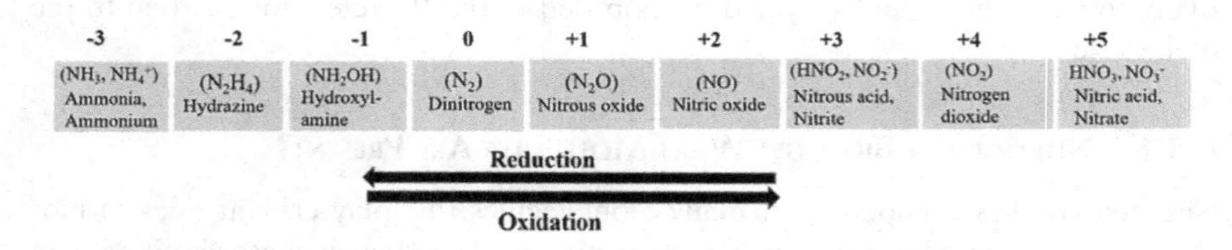


FIGURE 6.8 Different oxidation states of N that can range in oxidation number from –3 to +5 and, therefore, can exist in various chemical forms, for example, ammonia, dinitrogen, and nitric acid. These compounds play a significant role in the geochemical as well as life systems. The arrows indicate the reduction and oxidation capabilities of the N-bearing compounds.

### 6.3.2 Sources of N in Nature

Nitrogen is present in significant quantities on Earth with the atmosphere presently comprising 78% of $N_2$ (dinitrogen). However, despite the abundance of N in the form of $N_2$, it is regarded as a limiting nutrient for life. This is due to atmospheric $N_2$ bearing a triple bond that is extremely stable, and it requires a significant energy investment to break this bond, thus making this form of N relatively inaccessible for most life. Therefore, in order to be made available to biochemistry, N must be somehow 'fixed' to a more available and energetically accessible form (Todd, 2022).

Nitrogen can range in various oxidation states (–3 to +5) and various common forms including ammonia/ammonium ($NH_3$/$NH_4^+$), dinitrogen ($N_2$), nitrite ($NO_2^-$), or nitrate ($NO_3^-$) (Todd, 2022) (Figure 6.8). Of these, ammonia, nitrite, and nitrate are all easily accessible to biological systems. Both ammonia (as ammonium) and nitrate are present as solids on the surface of Earth as well. These compounds are present in solids as both compounds can be ionic (+1 for ammonium, −1 for nitrate). Nitrate is a constituent of fertilizer and guano, and ammonium is also a constituent of fertilizer, but can substitute for $Na^+$ and $K^+$ in some minerals and salts. An understanding the mineral and abiotic sources of nitrogen – in solid form – is the focus of the next section.

#### 6.3.2.1 Exogenous Sources of Nitrogen

In this section, we shall highlight the exogenous sources of N. Nitrogen is present in the interstellar medium (ISM, hereafter) (Cheung et al., 1968; Ho and Townes, 1983; Daniel et al., 2007), various star forming regions and molecular clouds (Ho et al., 1979; Gusten et al., 1981) as $NH_3$, and was presumably accreted to Earth in this form. Nitriles (compounds containing −C≡N group) have also been observed in both diffuse and dense interstellar regions (Thaddeus, 1972; Crutcher, 2012; Cernicharo et al., 2020), as components of dust grains (Jones, 2016) and molecular ices (such as HCN along with $NH_3$), and HCN is in protoplanetary disks (Bergner et al., 2019; Chapillon et al., 2012). A planet's N supply may possibly come from sources such as asteroids, or comets, or it may be a primary condensate (Epstein et al., 1987; Wyckoff et al., 1991; Owen et al., 2001; Kawakita et al., 2007; Marty et al., 2016 and 2017; Morbidelli et al., 2012). Both HCN and $NH_3$ could plausibly serve as major sources of N during the formation of the planets (Todd, 2022; Rice et al., 2018) though these would have changed over time.

Nitrogen has also been detected on planets' bodies besides our Earth. Venus' atmosphere bears more nitrogen than that of Earth (Todd, 2022; Lecuyer et al., 2000). The Martian atmosphere contains ~2% $N_2$ (Mahaffy et al., 2013; Owen et al., 1977) and various oxidized nitrogen compounds that have been detected in Martian sediments (Stern et al., 2015). $N_2$ has also been detected in the atmospheres of the icy outer planetary bodies such as Titan, Triton, and Pluto (Scherf et al., 2020). Pluto's surface may also composed of nitrogen ices (Lewis et al., 2021).

Various N species such as HCN, HNC, HNCO, $CH_3CN$, $HC_3N$, and $NH_2CHO$ (Mumma and Charnley, 2011; McKay and Roth, 2021; Todd and Oberg, 2020) have been detected in comets. $N_2$ has also been detected in situ in comet 67P/Churyumov–Gerasimenko by ROSINA/Rosetta (Rubin et al., 2015). Ammonium salts have also been detected on the surface of comet 67P/Churyumov–Gerasimenko (Poch et al., 2020). Nitrogen has also been found in various classes of meteorites. Various chondrites such as CI, CM, and CR chondrites can consist of around 500–2,000 ppm N. This N is mostly in the form of soluble and insoluble organics (Grady and Wright, 2003).

Interestingly, some nitrogen-containing organic heterocycles have also been detected in meteorites (Todd, 2022; Martins et al., 2020) such as pyridine-monocarboxylic acids in carbonaceous chondrites (Tagish Lake and Murchison) (Pizzarello et al., 2001, 2004; Pizzarello and Huang, 2005; Smith et al., 2014), diketopiperazine and hydantoins, and possibly purines, pyrimidines, triazines, pyridines, and quinolines in carbonaceous chondrites (Martins, 2018). Many meteorites also contain several amino acids (Cronin and Pizzarello, 1983; Glavin et al., 2020).

#### 6.3.2.2 Endogenous Nitrogen

There are many endogenous reservoirs of N, with one of the largest being Earth's atmosphere, where it exists as $N_2$. Nitrogen is also present in the oceans, is in sediment, and is minor constituent of some rocks.

Nitrogen is primarily found as $N_2$ today, but on early Earth it may have been in the form of ammonium/ammonia and nitriles. It is unlikely that nitrates would have been abundant on early Earth due to their formation under primarily oxidizing conditions. Reduced N compounds such as ammonia may not have persisted for long on early Earth, as ammonia readily photolyzes and is lost (Kuhn and Atreya, 1979). However, if ammonia was captured in rocks or in the ocean, it may have persisted till the origin of life.

### 6.3.3 The Nitrogen Cycle

Next, we set out to identify the transformations of nitrogen that take place on Earth. Assuming the largest reservoir of accessible nitrogen is the atmosphere as $N_2$, this molecule must be broken apart to make the nitrogen accessible to modern biology. Lightning is one of the important abiotic sources of N-fixation on Earth. Lightning is an electric discharge that, when it passes through an atmosphere composed of at least part $N_2$, it cleaves the triple bond and allows for the random reaction of N with other atmospheric gases. On modern Earth, this leads to NO compounds. On early Earth, this could have led to NC compounds. For example, the rate of formation of HCN and NO depends on the C/O ratio of the atmosphere. When C is more than O, HCN is favored, and when C is less than O, NO is favored (Chameides and Walker, 1981). The formation of the latter is also favored when lightning happens in an $N_2/CO_2$-dominated atmosphere (Mancinelli and McKay, 1988).

High-energy processes in general alter N chemistry. Meteoritic impacts result in high temperatures and pressures when the rock strikes the ground. This results in various chemical reactions that can fix $N_2$. Compounds like HCN and NO can be formed when C/O ratio is >1 (Fegley et al., 1986; Parkos et al., 2018), and NO may also form when C/O ratios are low (Kasting, 1992). Meteoritic impacts can also trigger the formation of high-temperature plasma in the impact plume which could also work as a potential site for the formation of significant N-bearing molecules such as amino acids and nucleobases (Miyakawa et al., 1998 and 1999). Similarly, hot volcanic vents also seem to facilitate the thermal fixation of atmospheric $N_2$ in various soluble forms such as various oxides of nitrogen (Mather et al., 2004; Martin et al., 2007; Brandes et al., 1998 and 2008). Syntheses of various amides, esters, and nitriles under simulated hydrothermal conditions suggest hydrothermal vents to be the significant source of dissolved nitrogenous compounds in the surrounding water (Rushdi and Simoneit, 2004).

Another endogenous, abiotic transformation of $N_2$ is promoted through photochemistry (Todd, 2022; Bickley and Vishwanathan, 1979). One example of such process is the generation of N atoms in the upper atmosphere, catalyzed by the strong UV radiation ($\lambda < 102$ nm) (Zahnle, 1986). These N atoms then form products such HCN and NO with the former in the more reducing environments and the latter in the more oxidizing environments. At even higher energy, cosmic rays and other high-energy particles formed by our Sun (Todd, 2022) can break apart $N_2$ (Nicolet, 1975; Calisto et al., 2011). Stars can also possibly release high-energy particles (solar particles and cosmic rays) from various sources such as stellar winds, coronal mass ejections, or stellar flares (Todd, 2022; Airapetian et al., 2019), which can catalyze the plausible prebiotic formation of various amino acids and nucleobases (Miyakawa et al., 2002).

Once the $N_2$ is broken apart to make new molecules, these NH, NC, and NO compounds slowly decompose under a variety of conditions and return to the atmosphere as $N_2$. To this end, the nitrogen that is used by life is cycled back into the atmosphere.

### 6.3.4 The Nitrogen Cycle on the Early Earth

The early Earth would have a primordial atmosphere through outgassing and accretion of the surrounding nebular gas (Rimmer and Shorttle, 2019; Hayashi et al., 1979). The atmosphere of early Earth was likely enriched in $H_2$, $NH_3$, $CH_4$, and CO (Zahnle et al., 2010), which would have transitioned into a secondary atmosphere (due to processes such as rainout and impacts). $N_2$ would have dominated this secondary atmosphere, along with $CO_2$ and $H_2O$ (Kasting, 1993). The precise atmospheric composition of Earth at the origin of life on the early Earth is not known, but an $N_2$-dominated atmosphere was plausible, with a range of partial pressures as suggested by Catling and Kasting (2017). Prior to the evolution of N-fixation, N must have been available through various geochemical abiotic processes such as lightning, photochemistry, or volcanism as discussed in Section 3.2 in order to afford chemically accessible forms of N (Todd, 2022; Rimmer and Shorttle, 2019; Fischer, 2008).

The early evolution of N-fixation on early Earth could be regarded as a significant step in the availability of nutrients and planet Earth's habitability. This event is thought to have come around ~3.5 Gyr ago (Todd, 2022; Nishizawa et al., 2014). On an early Earth with anoxic conditions, NO formed by abiotic processes would have been reduced by hydrogen forming the nitroxyl molecule (HNO) (Canfield et al., 2006; Kasting and Walker, 1981), which could lead to nitrites and nitrates (Mancinelli and McKay, 1988). Both nitrates and nitrites would be consumed by heterotrophic denitrifiers to oxidize organic matter and therefore releasing $N_2$ and possibly some amount of $NH_4^+$. Furthermore, via nitrites and nitrates, ammonia would be oxidized to $N_2$ through the *anammox process* (Dalsgaard et al., 2005) occurring mainly in anoxic sulfide-free zones (Dalsgaard et al., 2005). This N-cycle would potentially lack a biotic recycling route that is common to other element cycles, considering there is no known pathway from $NH_3$ to nitrates under anoxic conditions (Canfield et al., 2006).

Studies of nitrogen isotopes of kerogen extracted from the minimally altered shales Campbellrand–Malmani platform in South Africa, have shown that between the Palaeo – Archaean and about 2,670 million years ago, the $\delta^{15}N$ values of the kerogen rose by about 2‰. This increase could be attributed to the beginning of coupled nitrification and denitrification or anammox reactions in the surface oceans. These processes require free $O_2$. Then a second increase in N-isotopic composition around 2,520 million years ago suggests the instability of the N-cycle with loss of N that was fixed. These studies also suggest that the coupled processes of nitrification and denitrification were responsible for the loss of fixed inorganic N, leading to a deficiency that ultimately caused the overall lower availability of N for life. As a consequence, the growth of early microbes was likely hindered, thus delaying the formation and availability of free $O_2$ in the early Earth atmosphere (Godfrey and Falkowski, 2009).

#### 6.3.4.1 Effects of Ferruginous Conditions on the Nitrogen Cycle

Early Earth oceans would have been greatly impacted by the abundance of $Fe^{2+}$ dissolved in the oceans (Stüeken et al., 2016; Poulton and Canfield, 2011). The concentration of the $Fe^{2+}$ ions in the ancient Archean oceans may have been 40–120 μM (Canfield, 2005), compared to the concentration <1 nM today (Moore and Braucher, 2007). This suggests that the presence of ferrous iron could possibly have impacted the accumulation of N-bearing compounds. Iron could cause (a) the oxidation of $NH_4^+$ by $Fe^{3+}$ and (b) the reduction of $NO_2^-$ or $NO_3^-$ by $Fe^{2+}$ (Stüeken et al., 2016). Iron-driven oxidation of $NH_4^+$ to $NO_2^-$ or $NO_3^-$ is not plausible thermodynamically under realistic early Earth conditiONs with a pH ranging from 6 to 8 (Grotzinger and Kasting, 1993). However, a plausible reaction has been suggested by Stüeken et al., (2016) in which $NH_4^+$ can be oxidized to $N_2$ via the following reaction:

$$\mathbf{6Fe(OH)_3 + 2NH_4^+ + 10H^+ \rightarrow 6Fe^{2+} + N_2 + 18H_2O} \quad \text{(3)} \ \textbf{(Stüeken et al., 2016)}$$

The aforementioned reaction is active between pH values of 6 and 8 and can only be feasible in the early Archean oceans if the molar concentrations of $Fe^{2+}$ is 100 μM or less and $[NH_4^+]$ is 1 μM or higher (Stüeken et al., 2016). Due to the abundance of iron oxide-bearing banded iron formation in the Archean and Paleoproterozoic eras (Isley and Abbott, 1999), such processes could possibly have been a major sink of nitrogen in the early oceans and eventually released nitrogen to the atmosphere as $N_2$, prior to the onset of denitrification in the Paleoproterozoic era (Stüeken et al., 2016).

The reduction of nitrites and nitrates by $Fe^{2+}$ is in the reactions given next:

$$\mathbf{6Fe^{2+} + NO_2^- + 16H_2O \rightarrow 6Fe(OH)_3 + NH_4^+ + 10H^+} \quad \text{(4)} \ \textbf{(Stüeken et al., 2016)}$$

$$\mathbf{6Fe^{2+} + 2NO_2^- + 14H_2O \rightarrow 6Fe(OH)_3 + N_2 + 10H^+} \quad \text{(5)} \ \textbf{(Stüeken et al., 2016)}$$

For the reactions given here, with pH values between 6–8 with 1–100 μM $[Fe^{2+}]$ and 0.8 bar $N_2$, both reactions are thermodynamically feasible, even with an unrealistic high amount of 1 M $[NH_4^+]$ and as Little as 1 nM $[NO_2^-]$ (Stüeken et al., 2016). The implication of these reactions is that the denitrification of dissolved N by ferrous ions would have resulted in the concentrations of both nitrates and nitrites to be in extremely reduced amounts in the Precambrian deep ocean (Stüeken et al., 2016). Figure 6.9 gives a brief glimpse of plausible N-cycle on early Earth.

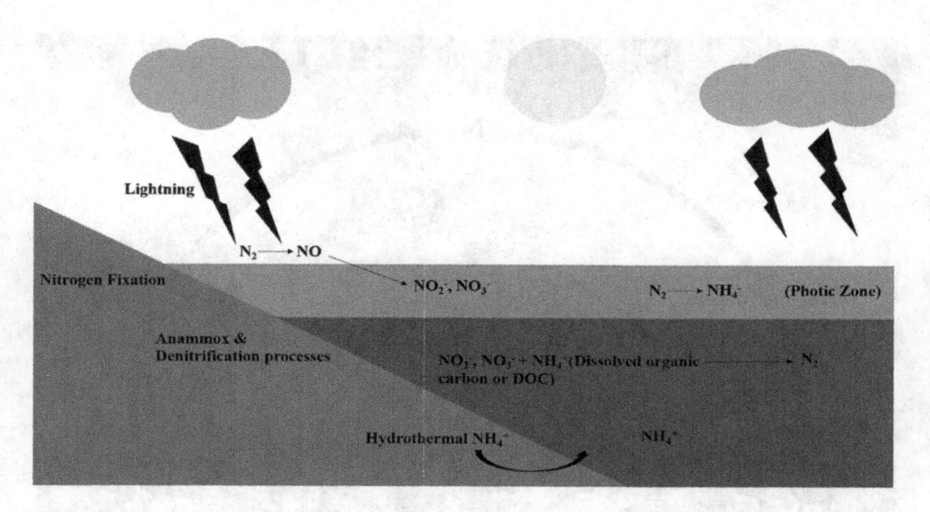


**FIGURE 6.9** The plausible early-Earth N-cycle. Nitric oxide (NO) would have been produced by high-energy events such as lightning and, subsequently, would have been transferred to the oceans, leading to the formation of nitrates $(NO_3)^-$ and nitrites $(NO_2)^-$. Both of these compounds would then be used in processes such as denitrification and the anammox reaction (Canfield et al., 2006).

### 6.3.5 Nitrogen Cycle on Earth Today

The nitrogen cycle by definition is the cyclic chain of events through which nitrogen (in various accessible forms) moves through living and nonliving matter, such as water, the atmosphere, and soil, and living organisms such as plants, animals, and microbes. It is a cycle that is separate from the water cycle or the rock cycle, but builds off of both.

Nitrogen-fixation transforms $N_2$ to $NH_3/NH_4^+$ or $NO_3^-$. In this process, atmospheric $N_2$ is converted into biologically usable N-compounds and is achieved either by natural processes such as lightning or by biological N-fixation. Various microorganisms fix nitrogen, including *Azotobacter*, as well as bacteria-forming symbiotic associations with plants such as *Azospirillum*, *Rhizobium*, and *Bradyrhizobium* and cyanobacteria. Second, nitrification transforms ammonia or ammonium into nitrates. This stage has two steps where first soil bacteria such as *Nitrosomonas* and *Nitrococcus* convert $NH_3$ to nitrites, and, second, another species of bacteria, *Nitrobacter*, helps the oxidation of nitrites to nitrates.

After fixing nitrogen and turning it into nitrate, assimilation occurs when $NH_3/NO_3^-$ is absorbed by plants roots. In this step, various N-bearing compounds such as ammonia and nitrates are absorbed through plants roots and then transferred to various parts to be utilized in forming functional biomolecules. Finally, ammonification occurs when organic-N bearing compounds are turned to $NH_3$. This source of excreted ammonia can then be used for nitrification and also denitrification, in which $NO_3^-$ is converted (reduced) into gaseous $N_2$ with the help of anaerobic bacteria. The nitrogen cycle is shown in Figure 6.10.

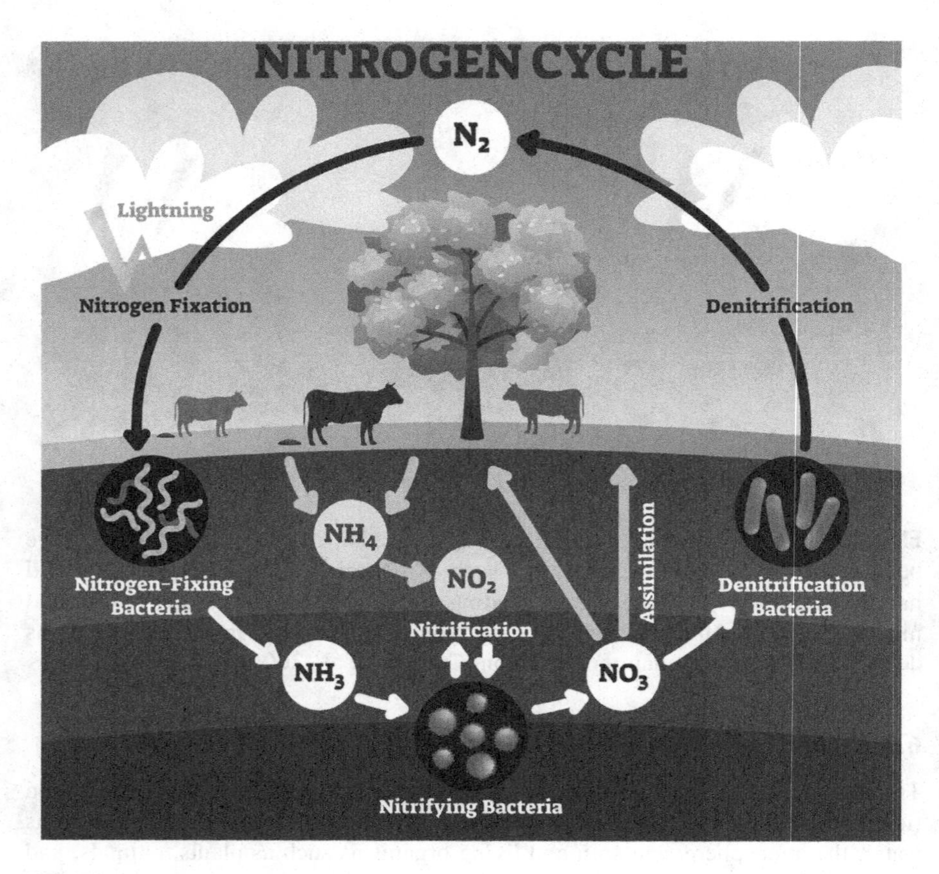


**FIGURE 6.10** Present-day nitrogen cycle.

The large-scale active nitrogen cycle today is a clear indicator of life's activity on the surface of Earth. Nitrogen is cycled into the living systems at a rate many times greater than abiotic sources. Life is a fantastic factory for fixing nitrogen. Intriguingly, in the time since the industrial age began, humans have taken over by developing actual factories for making ammonia out of air. Human industry presently surpasses the nitrogen-fixing capacity of the biosphere. Thus, the effects of life on the nitrogen cycle may also be a 'technosignature', an indicator of intelligent life active on another world.

## 6.4 CARBON

Our last system to investigate is the carbon cycle. In this subsection, we shall discuss the role of carbon geochemistry in forming major biochemical systems that would plausibly have kick-started life on the early Earth, and how modern life keeps carbon present in its environment.

### 6.4.1 Carbon Chemistry on the Earth

Carbon (C) is the most important element biologically due to its huge capacity to make covalent bonds. It is the fourth most abundant element in the universe, after hydrogen, helium, and oxygen. Carbon has a valency of 4 which can be satisfied by forming single bonds (think methane), double bonds (ethene), and triple bonds (acetylene). Carbon generally bonds with most other nonmetal elements very readily (H, S, O, N, and P). However, it also occurs in mineral form in rare carbides such as SiC in which it is reduced (−4 oxidation state).

Carbon is the foundation of organic chemistry, the subject dreaded by pre-meds across the globe. Broadly speaking, all organic compounds contain carbon bonded mainly to hydrogen and often with other atoms such as O, N, S, and the halogens. Organic compounds primarily composed of C and H atoms are known as hydrocarbons. However, not everything that bears carbon is considered organic: Some compounds such as $CaCO_3$ and $CO_2$ are not organic because these do not contain covalent C–H bonds. Carbon in carbides such as $Fe_3C$ and SiC is also not 'organic'.

Carbon is important to understanding the history of chemistry as well: The first synthesis of an organic compound from an inorganic compound was reported by a German scientist named Friedrich Wöhler who, in 1828, reported the formation of urea by combining cyanic acid and ammonium in vitro (Kinne-Saffran and Kinne, 1999). This demonstration showed that the chemistry of life was not separate from the chemistry of nonliving things, establishing that biochemistry is not inherently 'special'.

Life would not be possible without carbon. Carbon is essential to organic chemistry and is the element that allows the formation of biomolecules such as DNA and RNA, proteins, lipids, and many others. Carbon is also the main element of metabolism as both photosynthesis and cellular respiration concern the change of inorganic carbon ($CO_2$) into organic carbon and vice versa. All systems in biochemistry require C in some form, either as an active participant in the chemistry or as a catalyst in promoting a reaction. Even in inorganic form, carbon as carbonate is also used by various marine organisms such as molluscs to create hard shells composed of calcium carbonate, which protect their soft bodies.

### 6.4.2 Sources of Carbon on the Early Earth

C is found in space in almost all allotropic forms (diamond, and graphite, and fullerenes) (Cataldo, 2004). Carbon can get locked up in inorganic form, specifically in carbonates. Carbonates have been reported to be formed as a consequence of planetesimal alteration/differentiation via aqueous alteration. Some forms of C also occur as primary mineral phases in chondrules of type 3 chondrites such as cohenite ($Fe_3C$) and haxonite ($Fe_{23}C_6$, as a minor phase) (Hazen et al., 2008; Rubin, 1997; Brearley and Jones, 1998; Stroud et al., 2004; Ebel, 2006; Messenger et al., 2006; MacPherson, 2007). C-bearing micro and nano-mineral phases such as diamond/lonsdaleite (~2 nm) and graphite (allotropes of C) have also been identified from interstellar grains in various chondrites (Brearley and Jones, 1998; Nittler, 2003; Mostefaoui and Hoppe, 2004; Vollmer et al., 2007).

Various organic compounds are present in space as diffuse gas, as well as trapped in solid form in comets, asteroids, and meteorites (Ehrenfreund and Cami, 2010; Mumma and Charnley, 2011; Guelin and Cernicharo, 2022; Henning and Semenov, 2013; Walsh, 2014; Martins et al., 2020; Oró and Mills, 1989; Nakano et al., 2020). The intact delivery of various molecules of biological significance such as amino acids (Pierazzo and Chyba, 1999) and HCN (Todd and Oberg, 2020) from these solid sources has been suggested to have been important to the development of life (Pizzarello et al., 2004; Pizzarello and Huang, 2005; Smith et al., 2014; Cronin and Pizzarello, 1983; Pohorille, 2002; Mason, 1963). Organic molecules are ubiquitous in planetary atmospheres as well as on the surfaces of outer solar system moons (Cruikshank et al., 2005; Raulin, 2008; Lorenz et al., 2008; Kwok, 2016; Ehrenfreund et al., 2011). Organic compounds (including various derivatives of HCN) have also been detected in the cometary comae (Crovisier et al., 2009; Ehrenfreund and Cami, 2010). Atmospheric photochemical reactions produce several classes of organic compounds. The nature of these organic compounds is determined by the composition of the atmosphere and other factors such as redox state (Todd, 2022; Chyba and Sagan, 1992; Miller, 1993; Tian et al., 2011). Other potential sources of organic compounds are volcanism (Mukhin, 1976; Kolesnikov and Egorov, 1979; Liggins et al., 2022), serpentinization (Holm et al., 2015), and production at hydrothermal vents, particularly by a reaction called the *Fischer–Tropsch* synthesis. This geochemical route forms complex organics which are released into the surrounding waters (Nooner et al., 1976; Sleep et al., 2004).

The classic Miller–Urey spark-discharge experiment (Miller, 1953) and successive experiments have demonstrated the formation of organic compounds under simulated early Earth conditions (Johnson et al., 2008; Cooper et al., 2017). These experiments suggest the relevance of atmospheric conditions to the prebiotic organic synthesis generated by high-energy processes (Bada, 2013; Chyba and Sagan, 1992). It is now very comfortable to assume that early Earth was rife with a rich assortment of organic compounds, a selection of which would have been the prebiotic material feedstock for the development of life (Kobayashi et al., 1998). Such a justification is based on the analysis of meteorite organics, which show a huge diversity and a generally high abundance. Given how easily abiotic processes appear to have been able to have generated organic compounds relevant to modern life, early Earth was likely also rather enriched in such materials.

### 6.4.3 The Carbon Cycle

Carbon compounds move through the environment via the carbon cycle. The carbon cycle is based on the largest reservoir of carbon being solid carbonates, followed by carbon in the ocean, fossil fuels, soil carbon, and carbon in the atmosphere. Atmospheric carbon is especially familiar today as its excess is a main driver of anthropogenic climate change.

Carbon is found in the atmosphere in gaseous form and in trace amounts. These gases include $CO_2$ (presently dominant), as well as $CH_4$ and extremely small amounts of CO. Human activities such as industrialization and the burning of fossil fuels have led to a large increase in $CO_2$ emission. This source is on top of natural sources, such as the metabolism of both animals and plants, which also releases $CO_2$. $CO_2$ is also generated by biomass burning, and from the acidification of carbonates. $CO_2$ is chemically inert with respect to forming organic compounds, as it is wholly oxidized and hence must be reduced in order to form organic molecules. Ancient volcanism would have released $CO_2$ as well, though the release of C in the form of CO, $CO_2$, or $CH_4$ would have been dependent on the oxidation state of the upper mantle at the time of degassing event (Chang et al., 1983; Walker, 1985).

In the Hadean atmosphere, CO was probably the simplest source of carbon for reactions leading to organics as $CO_2$ is comparatively inert. In contrast, CO is capable of reacting with hydrogen and other organics to create prebiotically relevant molecules. CO is easy to generate abiotically by lightning, by impact shocks, or by photochemical reactions, provided there is some source of reducing power (Zahnle et al., 2010).

Methane ($CH_4$) is the most abundant hydrocarbon in nature and is another naturally occurring gaseous form of C. It is known to exist under low-pressure conditions under Earth's crust. Isotopically anomalous $CH_4$-rich gas deposits along with some other hydrocarbons have been discovered, which have been attributed to abiogenic processes such as serpentinization and the hydration of ultramafic rocks (Scott et al., 2004; Sherwood Lollar et al., 1993 and 2002; Horita and Berndt, 1999). Furthermore, $CH_4$ can also be formed from bicarbonates in the presence of nickel–iron alloys (Horita and Berndt, 1999) and through Fischer–Tropsch-type reactions forming hydrocarbons (Scott et al., 2004; Foustoukos and Seyfried, 2004; McCollom and Seewald, 2003). Studies have shown the formation of methane along with other hydrocarbons in Earth's interior at pressures between 5 and 11 GPa and temperatures from 500°C to 1,500°C, in the presence of FeO, $CaCO_3$-calcite, and water (Scott et al., 2004).

Living organisms are also a significant source of $CH_4$ (Saunois et al., 2020). $CH_4$ is released as a final product of the anoxic decomposition of the organic matter by methanogenic archaea (Ernst et al., 2022). Other organisms such as plants (Keppler et al., 2006), fungi (Lenhart et al., 2012), algae (Klintzsch et al., 2019), and cyanobacteria (Bižić et al., 2020) can also produce methane under oxic conditions. Recent studies have shown that it is formed by living organisms such as *Bacillus subtilis* and *Escherichia coli* by the consumption of free iron and reactive oxygen species (ROS). The latter is formed as a consequence of metabolic activity and is enhanced by oxidative stress (Ernst et al., 2022).

### 6.4.4 The Carbonate–Silicate Cycle

At some stage in the history of early Earth, the amount of C in the atmosphere was much greater than it is today. The high level of $CO_2$ would have impacted the average surface temperature of early Earth. This would have greatly enhanced the greenhouse effect, and Earth would have been warmer as a consequence (Kuhn and Kasting, 1983). However, this atmospheric reservoir has decreased over time as the continents have grown to capture C. This occurs through the carbonate–silicate cycle.

In the carbonate–silicate cycle, $CO_2$ reacts with rocks that contain divalent cations (mainly $Ca^{2+}$ and $Mg^{2+}$) to form carbonate minerals ($CaCO_3$, $CaMg(CO_3)_2$). The result is that the leftover rock becomes more silica-rich, which alters the chemistry of the crust. Additionally, the $CO_2$ in the atmosphere decreases over time as it gets incorporated into rock. This process is mediated by water and, in the absence of water, does not happen nearly as readily. For instance, on Venus, the carbonate–silicate cycle has been blocked, and all the planet's $CO_2$ is instead in its atmosphere, which bakes the surface of the planet and makes it uninhabitable.

The carbonate–silicate cycle is not one way (as per 'cycle'!), and as carbonate rocks are buried and subducted into the mantle, they release $CO_2$ which returns to the atmosphere. Figure 6.11 shows the carbon cycle on early Earth.

### 6.4.5 The Carbon Cycle Today

Carbon cycling is a biologically critical process. It involves the cyclic movement of the nutrient C through various stages where it is consumed and then finally given back to the environment. Simply speaking, C usually in the form of $CO_2$ (and carbonates for some marine algae) is taken by the chlorophyll of the plants where it is 'fixed' through photosynthesis and becomes a part of biochemistry and food chains. It is consumed and released back into the atmosphere by organisms during respiration as a by-product. Other geological processes such as weathering of the (specifically carbonate) rocks also releases $CO_2$ which can lock up in the form of bicarbonates and can also become part of oceans. $CO_2$ can also react with fresh rock, bonding with divalent cations to form carbonates. Human activities such as burning of fossil fuels due to rapid industrialization and deforestation have disturbed this natural cycling of carbon. Figure 6.12 shows carbon cycle on today's Earth.

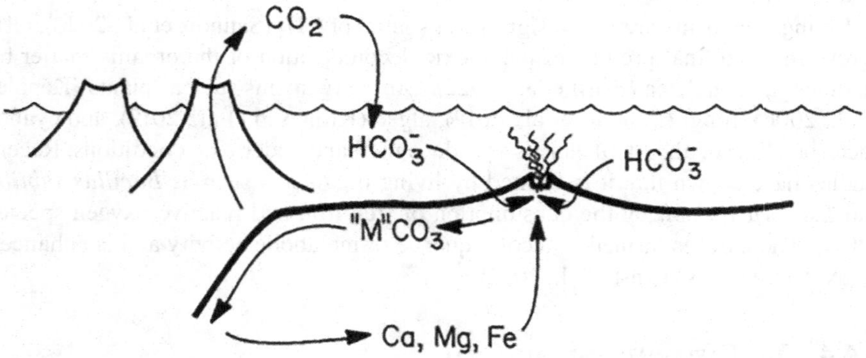

**FIGURE 6.11** Schematic illustration of plausible early-Earth C-cycle. Carbon dioxide ($CO_2$) is released by volcanic and metamorphic processes that cause the decarbonation of the weathered sea floor. It is then released from the ocean and atmosphere by hydrothermal interaction with sea. $CO_2$ then reacts with rocks, forming metal (M) carbonates such as $CaCO_3$ and $MgCO_3$.

*Source:* Reprinted with permission from (Walker, 1985)

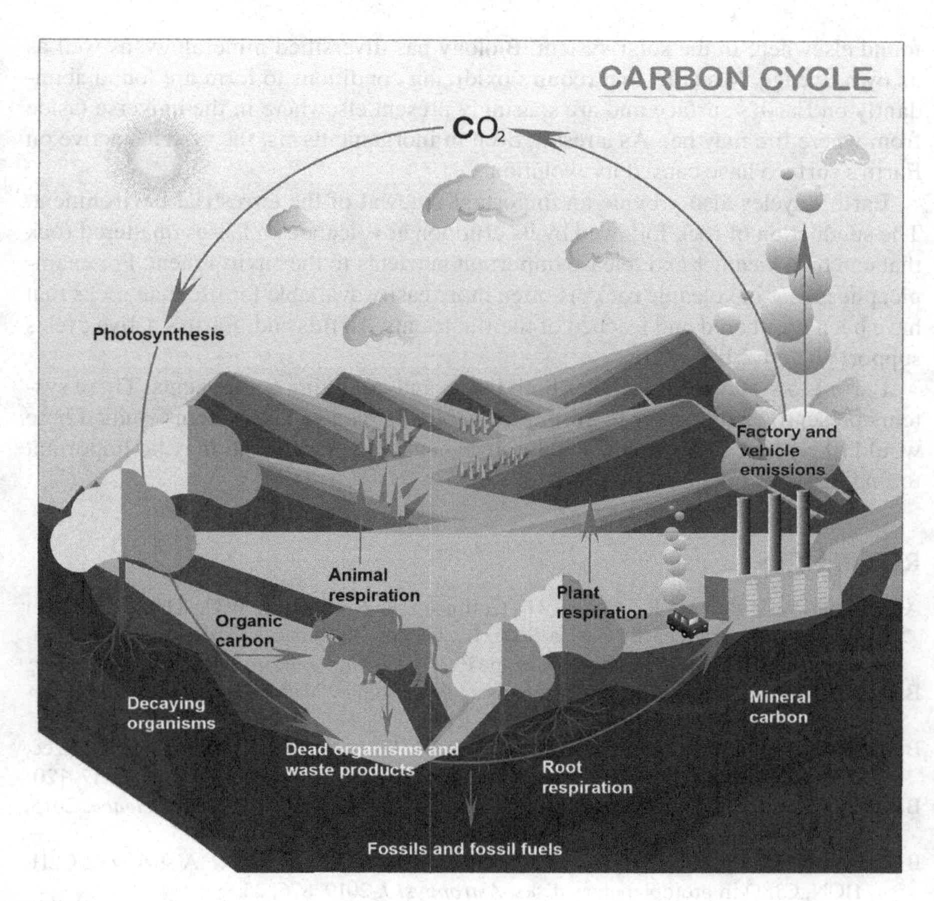


**FIGURE 6.12** Carbon cycle on Earth.

## 6.5 SUMMARY OF THE EFFECTS OF CYCLES ON PLANETS (E.G., HABITABILITY)

Earth is a system. Between the recycling of rocks, the movement of water, and the renewal of biologically critical elements, this active surface makes Earth habitable. To some extent, this change has come about because of the initial conditions of Earth's surface, which is temperate and bears important elements that are renewed through plate tectonics. To another extent, this cycling is caused by life: Life maintains these cycles, which in turn allows for its diversification and evolution.

These cycles exist because Earth's surface is active: These processes are driven by out-of-equilibrium processes such as solar heating, volcanism, and plate tectonics. Life acts on the top of these out-of-equilibrium processes to accumulate vital elements and actively recycle them. Earth's tight coupling of its cycles to biology has in turn impacted the composition of Earth's surface. As a result of Earth's biogeochemical cycling, there are many more minerals on Earth's surface than have been found elsewhere in the solar system. Biology has diversified mineralogy, as well as its own lineage. Minerals that require oxidizing conditions to form are found abundantly on Earth's surface and are sparingly present elsewhere in the universe (aside from where life may be). As a result, even in inorganic terms, the systems active on Earth's surface have caused its evolution.

Earth's cycles also provide an important renewal of the terrestrial environment. The subduction of rock followed by its eruption at volcanoes releases unaltered rock that can then weather and release important nutrients to the environment. For example, phosphate in volcanic rocks is often more easily available for life than rocks that have been weathered and leached of their nutrients. To this end, Earth's active cycles support an active biosphere.

Earth is a system that includes both living and nonliving components. These systems presently act together to promote a habitable planet. The systems studied here would likely be equally applicable to other habitable planets and may be important to finding other inhabited planets.

## REFERENCES

Airapetian, V.S.; Barnes, R.; Cohen, O.; Collinson, G.A.; Danchi, W.C.; Dong, C.F.; Del Genio, A.D.; France, K.; Garcia-Sage, K.; Glocer, A. Impact of space weather on climate and habitability of terrestrial-type exoplanets. *Int. J. Astrobiol.* 2019, 19, 139.

Bada, J.L. New insights into prebiotic chemistry from Stanley Miller's spark discharge experiments. *Chem. Soc. Rev.* 2013, 42, 2186–2196.

Bekker, A.; Holland, H.D.; Wang, P.-L.; Rumble III, D.; Stein, H.J.; Hannah, J.L.; Coetzee, L.L.; Beukes, N.J. Dating the rise of atmospheric oxygen. *Nature.* 2004, 427, 117–120.

Benitez-Nelson, C.; et al. The missing link in oceanic phosphorus cycling? *Science.* 2015, 348, 759–760.

Bergner, J.B.; Oberg, K.I.; Bergin, E.A.; Loomis, R.A.; Pegues, J.; Qi, C. A survey of C2H, HCN, C18O in protoplanetary disks. *Astrophys. J.* 2019, 876, 25.

Bickley, R.I.; Vishwanathan, V. Photocatalytically induced fixation of molecular nitrogen by near UV radiation. *Nature.* 1979, 280, 306–308.

Bindi, L.; Feng, T.; Pasek, M.A. Routes to reduction of phosphate by high-energy events. *Commun. Earth Environ.* 2023, 4, 70. https://doi.org/10.1038/s43247-023-00736-2

Bižić, M.; et al. Aquatic and terrestrial cyanobacteria produce methane. *Sci. Adv.* 2020, 6, eaax5343.

Bjerrum, C.J.; Canfield, D.E. Ocean productivity before about 1.9 Gyr ago limited by phosphorus adsorption onto iron oxides. *Nature.* 2002, 417, 159–162.

Brandes, J.A.; Boctor, N.Z.; Cody, G.D.; Cooper, B.A.; Hazen, R.M.; Yoder, H.S., Jr. Abiotic nitrogen reduction on the early Earth. *Nature.* 1998, 395, 365–367.

Brandes, J.A.; Hazen, R.M.; Yoder, H.S., Jr. Inorganic nitrogen reduction and stability under simulated hydrothermal conditions. *Astrobiology.* 2008, 8, 1113–1126.

Brearley, A.J.; Jones, R.H. Chondritic meteorites. In *Planetary Materials*, ed. Papike, J.J., Vol. 36. Reviews in Mineralogy and Geochemistry. Mineralogical Society of America: Chantilly, VA, 1998, pp. 3.1–3.398.

Bryant, D.E.; Kee, T.P. Direct evidence for the availability of reactive, water soluble phosphorus on the early Earth. H-Phosphinic acid from the Nantan meteorite. *Chem. Commun.* 2006, 22, 2344–2346.

Burr, D.M.; Drummond, S.A.; Cartwright, R.; Black, B.A.; Perron, J.T. Morphology of fluvial networks on Titan: Evidence for structural control. *ICARUS.* 2013, 226(1), 742–759.

Calisto, M.; Usoskin, I.; Rozanov, E.; Peter, T. Influence of Galactic Cosmic Rays on atmospheric composition and dynamics. *Atmos. Chem. Phys.* 2011, 11, 4547–4556.

Canfield, D.E. The early history of atmospheric oxygen: Homage to Robert M. Garrels. *Annu. Rev. Earth Planet. Sci.* 2005, 33, 1–36.

Canfield, D.E.; Rosing, M.T.; Bjerrum, C. Early anaerobic metabolisms. *Philos. Trans. R Soc. Lond. B Biol. Sci.* 2006, 361(1474), 1819–1834; discussion 1835–6. http://doi.org/10.1098/rstb.2006.1906.

Cataldo, F. From elemental carbon to complex macromolecular networks in space. In *Astrobiology: Future Perspectives*, eds. Ehrenfreund, P.; Becker, L.; Blank, J. Astrophysics and Space Science Library, Vol. 305. Kluwer Academic Publishers: Dordrecht, The Netherland, 2004, pp. 97–126.

Catling, D.C.; Kasting, J.F. The prebiotic and early postbiotic atmosphere. In *Atmospheric Evolution on Inhabited and Lifeless Worlds.* Cambridge University Press: Cambridge, 2017, Chapter 9, pp. 231–256.

Cernicharo, J.; Marcelino, N.; Pardo, J.R.; Agundez, M.; Tercero, B.; de Vicente, P.; Cabezas, C.; Bermudez, C. Interstellar nitrile anions: Detection of C3N- and C5N- in TMC-1. *Astron. Astrophys.* 2020, 641, L9.

Chameides, W.; Walker, J. Rates of fixation by lightning of carbon and nitrogen in possible primitive atmospheres. *Orig. Life.* 1981, 11, 291–302.

Chang, S.; DesMarais, D.; Mack, R.; Miller, S.L.; Strathearn, G.E. Prebiotic organic synthesis and the origin of life. In *Earth's Earliest Biosphere*, ed. Schopf, J.W. Princeton University Press: Princeton, 1983, p. 53.

Chapillon, E.; Guilloteau, S.; Dutrey, A.; Pletua, V.; Guelin, M. Chemistry in disks. VI. CN and HCN in protoplanetary disks. *Astron. Astrophys.* 2012, 537, A60.

Cheung, A.C.; Rank, D.M.; Townes, C.H.; Thornton, D.D.; Welch, W.J. Detection of $NH_3$ molecules in the interstellar medium by their microwave emission. *Phys. Rev. Lett.* 1968, 21, 1701.

Chyba, C.; Sagan, C. Endogenous production, exogenous delivery and impact-shock synthesis of organic molecules: An inventory for the origins of life. *Nature.* 1992, 355, 125–132.

Cockell, C.S. The origin and emergence of life under impact bombardment. *Philos. Trans. R. Soc. Lond. B Biol. Sci.* 2006, 361, 1845–1856.

Cohn, M. A study of oxidative phosphorylation with O18-labeled inorganic phosphate. *J. Biol. Chem.* 1953, 201, 735–750.

Cooper, G.J.; Surman, A.J.; McIver, J.; Colon-Santos, S.M.; Gromski, P.S.; Buchwald, S.; Suarez Marina, I.; Cronin, L. Miller-Urey spark-discharge experiments in the deuterium world. *Angew. Chem.* 2017, 56, 8079–8082.

Cooper, G.W.; Onwo, W.M.; Cronin, J.R. Alkyl phosphonic-acids and sulfonic-acids in the Murchison meteorite. *Geochim. Cosmochim. Acta.* 1992, 56, 4109–4115.

Cronin, J.R.; Pizzarello, S. Amino acids in meteorites. *Adv. Space Res.* 1983, 3, 5–18.

Crovisier, J.; Biver, N.; Bockelée-Morvan, D.; Boissier, J.; Colom, P.; Dariusz, C. The chemical diversity of comets: Synergies between space exploration and ground-based radio observations. *Earth, Moon, and Planets.* 2009, 105, 267–272.

Cruikshank, D.; Imanaka, H.; Dalle, O.; Cristina, M. Tholins as coloring agents on outer Solar system bodies. *Adv. Space Res.* 2005, 36, 178–183.

Crutcher, R.M. Magnetic fields in molecular clouds. *Annu. Rev. Astron. Astrophys.* 2012, 50, 29.

Da Costa, F.P.; Grinfeld, M.; Wattis, J.A. A hierarchical cluster system based on Horton – Strahler rules for river networks. *Stud. Appl. Math.* 2002, 109(3), 163–204.

Dalsgaard, T.; Thamdrup, B.; Canfield, D.E. Anaerobic ammonium oxidation (anammox) in the marine environment. *Res. Microbiol.* 2005, 156, 457–464.

Daniel, F.; Cernicharo, J.; Roueff, E.; Gerin, M.; Dubernet, M.L. The excitation of $N_2H^+$ in interstellar molecular clouds. II. Observations. *Astrophys. J.* 2007, 667, 980–1001.

Ebel, D.S. Condensation of rocky material in astrophysical environments. In *Meteorites and the Early Solar System II*, eds. Lauretta, D.; McSween Jr, H.Y. Univ Ariz Press: Tucson, 2006, pp. 253–278.

Ehrenfreund, P.; Cami, J. Cosmic carbon chemistry: From the interstellar medium to the early Earth. *Cold Spring Harb. Perspect. Biol.* 2010, 2(12), a002097.

Ehrenfreund, P.; Spaans, M.; Holm, N.G. The evolution of organic matter in space. *Philos. Trans. R. Soc. A Math. Phys. Eng. Sci.* 2011, 369, 538–554.

Epstein, S.; Krishnamurthy, R.V.; Cronin, J.R.; Pizzarello, S.; Yuen, G.U. Unusual stable isotope ratios in amino acid and carboxylic acid extracts from the Murchison meteorite. *Nature.* 1987, 326, 477–479.

Ernst, L.; Steinfeld, B.; Barayeu, U.; et al. Methane formation driven by reactive oxygen species across all living organisms. *Nature.* 2022, 603, 482–487.

Farquhar, J.; Bao, H.M.; Thiemens, M. Atmospheric influence of Earth's earliest sulfur cycle. *Science.* 2000, 289, 756–758.

Fegley, B.; Prinn, R.G.; Hartman, H.; Watkins, G.H. Chemical effects of large impacts on the Earth's primitive atmosphere. *Nature.* 1986, 319, 305–308.

Fernández-García, C.; Coggins, A.J.; Powner, M.W. A Chemist's perspective on the role of phosphorus at the origins of life. *Life.* 2017, 7, 31.

Fischer, T.P. Fluxes of volatiles ($H_2O$, $CO_2$, $N_2$, Cl, F) from arc volcanoes. *Geochem. J.* 2008, 42, 21–38.

Forsythe, J.G.; Yu, S.S.; Mamajanov, I.; Grover, M.A.; Krishnamurthy, R.; Fernández, F.M.; Hud, N.V. Ester-mediated amide bond formation driven by wet – dry cycles: A possible path to polypeptides on the prebiotic earth. *Angew. Chem. Int. Ed.* 2015, 54(34), 9871–9875.

Foustoukos, D.I.; Seyfried, W.E. Hydrocarbons in hydrothermal vent fluids: The role of chromium-bearing catalysts. *Science.* 2004, 304, 1002–1005.

Gajewski, E.; Steckler, D.K.; Goldberg, R.N. Thermodynamics of the hydrolysis of adenosine 5′-triphosphate to adenosine 5′-diphosphate. *J. Biol. Chem.* 1986, 261, 12733–12737.

Galofre, A.G.; Bahia, R.S.; Jellinek, A.M.; Whipple, K.X.; Gallo, R. Did Martian valley networks substantially modify the landscape? *Earth Planet. Sci. Lett.* 2020, 547, 116482.

Gibard, C.; Gorrell, I.B.; Jiménez, E.I.; Kee, T.P.; Pasek, M.A.; Krishnamurthy, R. Geochemical sources and availability of amidophosphates on the early earth. *Angew. Chem.* 2019, 131, 8235–8239.

Glavin, D.P.; Burton, A.S.; Elsila, J.E.; Aponte, J.C.; Dworkin, J.P. The search for chiral asymmetry as a potential biosignature in our solar system. *Chem. Rev.* 2020, 120, 4660–4689.

Godfrey, L.; Falkowski, P. The cycling and redox state of nitrogen in the Archaean ocean. *Nat. Geosci.* 2009, 2, 725–729. https://doi.org/10.1038/ngeo633

Graaf, R.M.D.; Schwartz, A.W. Reduction and activation of phosphate on the primitive earth. *Orig. Life Evol. Biosph.* 2000, 30, 405–410.

Grady, M.M.; Wright, I.P. Elemental and isotopic abundances of carbon and nitrogen meteorites. *Space Sci. Rev.* 2003, 106, 231–248.

Grotzinger, J.P.; Kasting, J.F. New constraints on Precambrian ocean composition. *J. Geol.* 1993, 101, 235–243.

Guelin, M.; Cernicharo, J. Organic molecules in interstellar space: Latest advances. *Front. Astron. Space Sci.* 2022, 9, 787567.

Gulick, A. Phosphorus as a factor in the origin of life. *Amer. Sci.* 1955, 43, 479–489.

Gull, M. Prebiotic phosphorylation reactions on the early earth. *Challenges*. 2014, 5, 193–212.

Gull, M.; Feng, T.; Bracegirdle, J.; Abbott-Lyon, H.; Pasek, M.A. Organophosphorus compound formation through the oxidation of reduced oxidation state phosphorus compounds on the hadean earth. *J. Mol. Evol.* 2023a, 91(1), 60–75.

Gull, M.; Feng, T.; Cruz, H.A.; Krishnamurthy, R.; Pasek, M.A. Prebiotic chemistry of phosphite: Mild thermal routes to form condensed-p energy currency molecules leading up to the formation of organophosphorus compounds. *Life*. 2023b, 13, 920.

Gull, M.; Feng, T.; Pasek, M.A. Results of an Eight-year extraction of phosphorus minerals within the Seymchan meteorite. *Life*. 2022, 12, 1591. https://doi.org/10.3390/life12101591.

Gull, M.; Pasek, M.A. The role of glycerol and its derivatives in the biochemistry of living organisms, and their prebiotic origin and significance in the evolution of life. *Catalysts*. 2021, 11, 86.

Gusten, R.; Walmsley, C.M.; Pauls, T. Ammonia in the neighborhood of the galactic center. *Astron. Astrophys.* 1981, 103, 197–206.

Hargreaves, W.R.; Mulvihill, S.J.; Deamer, D.W. Synthesis of phospholipids and membranes in prebiotic conditions. *Nature*. 1977, 266, 78–80.

Hayashi, C.; Nakazawa, K.; Mizuno, H. Earth's melting due to the blanketing effect of the primordial dense atmosphere. *Earth Planet. Sci. Lett.* 1979, 43, 22–28.

Hazen, R.M. Paleomineralogy of the Hadean eon: A preliminary species list. *Am. J. Sci.* 2013, 313(9), 807–843.

Hazen, R.M.; Papineau, D.; Bleeker, W.; Downs, R.; Ferry, J.; Coy, T.; Sverjrjensky, D.; Yang, H. Mineral evolution. *Am. Min.* 2008, 93, 1693–1720.

Henning, T.; Semenov, D. Chemistry in protoplanetary disks. *Chem. Rev.* 2013, 113, 9016–9042.

Herschy, B.; Chang, S.J.; Blake, R.; et al. Archean phosphorus liberation induced by iron redox geochemistry. *Nat. Commun.* 2018, 9, 1346.

Ho, P.T.P.; Barrett, A.H.; Myers, P.C.; Matsakis, D.N.; Cheung, A.L.; Chui, M.F.; Townes, C.H.; Yngvesson, K.S. Ammonia observations of the Orion molecular cloud. *Astrophys J.* 1979, 234, 912–921.

Ho, P.T.P.; Townes, C.H. Interstellar ammonia. *Annu. Rev. Astron. Astrophys.* 1983, 21, 239.

Holm, N.G.; Oze, C.; Mousis, O.; Waite, J.H.; Guilbert-Lepoutre, A. Serpentinization and the formation of H2 and CH4 on celestial bodies (Planets, Moons, Comets). *Astrobiology*. 2015, 15, 587–600.

Horita, J.; Berndt, M.E. Abiogenic methane formation and isotopic fractionation under hydrothermal conditions. *Science*. 1999, 285, 1055–1057.

Horsfield, K. Some mathematical properties of branching trees with application to the respiratory system. *Bull. Math. Biol.* 1976, 38(3), 305–315.

Isley, A.E.; Abbott, D.H. Plume-related mafic volcanism and the deposition of banded1440 iron formation. *J. Geophys. Res. Solid Earth*. 1999, 104(B7), 15461–15477.

Johnson, A.P.; Cleaves, H.J.; Dworkin, J.P.; Glavin, D.P.; Lazcano, A.; Bada, J.L. The Miller volcanic spark discharge experiment. *Science*. 2008, 322, 404.

Jones, A.P. Dust evolution, a global view I. Nanoparticles, nascence, nitrogen and natural selection . . . joining the dots. *R. Soc. Open Sci.* 2016, 3, 160221.

Jones, C.; Nomosatryo, S.; Crowe, S.A.; Bjerrum, C.J.; Canfield, D.E. Iron oxides, divalent cations, silica, and the early earth phosphorus crisis. *Geology*. 2015, 43, 135–138.

Kakegawa, T.; Noda, M.; Nannri, H. Geochemical cycles of bio-essential elements on the early earth and their relationships to origin of life. *Resour. Geol.* 2002, 52(2), 83–89.

Karl, D.M. Microbially mediated transformations of phosphorus in the sea: New views of an old cycle. *Annu. Rev. Mar. Sci.* 2014, 6, 279.

Kasting, J.F. Bolide impacts and the oxidation state of carbon in the Earth's early atmosphere. *Orig. Life Evol. B.* 1992, 20, 199–231.

Kasting, J.F. Earth's early atmosphere. *Science.* 1993, 259, 920–926.

Kasting, J.F.; Walker, J.C.G. Limits on oxygen concentration in the prebiotic atmosphere and the rate of abiotic fixation of nitrogen. *J. Geophys. Res. Oceans Atmos.* 1981, 86, 1147–1158.

Kawakita, H.; Jehin, E.; Manfroid, J.; Hutsemekers, D. Nuclear spin temperature of ammonia in Comet 9P/Tempel 1 before and after the Deep Impact event. *ICARUS.* 2007, 191, 513–516.

Kee, T.P.; Bryant, D.E.; Herschy, B.; Marriott, K.E.; Cosgrove, N.E.; Pasek, M.A.; Atlas, Z.D.; Cousins, C.R. Phosphate activation via reduced oxidation state phosphorus (P). Mild routes to condensed-P energy currency molecules. *Life.* 2013, 3, 386–402.

Kee, T.P.; Monnard, P.A. Chemical systems, chemical contiguity and the emergence of life. *Beilstein J. Org. Chem.* 2017, 13, 1551–1563.

Keppler, F.; Hamilton, J.T.G.; Braß, M.; Röckmann, T. Methane emissions from terrestrial plants under aerobic conditions. *Nature.* 2006, 439, 187–191.

Kinne-Saffran, E.; Kinne, R.K. Vitalism and synthesis of urea. From Friedrich Wöhler to Hans A. Krebs. *Am. J. Nephrol.* 1999, 19(2), 290–294. http://doi.org/10.1159/000013463. PMID: 10213830.

Kipp, M.A.; Eva, E. Stüeken. Biomass recycling and Earth's early phosphorus cycle. *Sci. Adv.* 2017, 3. http://doi.org/10.1126/sciadv.aao4795.

Klintzsch, T.; et al. Methane production by three widespread marine phytoplankton species: Release rates, precursor compounds, and potential relevance for the environment. *Biogeosciences.* 2019, 16, 4129–4144.

Kobayashi, K.; Kasamatsu, T.; Kaneko, T.; Saito, T. Production of organic compounds in interstellar space. In *Exobiology: Matter, Energy, and Information in the Origin and Evolution of Life in the Universe*, eds. Chela-Flores, J.; Raulin, F., 1st ed. Springer: Dordrecht, 1998, pp. 213–216.

Kolesnikov, M.P.; Egorov, A. Metalloporphyrins and molecular complexes of amino acids with porphyrins in juvenile volcanic ash. *Orig. Life.* 1979, 9, 267–277.

Krishnamurthy, R.; Arrhenius, G.; Eschenmoser, A. Formation of glycolaldehyde phosphate from glycolaldehyde in aqueous solution. *Orig. Life Evol. Biosph.* 1999, 29, 333–354.

Kuhn, W.R.; Atreya, S.K. Ammonia photolysis and the greenhouse effect in the primordial atmosphere of the Earth. *ICARUS.* 1979, 37(1), 207–213.

Kuhn, W.R.; Kasting, J.F. Effects of increased $CO_2$ concentrations on surface temperature of the early Earth. *Nature.* 1983, 301, 53.

Kwok, S. Complex organics in space from solar system to distant galaxies. *Astron. Astrophys. Rev.* 2016, 24, 8.

Lecuyer, C.; Simon, L.; Guyot, F. Comparison of carbon, nitrogen, and water budgets on Venus and the Earth. *Earth Planet. Sci. Lett.* 2000, 181, 33–40.

Lenhart, K.; et al. Evidence for methane production by saprotrophic fungi. *Nat. Commun.* 2012, 3, 1046.

Lewis, B.L.; Stansberry, J.A.; Holler, B.J.; Grundy, W.M.; Schmitt, B.; Protopapa, S.; Lisse, C.; Alan Stern, S.; Young, L.; Weaver, H.A. Distribution and energy balance of Pluto's nitrogen ice, as seen by New Horizons in 2015. *ICARUS.* 2021, 356, 113633.

Liggins, P.; Jordan, S.; Rimmer, P.B.; Shorttle, O. Growth and evolution of secondary volcanic atmospheres: I. Identifying the geological character of hot rocky planets. *J. Geophys. Res.-Planet.* 2022, 127, e2021JE007123.

Lindenmayer, A. Mathematical models for cellular interactions in development II. Simple and branching filaments with two-sided inputs. *J. Theor. Biol.* 1968, 18(3), 300–315.

Lohrmann, L.; Orgel, L.E. Urea-inorganic phosphate mixtures as prebiotic phosphorylating agents. *Science*. 1971, 171, 490–494.

Lorenz, R.D.; Mitchell, K.L.; Kirk, R.L.; Hayes, A.G.; Aharonson, O.; Zebker, H.A.; Paillou, P.; Radebaugh, J.; Lunine, J.I.; Janssen, M.A.; et al. Titan's inventory of organic surface materials. *Geophys Res. Lett*. 2008, 35, L02206.

Macià, E.; Hernández, M.V.; Oró, J. Primary sources of phosphorus and phosphates in chemical evolution. *Orig. Life Evol. Biosph*. 1997, 27, 459–480.

MacPherson, G.J. Calcium-aluminum-rich inclusions in chondritic meteorites. In *Treatise on Geochemistry*, eds. Holland, H.D.; Turekian, K.K., Vol. 1. Elsevier: Amsterdam, 2007, pp. 201–246.

Mahaffy, P.R.; Webster, C.R.; Atreya, S.K.; Franz, H.; Wong, M.; Conrad, P.G.; Harpold, D.; Jones, J.J.; Leshin, L.A.; Manning, H.; et al. Abundance and isotopic composition of gases in the Martian atmosphere from the Curiosity Rover. *Science*. 2013, 341, 263–266.

Mancinelli, R.L.; McKay, C.P. The evolution of nitrogen cycling. *Orig. Life Evol. Biosph*. 1988, 18, 311–25.

Martin, R.S.; Mather, T.A.; Pyle, D.M. Volcanic emissions and the early Earth. *Geochim. Cosmochim. Acta*. 2007, 71, 3673–3685.

Martins, Z. The nitrogen heterocycle content of meteorites and their significance for the origin of life. *Life*. 2018, 8, 28.

Martins, Z.; Chan, Q.H.S.; Bonal, L.; King, A.; Yabuta, H. Organic matter in the solar system – implications for future on-site and sample return missions. *Space Sci. Rev*. 2020, 216, 54.

Marty, B.; Altwegg, K.; Balsiger, H.; Bar-Nun, A.; Bekaert, D.V.; Berthelier, J.-J.; Bieler, A.; Briois, C.; Calmonte, U.; Combi, M.; et al. Xenon isotopes in 67P/Churyumov-Gerasimenko show that comets contributed to Earth's atmosphere. *Science*. 2017, 356, 1069–1072.

Marty, B.; Avice, G.; Sano, Y.; Altwegg, K.; Balsiger, H.; Hassig, M.; Morbidelli, A.; Mousis, O.; Rubin, M. Origins of volatile elements (H, C, N, noble gases) on Earth and Mars in light of recent results from the ROSETTA cometary mission. *Earth Planet Sc. Lett*. 2016, 441, 91–102.

Mason, B. Organic matter from space. *Sci. Am*. 1963, 208, 43–49.

Mather, T.; Pyle, D.; Allen, A. Volcanic source for fixed nitrogen in the early Earth atmosphere. *Geology*. 2004, 32, 905–908.

McCollom, T.M.; Seewald, J.S. Geochim. *Cosmochim. Acta*. 2003, 67, 3625–3644.

McKay, A.J.; Roth, N.X. Organic matter in cometary environments. *Life*. 2021, 11, 37.

Messenger, S.; Sandford, S.; Brownlee, D. The population of starting materials available for solar system construction. In *Meteorites and the Early Solar System II*, eds. Lauretta, D.S.; McSween Jr., H.Y. University of Arizona Press: Tucson, 2006, pp. 187–207.

Miller, S.L. A production of amino acids under possible primitive Earth conditions. *Science*. 1953, 117, 528–529.

Miller, S.L. The prebiotic synthesis of organic compounds on the early Earth. In *Organic Geochemistry*, eds. Engel, M.H.; Mack, S.A., Vol. 11. Springer: Boston, MA, 1993, pp. 625–637.

Miyakawa, S.; Murasawa, K.-I.; Kobayashi, K.; Sawaoka, A.B. Cytosine and uracil synthesis by quenching with high-temperature plasma. *J. Am. Chem. Soc*. 1999, 121, 8144–8145.

Miyakawa, S.; Tamura, H.; Sawaoka, A.B. Amino acid synthesis from an amorphous substance composed on carbon, nitrogen, and oxygen. *Appl. Phys. Lett*. 1998, 72, 990.

Miyakawa, S.; Yamanashi, H.; Kobayashi, K.; Cleaves, H.J.; Miller, S.L. Prebiotic synthesis from CO atmospheres: Implications for the origins of life. *Proc. Natl. Acad. Sci. USA*. 2002, 99, 14628–14631.

Molden, R.C.; Goya, J.; Khan, Z.; Garcia, B.A. Stable isotope labeling of phosphoproteins for large-scale phosphorylation rate determination. *Mol. Cell. Proteomics.* 2014, 13, 1106–1118.

Moore, J.K.; Braucher, O. Observations of dissolved iron concentrations in the World Ocean: Implications and constraints for ocean biogeochemical models. *Biogeosciences Discuss.* 2007, 4(2), 1241–1277.

Morbidelli, A.; Lunine, J.I.; O'Brien, D.P.; Raymond, S.N.; Walsh, K.J. Building terrestrial planets. *Annu. Rev. Earth Planet. Sci.* 2012, 40, 251–275.

Mostefaoui, S.; Hoppe, P. Discovery of abundant in situ silicate and spinel grains from red giant stars in a primitive meteorite. *Astrophys. J.* 2004, 613, L149–L152.

Mukhin, L.M. Volcanic processes and synthesis of simple organic compounds on primitive Earth. *Orig. Life Evol. B* 1976, 7, 355–368.

Mumma, M.J.; Charnley, S.B. The chemical composition of comets – Emerging taxonomies and natal heritage. *Ann. Rev. Astron. Astrophys.* 2011, 49, 471–524.

Nakano, H.; Hirakawa, N.; Matsubara, Y.; Yamashita, S.; Okuchi, T.; Asahina, K.; Tanaka, R.; Suzuki, N.; Naraoka, H.; Takano, Y.; et al. Precometary organic matter: A hidden reservoir of water inside the snow line. *Sci. Rep.* 2020, 10, 7755.

Nicolet, M. On the production of nitric oxide by cosmic rays in the mesosphere and stratosphere. *Planet Space Sci.* 1975, 23, 637–649.

Nishizawa, M.; Miyazaki, J.; Makabe, A.; Koba, K.; Takai, K. Physiological and isotopic characteristics of nitrogen fixation by hyperthermophilic methanogens: Key insights into nitrogen anabolism of the microbial communities in Archean hydrothermal systems. *Geochim. Cosmochim. Acta.* 2014, 138, 117–135.

Nittler, L.R. Presolar stardust in meteorites: Recent advances and scientific frontiers. *Earth Planet. Sci. Lett.* 2003, 209, 259–273.

Nooner, D.W.; Gibert, J.M.; Gelp, E.; Oro, J. Closed system Fischer-Tropsch synthesis over meteoritic iron, iron ore, and iron-nickel alloy. *Geochim. Cosmochim. Acta.* 1976, 40, 915–924.

Oró, J.; Mills, T. Chemical evolution of primitive solar system bodies. *Adv. Space Res.* 1989, 9, 105–120.

Osinski, G.R.; Cockell, C.S.; Pontefract, A.; Sapers, H.M. The role of meteorite impacts in the origin of life. *Astrobiology.* 2020, 20, 1121–1149.

Owen, T.; Biemann, K.; Rushneck, D.R.; Biller, J.E.; Haworth, D.W.; Lafleur, A.L. The composition of the atmosphere at the surface of Mars. *J. Geophys. Res.* 1977, 82, 4635–4639.

Owen, T.; Mahaffy, P.R.; Niemann, H.B.; Atreya, S.; Wong, M. Protosolar nitrogen. *Astrophys. J.* 2001, 553, L77–L79.

Parkos, D.; Pikus, A.; Alexeenko, A.; Melosh, H.J. HCN production via impact ejecta reentry during the late heavy bombardment. *J. Geophys. Res.* 2018, 123, 892–909.

Pasek, M.A. Rethinking early earth phosphorus geochemistry. *Proc. Natl. Acad. Sci. USA.* 2008, 105, 853–858.

Pasek, M.A. Thermodynamics of prebiotic phosphorylation. *Chem. Rev.* 2020, 120(11), 4690–4706. http://doi.org/10.1021/acs.chemrev.9b00492.

Pasek, M.A.; Block, K. Lightning-induced reduction of phosphorus oxidation state. *Nat. Geosci.* 2009, 2(8), 553.

Pasek, M.A.; Dworkin, J.; Lauretta, D.S. A radical pathway for phosphorylation during schreibersite corrosion with implications for the origin of life. *Geochim. Cosmochim. Acta.* 2007, 71, 1721–1736.

Pasek, M.A.; et al. Serpentinization as a route to liberating phosphorus on habitable worlds. *Geochim. Cosmochim. Acta.* 2022, 336, 332–340.

Pasek, M.A.; Gull, M.; Herschy, B. Phosphorylation on the early earth. *Chem. Geol.* 2017, 475, 149–170.

Pasek, M.A.; Harnmeijer, J.P.; Buick, R.; Gull, M.; Atlas, Z. Evidence for reactive reduced phosphorus species in the early Archean ocean. *Proc. Natl. Acad. Sci. USA*. 2013, 110, 10089–10094.

Pasek, M.A.; Kee, T.P. On the origin of phosphorylated biomolecules. In *Origins of Life: The Primal Self-Organization*, eds. Egel, R.; Lankenau, D.-H.; Mulkidjanian, A.Y. Springer-Verlag: Berlin/Heidelberg, Germany, 2011, pp. 57–84.

Pasek, M.A.; Kee, T.P.; Bryant, D.E.; Pavlov, A.A.; Lunine, J.I. Production of potentially prebiotic condensed phosphates by phosphorus redox chemistry. *Angew. Chem. Int. Ed. Engl.* 2008, 47, 7918–7920.

Pasek, M.A.; Lauretta, D.S. Aqueous corrosion of phosphide minerals from iron meteorites: A highly reactive source of prebiotic phosphorus on the surface of the early Earth. *Astrobiology*. 2005, 5, 515–535.

Pasek, M.A.; Lauretta, D.S. Extraterrestrial flux of potentially prebiotic C, N, and P to the early Earth. *Orig. Life Evol. Biosph.* 2008, 38, 5–21.

Pereto, J.; Bada, J.L.; Lazcano, A. Charles Darwin and the origin of life. *Orig. Life Evol. Biosph.* 2009, 39, 395–406.

Pierazzo, E.; Chyba, C.F. Amino acid survival in large cometary impacts. *Meteorit. Planet. Sci.* 1999, 34, 909–918.

Pirim, C.; Pasek, M.A.; Sokolov, D.A.; Sidorov, A.N.; Gann, R.D.; Orlando, T.M. Investigation of schreibersite and intrinsic oxidation products from Sikhote-Alin, Seymchan, and Odessa meteorites and $Fe_3P$ and $Fe_2NiP$ synthetic surrogates. *Geochim. Cosmochim. Acta*. 2014, 140, 259–274.

Pizzarello, S.; Huang, Y. The deuterium enrichment of individual amino acids in carbonaceous meteorites: A case for the presolar distribution of biomolecule precursors. *Geochim. Cosmochim. Acta*. 2005, 69, 599–605.

Pizzarello, S.; Huang, Y.; Becker, L.; Poreda, R.J.; Nieman, R.A.; Cooper, G.; Williams, M. The organic content of the Tagish Lake meteorite. *Science*. 2001, 293, 2236–2239.

Pizzarello, S.; Huang, Y.; Fuller, M. The carbon isotopic distribution of Murchison amino acids. *Geochim. Cosmochim. Acta*. 2004, 68, 4963–4969.

Planavsky, N.J.; Asael, D.; Hofmann, A.; Reinhard, C.T.; Lalonde, S.V.; Knudsen, A.; Wang, X.; Ossa, F.; Pecoits, E.; Smith, A.J.B.; Beukes, N.J.; Bekker, A.; Johnson, T.M.; Konhauser, K.O.; Lyons, T.W.; Rouxel, O.J. Evidence for oxygenic photosynthesis half a billion years before the Great Oxidation Event. *Nat. Geosci.* 2014, 7, 283–286.

Planavsky, N.J.; Rouxel, O.J.; Bekker, A.; Lalonde, S.V.; Konhauser, K.O.; Reinhard, C.T.; Lyons, T.W. The evolution of the marine phosphate reservoir. *Nature*. 2010, 467, 1088–1090.

Poch, O.; Istiqomah, I.; Quirico, E.; Beck, P.; Schmitt, B.; Theulé, P.; Faure, A.; Hily-Blant, P.; Bonal, L.; Raponi, A.; et al. Ammonium salts are a reservoir of nitrogen on a cometary nucleus and possibly on some asteroids. *Science*. 2020, 367, eaaw7462.

Pohorille, A. From organic molecules in space to the origins of life and back. *Adv. Space Res.* 2002, 30, 1509–1520.

Ponnamperuma, C.; Chang, S. The role of phosphates in chemical evolution. In *Chemical Evolution and the Origin of Life; Proceedings of the Third International Conference, Pont-a-Mousson, France*. North-Holland Publishing Co.: Amsterdam, 1971, pp. 216–223.

Poulton, S.W.; Canfield, D.E. Ferruginous conditions: A dominant feature of the ocean through Earth's history. *Elements*. 2011, 7, 107–112.

Powner, M.W.; Gerland, B.; Sutherland, J.D. Synthesis of activated pyrimidine ribonucleotides in prebiotically plausible conditions. *Nature*. 2009, 459(7244), 239–242.

Quimby, O.T.; Flautt, T.J. Ammonolyse des Trimetaphosphats. *Z. Anorg. Allg. Chem.* 1958, 296, 220–228.

Raulin, F. Astrobiology and habitability of Titan. *Space Sci. Rev.* 2008, 135, 37–48.

Reimann, R.; Zubay, G. Nucleoside phosphorylation: A feasible step in the prebiotic pathway to RNA. *Orig. Life Evol. Biosph.* 1999, 29(3), 229–247.

Reinhard, C.T.; Planavsky, N.J.; Gill, B.C.; Ozaki, K.; Robbins, L.J.; Lyons, T.W.; Fischer, W.W.; Wang, C.; Cole, D.B.; Konhauser, K.O. Evolution of the global phosphorus cycle. *Nature.* 2017, 541, 386–389.

Rice, T.S.; Bergin, E.A.; Jorgenson, J.K.; Wampfler, S.F. Exploring the origins of Earth's nitrogen: Astronomical observations of nitrogen-bearing organics in protostellar environments. *Astrophys. J.* 2018, 866, 156.

Rimmer, P.B.; Shorttle, O. Origin of life's building blocks in carbon- and nitrogen-rich surface hydrothermal vents. *Life.* 2019, 9, 12.

Ritson, D.J.; Mojzsis, S.J.; Sutherland, J. Supply of phosphate to early Earth by photogeochemistry after meteoritic weathering. *Nat. Geosci.* 2020, 13, 344–348.

Rubin, A.E. Mineralogy of meteorite groups. *Meteorit. Planet. Sci.* 1997, 32, 231–247.

Rubin, M.; Altwegg, K.; Balsiger, H.; Bar-Nun, A.; Berthelier, J.-J.; Bieler, A.; Bochsler, P.; Briois, C.; Calmonte, U.; Combi, M.; et al. Molecular nitrogen in comet 67P/Churyumov-Gerasimenko indicates a low formation temperature. *Science.* 2015, 348, 232–235.

Ruiz-Mirazo, K.; Briones, C.; de la Escosura, A. Prebiotic systems chemistry: New perspectives for the origins of life. *Chem. Rev.* 2014, 114, 285–366.

Rushdi, A.I.; Simoneit, B.R. Condensation reactions and formation of amides, esters, and nitriles under hydrothermal conditions. *Astrobiology.* 2004, 4(2), 211–224.

Saunois, M. et al. The global methane budget 2000–2017. *Earth Syst. Sci. Data.* 2020, 12, 1561–1623.

Scherf, M.; Lammer, H.; Erkaev, N.V.; Mandt, K.E.; Thaller, S.E.; Marty, B. Nitrogen atmospheres of the icy bodies in the solar system. *Space Sci. Rev.* 2020, 216, 123.

Schwartz, A.W. Prebiotic phosphorylation-nucleotide synthesis with apatite. *Biochim. Biophys. Acta (BBA)-Nucleic Acids Protein Synthesis.* 1972, 281(4), 477–480.

Scott, H.P.; Hemley, R.J.; Mao, H.K.; Herschbach, D.R.; Fried, L.E.; Howard, W.M.; Bastea, S. Generation of methane in the Earth's mantle: In situ high pressure-temperature measurements of carbonate reduction. *Proc. Natl. Acad. Sci. USA.* 2004, 101(39), 14023–14026.

Sherwood Lollar, B.; Frape, S.K.; Weise, S.M.; Fritz, P.; Macko, S.A.; Welhan, J.A. Abiogenic methanogenesis in crystalline rocks. *Geochim. Cosmochim. Acta.* 1993, 57, 5087–5097.

Sherwood Lollar, B.; Westgate, T.D.; Ward, J.A.; Slater, G.F.; Lacrampe-Couloume, G. Abiogenic formation of alkanes in the Earth's crust as a minor source for global hydrocarbon reservoirs. *Nature.* 2002, 416, 522–524.

Sleep, N.H.; Meibom, A.; Fridriksson, T.; Coleman, R.G.; Bird, D.K. H2-rich fluids from serpentinization: Geochemical and biotic implications. *Proc. Natl. Acad. Sci. USA.* 2004, 101, 12818–12823.

Smith, K.E.; Callahan, M.P.; Gerakines, P.A.; Dworkin, J.P.; House, C.H. Investigation of pyridine carboxylic acids in CM2 carbonaceous chondrites: Potential precursor molecules for ancient coenzymes. *Geochim. Cosmochim. Acta.* 2014, 136, 1–12.

Srinivasan, V.; Morowitz, H.J. Analysis of the intermediary metabolism of a reductive chemoautotroph. *Biol. Bull.* 2009, 217, 222–232.

Stern, J.C.; Sutter, B.; Freissinet, C.; Navarro-González, R.; McKay, C.P.; Archer, P.D.; Buch, A.; Brunner, A.E.; Coll, P.; Eigenbrode, J.L.; et al. Evidence for indigenous nitrogen in sedimentary and aeolian deposits from the Curiosity rover investigations at Gale Crater, Mars. *Proc. Natl. Acad. Sci. USA.* 2015, 112, 4245–4250.

Stroud, R.M.; Nittler, L.R.; Alexander, C.M.O'D. Polymorphism in presolar $Al_2O_3$ grains from asymptotic giant branch stars. *Science*. 2004, 305, 1455–1457.

Stüeken, E.E.; Catling, D.C.; Buick, R. Contributions to late Archaean sulphur cycling by life on land. *Nat. Geosci.* 2012, 5, 722–725.

Stüeken, E.E.; Kipp, M.A.; Koehler, M.C.; Buick, R. The evolution of Earth's biogeochemical nitrogen cycle. *Earth Sci. Rev.* 2016, 160, 220–239.

Tarelli, E.; Wheeler, S.F. Formation of esters, especially phosphate-esters, under dry conditions and mild pH. *Chem. Ind.* 1993, 5, 164–165.

Thaddeus, P. The short-wavelength spectrum of the microwave background. *Annu. Rev. Astron. Astrophys.* 1972, 10, 305–334.

Tian, F.; Kasting, J.F.; Zahnle, K. Revisiting HCN formation in Earth's early atmosphere. *Earth Planet Sc. Lett.* 2011, 308, 417–423.

Todd, Z.R. Sources of Nitrogen-, Sulfur-, and phosphorus-containing feedstocks for prebiotic chemistry in the planetary environment. *Life*. 2022, 12, 1268.

Todd, Z.R.; Oberg, K.I. Cometary delivery of HCN to the early Earth. *Astrobiology*. 2020, 20, 1109–1120.

Toner, J.D.; Catling, D.C. A carbonate-rich lake solution to the phosphate problem of the origin of life. *Proc. Natl. Acad. Sci. USA*. 2020,117, 883–888.

Tyrrell, T. The relative influences of nitrogen and phosphorus on oceanic primary production. *Nature*. 1999, 400, 525–531.

Van Mooy, B.A.S.; Krupke, A.; Dyhrman, S.T.; Fredricks, H.F.; Frischkorn, K.R.; Ossolinski, J.E.; Repeta, D.J.; Rouco, M.; Seewald, J.D.; Sylva, S.P. Phosphorus cycling. Major role of planktonic phosphate reduction in the marine phosphorus redox cycle. *Science*. 2015, 348, 783.

Vollmer, C.; Hoppe, P.; Brenker, F.E.; Holzapfel, C. Stellar $MgSiO_3$ perovskite: A shock-transformed silicate found in a meteorite. *Astrophys. J.* 2007, 666, L49–L52.

Walker, J.C.G. Carbon dioxide on the early earth. *Orig. Life Evol. Biosph.* 1985, 16, 117–127.

Walsh, C. Complex organic molecules in protoplanetary disks. *Astron. Astrophys.* 2014, 563, A33.

Wyckoff, S.; Tegler, S.C.; Engel, L. Nitrogen abundance in Comet Halley. *Astrophys. J.* 1991, 367, 641.

Young C.L.; Ingall E.D. Marine dissolved organic phosphorus composition: Insights from samples recovered using combined electrodialysis/reverse osmosis. *Aquat. Geochem.* 2010, 16, 563–574.

Zahnle, K.J. Photochemistry of methane and the formation of hydrocyanic acid (HCN) in the Earth's early atmosphere. *J. Geophys. Res.-Atmos.* 1986, 91, 2819–2834.

Zahnle, K.J.; Schaefer, L.; Fegley, B. Earth's earliest atmospheres. *Cold Spring Harb. Perspect. Biol.* 2010, 2(10), a004895.

# 7 Asymmetric Autocatalysis and the Origins of Homochirality

*Kenso Soai, Tsuneomi Kawasaki and Arimasa Matsumoto*

## 7.1 IMPLICATIONS OF HOMOCHIRALITY AND SELF-REPLICATION FOR THE ORIGIN OF LIFE

One of the most characteristic features of life is the homochirality of biomolecules, such as seen in L-amino acids and D-sugars. This is not a mere aspect of life but has been considered to be a prerequisite for the emergence of life. In the introduction, we explain the concept and implications of chirality. Shapes of left and right hands are mirror images and are not superimposable. In the same manner, L- and D-amino acids are mirror images and are not superimposable, i.e., they are chiral and enantiomers (Figure 7.1). In general, when a carbon atom has four different substituents or atoms, the carbon atom becomes a stereogenic (chiral) center and is termed an asymmetric carbon atom (Eliel and Wilen 1994). Other than amino acids, there are many biological chiral compounds such as D- and L-sugars. One of the most striking aspects of life is that all living creatures on Earth are composed of essentially L-amino acids and D-sugars. This aspect is often called homochirality of life. Thus, the question when and how did biomolecules reach the stage of homochirality is closely related to the question of the origin of life and astrobiology.

Homochirality of the components is essential for life. Why? Let's consider the situation of shaking hands. The normal manner of shaking hands is that two people use their same (right) hands; however, if one uses the left hand and the other the right, the situation of linking hands bears different meaning from normal shaking hands. Peptides and proteins are formed from linking amino acids by peptide bonds. In dipeptides, for example, L-alanyl-L-alanine and D-alanyl-L-alanine are not enantiomers but diastereomers. They have different properties such as melting points. Natural peptides and proteins are composed of L-amino acids by forming peptide bonds. If D-amino acids are irregularly incorporated into peptides and proteins, the conformation and physical properties of the protein change and the enzymes would not be able to operate their functions. D-Deoxyribose is a component of DNA and forms a helix. If L-deoxyribose, the enantiomer of D-deoxyribose, is irregularly incorporated in DNA, the natural structure of double helix would not be formed, and the gene information would not be transferred to the next generation. Indeed, Eschenmoser described that the reaction of the self-assembling by ligative oligomerization of tetrameric D-pyranosyl-RNA (analogues of RNA) is slower by at least two orders of magnitude when one of the components


DOI: 10.1201/9781003294276-7

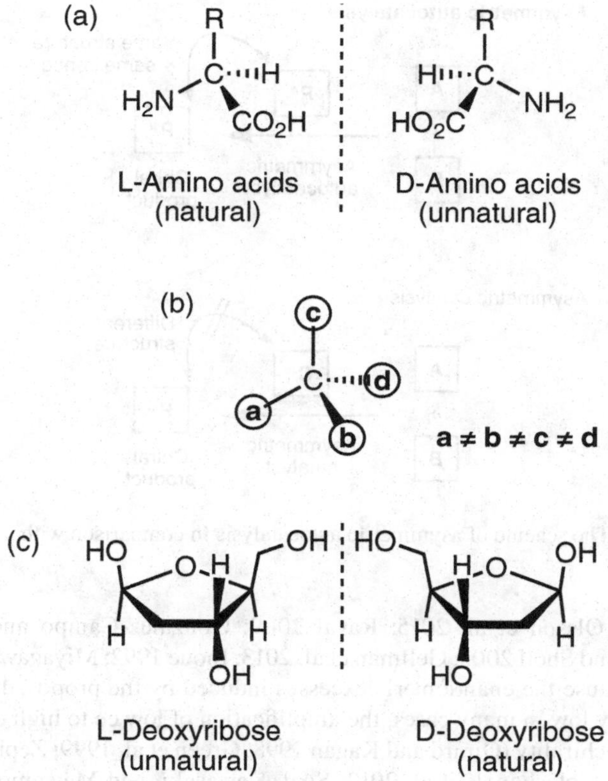

**FIGURE 7.1** (a) Structures of L- and D-amino acids (R = $CH_3$: alanine). (b) Structure of asymmetric carbon, i.e., C (carbon) with four different substituents (a ≠ b ≠ c ≠ d). (c) Structures of D-deoxyribose (natural) and L-deoxyribose (unnatural) in nucleosides.

is substituted by the L-enantiomer (Bolli, Micura, and Eschenmoser 1997). Thus, the homochirality of components is essential for living creatures and the origin of life.

Then, how did biomolecules become highly enantioenriched? How was the first chiral organic compound formed? Usual chemical reactions using achiral reactants and without the intervention of any chiral factor afford a mixture with an equal ratio of D- and L-products, i.e., racemates. (However, it should be noted that this statement is not a scientific truth because asymmetric autocatalysis with the amplification of enantiomeric excess (ee), the Soai reaction, is capable of achieving absolute asymmetric synthesis which affords enantioenriched chiral compounds without the intervention of any chiral factor. This will be described later in this chapter). Thus, the origin of the homochirality of organic compounds has been a puzzle for the chemical origin of life for over 170 years, ever since the discovery of molecular chirality by Pasteur in 1848 (Gal 2008).

Several theories have been proposed as the origins of chirality of organic compounds (Weissbuch et al. 1984; Kondepudi, Kaufman, and Singh 1990; Siegel 1998; Feringa and van Delden 1999; Ribó et al. 2001; Mislow 2003; Pizzarello and Weber 2004; Breslow and Cheng 2010; Weissbuch and Lahav 2011; Ernst 2012; Saito and

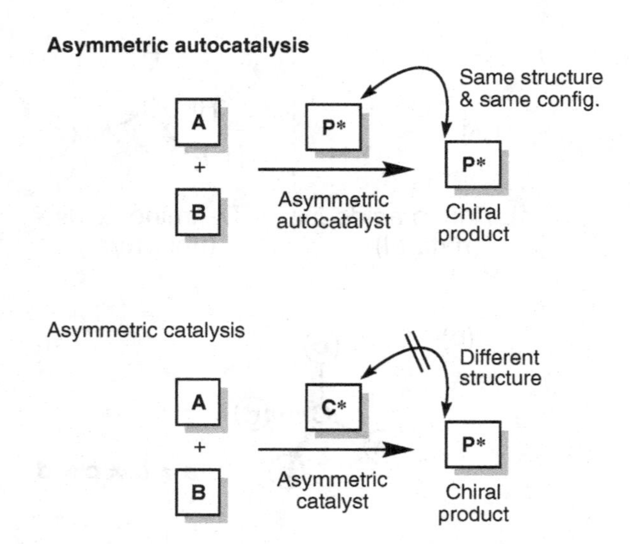


**FIGURE 7.2** The scheme of asymmetric autocatalysis in comparison with usual asymmetric catalysis.

Hyuga 2013; Olsson et al. 2015; Raval 2009; Gonzalez-Campo and Amabilino 2013; Hazen and Sholl 2003; Gellman et al. 2013; Inoue 1992; Miyagawa et al. 2017). However, because the enantiomeric excesses induced by the proposed mechanisms have been very low in many cases, the amplification of low ee to high ee is required to reach homochirality (Girard and Kagan 1998; Green et al. 1999; Zepik et al. 2002; Viedma 2005; Soloshonok et al. 2012; Soai, Kawasaki, and Matsumoto 2014a). In this chapter, we describe asymmetric autocatalysis and the origin of homochirality. Asymmetric autocatalysis is a reaction in which a chiral product acts as a chiral catalyst for its own production (Figure 7.2) (Soai et al. 1995; Shibata, Yonekubo, and Soai 1999; Sato et al. 2003a; Soai, Kawasaki, and Matsumoto 2014b; Soai, Matsumoto, and Kawasaki 2017; Soai, Shibata, and Sato 2000; Soai and Kawasaki 2008; Avalos et al. 2000; Bissette and Fletcher 2013; Todd 2002). The process is a catalytic self-amplification of a chiral compound. Soai and co-workers discovered asymmetric autocatalysis of pyrimidyl alkanol in the addition of diisopropylzinc ($i$-$Pr_2Zn$) to pyrimidine-5-carbaldehyde. During the reaction, the amount of pyrimidyl alkanol increases and its ee amplifies from extremely low to >99.5% ee (this is named as the Soai reaction). In the Soai reaction, various chiral factors, such as circularly polarized light (CPL) and chiral inorganic crystals, trigger asymmetric autocatalysis and afford the highly enantioenriched chiral product with absolute configurations corresponding to those of the chiral triggers. In addition, spontaneous absolute asymmetric synthesis without the intervention of any chiral factor has been achieved by the Soai reaction, i.e., the reaction between pyrimidine-5-carbaldehyde and $i$-$Pr_2Zn$.

Automultiplication, i.e., self-replication, at the individual and cellular levels is another characteristic feature of life. Self-replication at the molecular level has been examined with organic compounds without producing any chiral stereogenic center (Sievers and von Kiedrowski 1994; Ashkenasy et al. 2004; Wintner, Conn, and

Rebek 1994; Lee et al. 1996; Ashkenasy et al. 2017). In sharp contrast, asymmetric autocatalysis of pyrimidyl alkanol, i.e., the Soai reaction, is unique in that the process is a catalytic self-multiplication of a chiral compound.

Concerned with the theories of the origins of homochirality, several mechanisms other than asymmetric autocatalysis have been proposed, including crystallization with racemization, self-disproportionation by distillation, crystallization with abrasion (Viedma ripening), and amino acids in meteorites. These are beyond the scope of this chapter, and readers are encouraged to consult the corresponding articles and reviews.

## 7.2 ASYMMETRIC AUTOCATALYSIS OF 5-PYRIMIDYL ALKANOL WITH AMPLIFICATION OF ENANTIOMERIC EXCESS IN THE ADDITION OF DIISOPROPYLZINC TO PYRIMIDINE-5-CARBALDEHYDE: THE SOAI REACTION

In 1953, Frank proposed a scheme of asymmetric autocatalysis without mentioning any actual chemical structure of compounds (Frank 1953). In 1990, Soai et al. reported first asymmetric autocatalysis of 3-pyridyl alkanol in the reaction between pyridine-3-carbaldehyde and $i$-$Pr_2Zn$ (Soai, Niwa, and Hori 1990). They also found in 1995, for the first time, the asymmetric autocatalysis with the amplification of ee as a real chemical reaction in the enantioselective addition of $i$-$Pr_2Zn$ to pyrimidine-5-carbaldehyde (Soai et al. 1995). After further investigation of the substituent effect, i.e., by the introduction of a 2-alkynyl group at the 2-position of the pyrimidine ring (Shibata, Yonekubo, and Soai 1999), the isopropylzinc alkoxide of 2-alkynyl-5-pyrimidyl alkanol **1** (Matsumoto et al. 2015a; Matsumoto et al. 2016a) formed in situ acted as a practically perfect asymmetric autocatalyst. It has been demonstrated that 5-pyrimidyl alkanol **1** shows remarkable amplification of ee in the addition reaction of $i$-$Pr_2Zn$ to aldehyde **2** (Figure 7.3). An initial very slight enantiomeric imbalance, as low as ca. 0.00005% ee, was

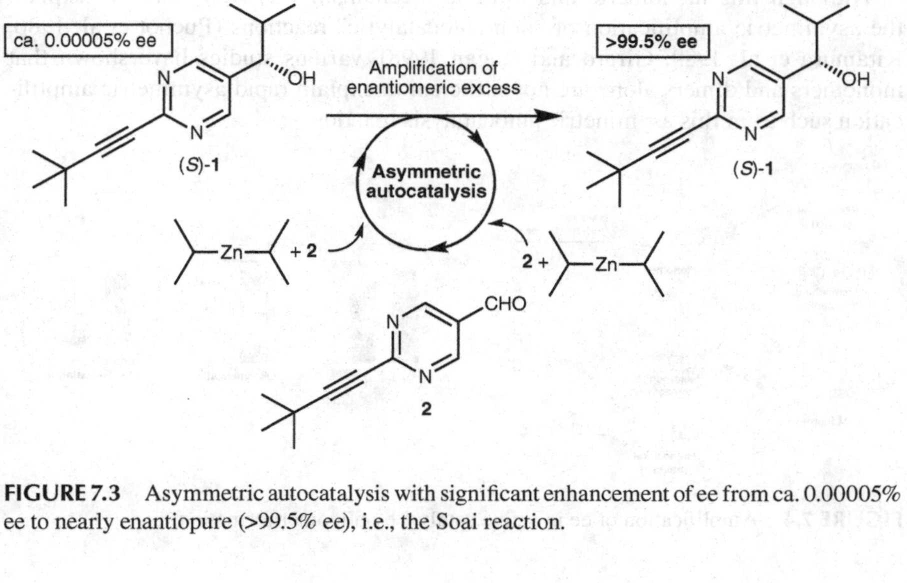

**FIGURE 7.3** Asymmetric autocatalysis with significant enhancement of ee from ca. 0.00005% ee to nearly enantiopure (>99.5% ee), i.e., the Soai reaction.

significantly amplified to greater than 99.5% ee in only three consecutive asymmetric autocatalyses (Sato et al. 2003a). During these three consecutive asymmetric autocatalyses, the initial very slight excess of the (*S*)-enantiomer **1** was automultiplied by a factor of ca. 630,000, whereas the slightly minor initial (*R*)-**1** was automultiplied by a factor of less than 1,000. Therefore, the significant asymmetric amplification of pyrimidyl alkanol **1** has been realized during its formation without the intervention of any other chiral auxiliary. Implication of these results is that there is an actual chemical reaction in which very slight enantiomeric imbalance is enhanced to near enantiopure by the mechanism of asymmetric autocatalysis (Soai 2019: Soai, Kawasaki, and Matsumoto 2019; Soai 2022; Soai, Kawasaki, and Matsumoto 2022).

## 7.3 MECHANISM OF THE ASYMMETRIC AUTOCATALYSIS: THE SOAI REACTION

This asymmetric autocatalytic reaction is very attractive not simply as a new and efficient method of asymmetric synthesis, but also as a model of how the initial molecular chirality bias can actually be amplified to achieve homochirality. Therefore, many researchers have been interested in studying the mechanism of this reaction, which has been elucidated from a wide variety of perspectives, including the capture of reaction rates and intermediates by analysis, theoretical studies based on DFT calculations, and mathematical modeling. Here, we would like to introduce some of the findings on this amplification reaction.

The simplest model to the amplification is the mechanism by which the formation of an inactive dimer enhances the ee of the active catalyst. Or it might be the idea that the active species of the monomer is the dimer is the actual catalytically active species (Figure 7.4). In both cases, the *R*, *R* or *S*, *S* dimer with the same chirality and the *R*, *S* dimer with a different chirality are diastereomeric, and the difference in stability and reactivity creates a nonlinear relationship between the original monomer and the actual active species.

Although this monomeric and dimeric mechanism was well suited to explain the asymmetric amplification of "non-autocatalytic" reactions (Puchot et al. 1986; Kitamura et al. 1989; Girard and Kagan 1998), various studies have shown that monomers and dimers alone are not sufficient to explain rapid asymmetric amplification such as in this asymmetric autocatalysis reaction.

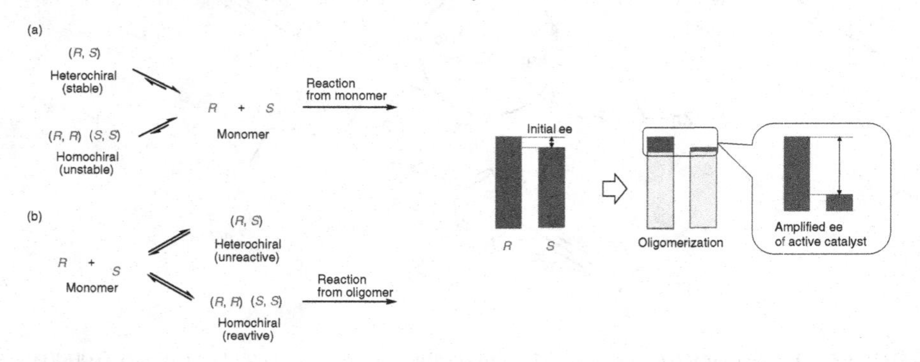


**FIGURE 7.4** Amplification of ee in active catalyst by oligomer formation.

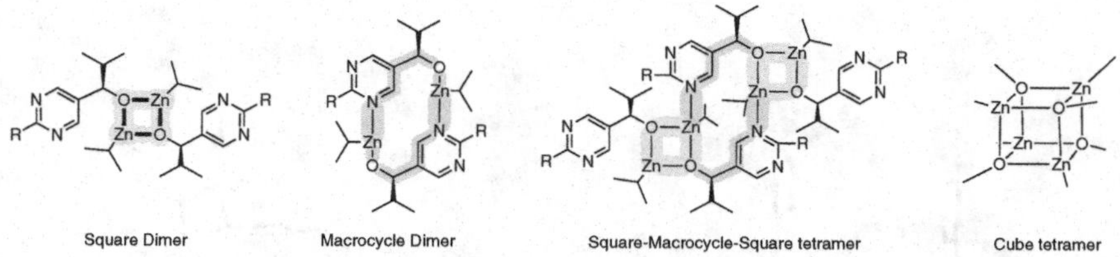

**FIGURE 7.5** Various possible aggregation forms of zinc alkoxide.

It has been proposed that the reaction involves the formation of a high aggregate due to a combination of association by Zn-O-Zn-O square of the zinc alkoxide and macrocyclic structure by zinc coordination of the nitrogen atom of the pyrimidine at the γ-position of the alcohol (Figure 7.5).

Based on measured data of initial ee changes and detailed reaction kinetics using reaction calorimetry, Blackmond et al. suggested that there is little difference in stability between the heterochiral and homochiral dimers and that the monomer is unlikely to be the active species (Blackmond et al. 2001; Sato et al. 2001a; Sato et al. 2003b). Studies using NMR and DFT calculations by Brown and Gridnev et al. also confirmed that there is little difference in the stability of the dimers (Gridnev, Serafimov, and Brown 2004; Gridnev and Brown 2004).

Subsequently, more detailed kinetic studies have shown that the reaction rate is almost independent of the concentration of the zinc reagent and is close to second order for aldehydes (Buono and Blackmond 2003). This rate trend has also been confirmed in a recent analysis by Trapp et al. using a significant amount of HPLC and MS data (Trapp et al. 2020). Based on the second-order data for this aldehyde, Schiaffino and Ercolani proposed a cycle in which the dimer reacts with two molecules of the aldehyde to form a tetramer (Schiaffino and Ercolani 2008). Micheau et al., through the analysis of kinetic model of asymmetric autocatalysis, demonstrated that higher order aggregate of catalytic species results in increased asymmetric selectivity, as if undergoing "proofreading" (Micheau et al. 2012). The tetrameric structure proposed at this time differs from that currently considered, and Gridnev considered several multimeric structures, with the Square-Macrocycle-Square type tetramer being the most stable tetramer. Athavale et al. showed that a nonreactive tetramer with cubic structure was formed on a substrate with pyridine instead of pyrimidine, suggesting that the coordination of excess diisopropylzinc to the pyrimidine nitrogen is responsible for the stability of the tetramer structure (Athavale et al. 2020). Several tetrameric and multimeric structures have been calculated by Gridnev, and the Square-Macrocycle-Square type tetramer is considered as a stable tetramer (Gridnev and Vorobiev 2012).

We have determined the crystal structure of tetramers by X-ray crystallography and found that the conformation of the enantiopure tetramer and racemic tetramer is very different (Figure 7.6). We have also confirmed that oligomers form in the absence of excess zinc reagent (Matsumoto et al. 2015a; Matsumoto et al. 2016a; Matsumoto, Kawasaki and Soai 2022).

Recent NMR and MS studies have confirmed the formation of acetal structures as reaction intermediates (Gehring et al. 2012; Trapp et al. 2020), which are also thought

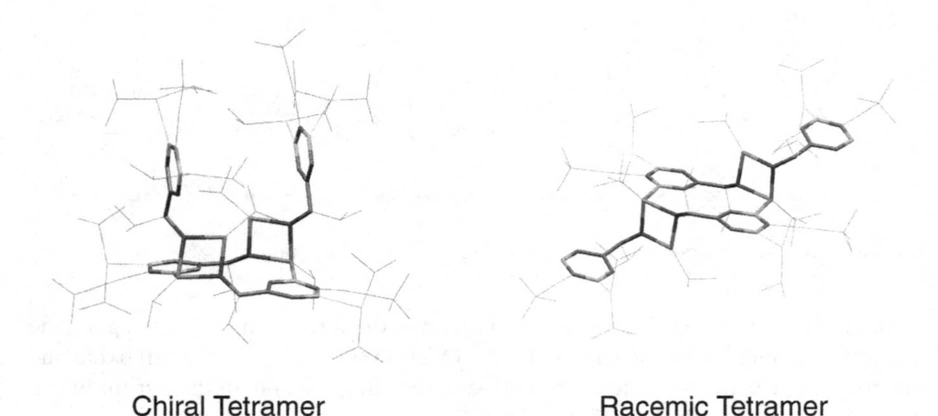


**FIGURE 7.6** Tetramer structure revealed by X-ray diffraction analysis.

to be the active species of the catalyst. In any case, it is clear from DOSY NMR analysis (Quaranta et al. 2010) and the temperature and concentration dependence of solution CD spectra (Matsumoto et al. 2021) that such tetramers and higher multimeric forms are formed in the equilibrium of the catalyst precursors in solution, and it is important to note that various equilibrium relationships of association formation are involved in the rapid asymmetric amplification of this reaction. The equilibrium relationships of the various aggregation formations are involved in the rapid asymmetric amplification of this reaction.

## 7.4 ASYMMETRIC AUTOCATALYSIS TRIGGERED BY CHIRAL COMPOUNDS

Asymmetric autocatalysis of 5-pyrimidyl alkanol **1** is capable of enhancing the slight enantiomeric imbalance to near enantiopure. Therefore, if the initial chiral imbalance of the autocatalyst **1** could be introduced by chiral compounds other than the autocatalyst **1**, it can be expected to obtain pyrimidyl alkanol **1** with detectable ee by the subsequent significant amplification of ee during autocatalytic formation of **1**. The absolute configuration of the product **1** should be controlled by the absolute handedness of the originally used external chiral factor; thus, it has been found that various chiral compounds, even those whose asymmetric induction power is assumed to be weak, can act as chiral initiators of asymmetric autocatalysis (Soai, Kawasaki, and Matsumoto 2014a).

Meanwhile, propylene oxide, i.e., 2-methyloxirane, is known as the first interstellar chiral compound and was found in the Sagittarius B2 star formation region (McGuire et al. 2016). However, its absolute configuration has not been determined. (*R*)- and (*S*)-(+)-Propylene oxide with low-to-high ee was employed as a chiral trigger in the enantioselective addition of *i*-$Pr_2Zn$ to pyrimidine-5-carbaldehyde **2** (Figure 7.7) (Kawasaki et al. 2004). In the presence of (*R*)-(+)-propylene oxide with 97% ee as a chiral trigger, the addition of *i*-$Pr_2Zn$ to pyrimidine-5-carbaldehyde **2** was examined. (*S*)-5-Pyrimidyl alkanol **1** with 96% ee was obtained after the asymmetric autocatalysis with an amplification of ee. In sharp contrast, in the presence of (*S*)-(–)-propylene oxide with 97% ee,

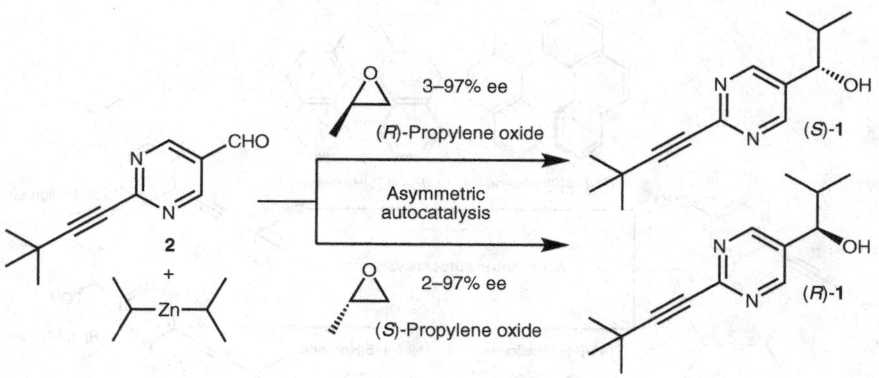

**FIGURE 7.7** Asymmetric autocatalysis triggered by propylene oxide.

oppositely configured (*R*)–**1** with 94% ee was formed. Even when propylene oxide with low (2–3%) ee was utilized, the same stereochemical relationship could be observed, i.e., the (*R*)-(+)-enantiomer with only 3% ee and (*S*)-(–)-isomer enantiomer with only 2% ee were found to serve as chiral initiators in the asymmetric autocatalysis, producing (*S*) –**1** and (*R*)–**1** with high ee, respectively (Kawasaki et al. 2004). Thus, the chiral interstellar chiral compound propylene oxide, acting as the source of chirality, is correlated with highly enantioenriched 5-pyrimidyl alkanol **1** via the asymmetric autocatalysis with amplification of ee. In other words, the Soai reaction is capable of detecting the absolute configuration of chiral compound with low ee.

It should also be mentioned that enantioenrichments are often observed in extraterrestrial amino acids detected in meteorites (Cronin and Pizzarello 1997). These show that space is a candidate for the place of origin of chirality. In fact, the Rosetta spacecraft of the ESA with a probe was launched to perform the in-situ analysis of chirality of amino acids on a comet (Myrgorodska et al. 2015).

The existence of polycyclic aromatic hydrocarbons (PAHs) in space has been discussed (Tielens 2005). Fullerenes, for example, have been detected in meteorites (Becker et al. 1994) and the planetary nebula (Cami et al. 2010). Helicenes are chiral with right- and left-handed helical structures, and the asymmetric photosynthesis of [6]helicene has been reported by irradiation with CPL to racemic [6]helicene followed by oxidation (Kagan et al. 1971). We have reported that chiral [6]helicene and [5]helicene acted as highly efficient chiral triggers in the Soai reaction (Figure 7.8) (Sato et al. 2001b). It should be noted that the asymmetric induction using chiral hydrocarbons without heteroatoms as a chiral catalyst or chiral ligand has been a challenging theme from the standpoint of organic synthesis.

When pyrimidine-5-carbaldehyde **2** was treated with *i*-$Pr_2Zn$ in the presence of (*P*)-(+)-[6]helicene, after the amplification of ee by asymmetric autocatalysis, (*S*)-pyrimidyl alkanol **1** with 95% ee was formed. In contrast, the opposite (*M*)-(–)-[6] helicene induced the formation of (*R*)-**1** with 93% ee. Even when the very low ee values of (P)-(+)-[6]helicene of ca. 0.13% ee and (*M*)-(–)-[6]helicene of ca. 0.54% ee were employed as chiral trigger, asymmetric autocatalysis afforded (*S*)- and (*R*)–**1** with high ee with the same stereochemical relationships, respectively. (*P*)-(+)- and

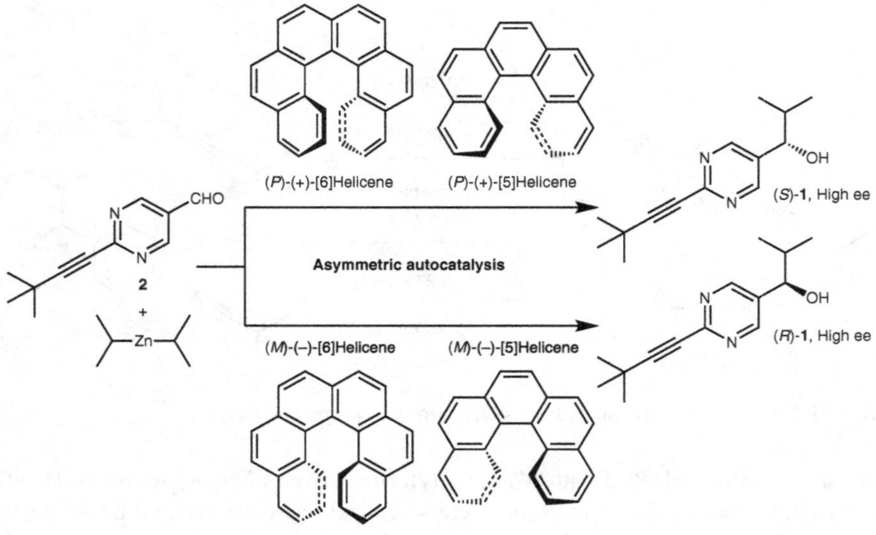

(P)-(+)-[6]Helicene (P)-(+)-[5]Helicene

CHO

N

N

2

+

Zn

Asymmetric autocatalysis

(M)-(−)-[6]Helicene (M)-(−)-[5]Helicene

N

N

OH

(S)-1, High ee

N

N

OH

(R)-1, High ee

**FIGURE 7.8** Asymmetric autocatalysis triggered by [6] and [5]helicenes with chiral helicity.

(*M*)-(–)-[5]helicenes also gave (*S*)- and (*R*) **–1**, respectively. As described, [6] and [5]helicenes are shown to act as the origin of chirality in conjunction with asymmetric autocatalysis. 1,1¢-Binaphthyl (Sato et al. 2002) and finite single-wall carbon nanotube molecules (Hitosugi et al. 2014) can also be employed as chiral triggers of asymmetric autocatalysis.

In 1953, Miller (Miller 1953) demonstrated that electric discharges under primitive Earth conditions produce racemic amino acids (Miller and Urey 1959). Formation of amino acids and asymmetric induction by irradiation with CPL (Meierhenrich et al. 2005) have also been reported under simulated interstellar conditions (Bernstein et al. 2002; Munoz Caro et al. 2002). Furthermore, enantiomeric enrichment of amino acids was detected in meteorite samples (Cronin and Pizzarello 1997; Engel and Macko 1997). Thus, asymmetric autocatalysis with the amplification of ee triggered by the meteoritic amino acids becomes an important approach toward understanding the origin of biological homochirality, such as in L-amino acids.

We have reported the asymmetric autocatalysis triggered by enantioenriched proteinogenic amino acids (Shibata et al. 1998; Sato et al. 2007). When asymmetric autocatalysis was triggered by L-alanine with high ee, (*S*)-5-pyrimidyl alkanol **1** with 92% ee was formed, whereas D-alanine afforded (*R*)-alkanol **1** with 90% ee (Figure 7.9). When L-alanine with ca. 10% ee was used as the chiral trigger of asymmetric autocatalysis, the (*S*)-alkanol **1** with 94% ee was formed. Even when the ee of L-alanine was very low such as ca. 1% and ca. 0.1% ee, (*S*)–**1** with high ee was formed. In contrast, asymmetric autocatalysis triggered by D-alanine with low ee (ca. 10%, 1%, and 0.1% ee) afforded oppositely configured (*R*)-alkanol **2**. Thus, even if the enantiomeric excess of alanine is as low as ca. 0.1% ee, it can work as the origin of chirality in asymmetric autocatalysis to give an increased amount of highly enantioenriched alkanol **1**.

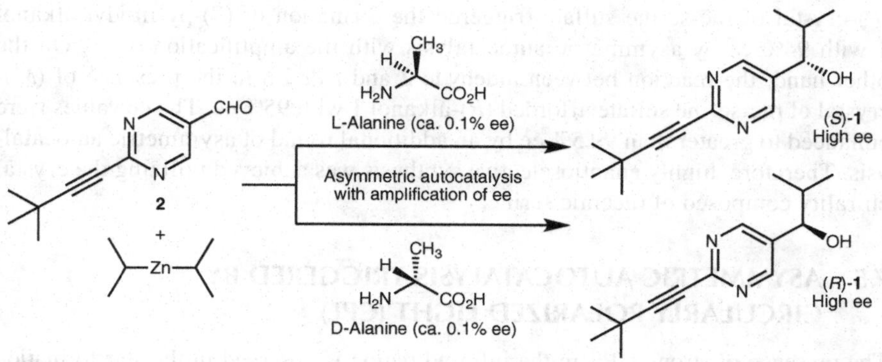

**FIGURE 7.9** Asymmetric autocatalysis triggered by L- and D-alanine and the formation of (*S*)- and (*R*)-5-pyrimidyl alkanols with high ee.

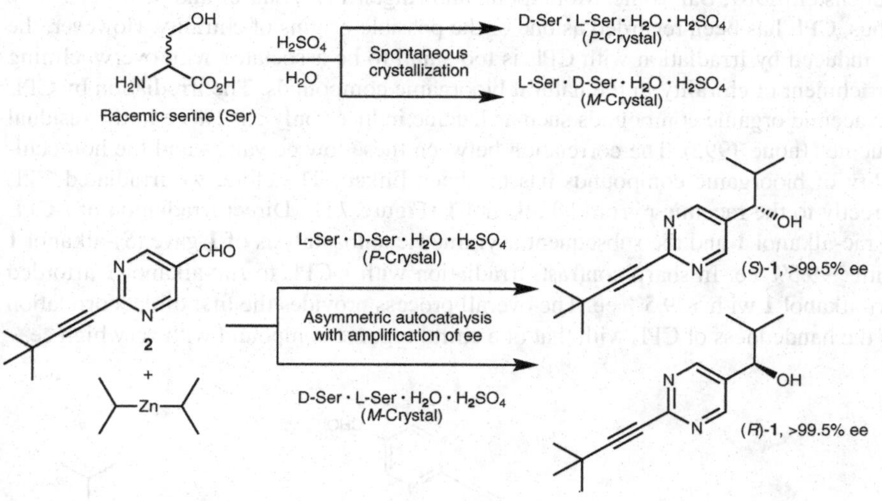

**FIGURE 7.10** Asymmetric autocatalysis triggered by chiral crystal composed of racemic serine.

Serine has been regarded as one of the prebiotic molecules (Ring et al. 1972), and a chiral effect in the formation of its aggregate has been reported (Nanita and Cooks 2006). It is also known that the crystallization of racemic serine from aqueous sulfuric acid affords chiral crystals of DL-diserinium sulfate monohydrate, which contain an equimolar amount of D- and L-serine with (*P*)- or (*M*)-crystal chirality. This means that the crystal is chiral that contains equimolar amount of D- and L-serine. We reported that the chiral crystals formed from racemic serine trigger asymmetric autocatalysis (Figure 7.10) (Kawasaki et al. 2011a). Thus, the enantioselective addition of $i$-$Pr_2Zn$ to aldehyde **2** was conducted in the presence of enantiomorphs of racemic serine as a heterogeneous chiral trigger. As a result, the

(*P*)-crystal of rac-serine sulfate triggered the formation of (*S*)-pyrimidyl alkanol **1** with 94% ee by asymmetric autocatalysis with the amplification of ee. On the other hand, the reaction between aldehyde **2** and $i$-$Pr_2Zn$ in the presence of (*M*)-crystal of rac-serine sulfate afforded (*R*)-alkanol **1** with 95% ee. The ee values were enhanced to greater than 99.5% ee by an additional round of asymmetric autocatalysis. Therefore, highly enantioselective synthesis was achieved utilizing the crystal chirality composed of racemic serine.

## 7.5 ASYMMETRIC AUTOCATALYSIS TRIGGERED BY CIRCULARLY POLARIZED LIGHT (CPL)

The presence of strong CPL in the infrared region is observed in the star formation region of the Orion nebula (Bailey et al. 1998). As reported previously, left (*l*)- and right (*r*)-handed CPL is a chiral physical force to induce ee in chiral compounds by asymmetric photolysis or asymmetric photoequilibrium (Kagan et al. 1971; Bonner and Rubenstein 1987; Balavoine, Moradpour, and Kagan 1974; Suarez and Schuster 1995). Thus, CPL has been regarded as one of the possible origins of chirality. However, the ee induced by irradiation with CPL is too small to be associated with overwhelming enrichment of chirality in the natural bioorganic compounds. The irradiation by CPL to racemic organic compounds such as leucine induces only ca. 2% ee in the residual leucine (Inoue 1992). The correlation between these low ee values and the homochirality of bioorganic compounds has not been linked. Therefore, we irradiated CPL directly to the racemic pyrimidyl alkanol **1** (Figure 7.11). Direct irradiation of *l*-CPL to rac-alkanol **1** and the subsequent asymmetric autocatalysis of **1** gave (*S*)-alkanol **1** with >99.5% ee. In sharp contrast, irradiation with *r*-CPL to *rac*-alkanol **1** afforded (*R*)-alkanol **1** with >99.5% ee. The overall process provides the first direct correlation of the handedness of CPL with that of a chiral organic compound with very high ee.

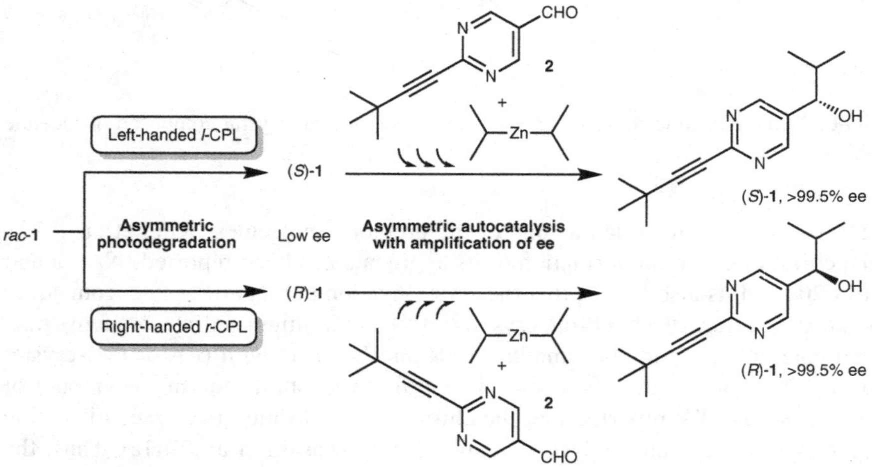

**FIGURE 7.11** Direct irradiation of CPL triggers asymmetric autocatalysis.

## 7.6 ISOTOPICALLY CHIRAL COMPOUNDS AND ISOTOPOMERS OF METEORITIC AMINO ACIDS AS CHIRAL TRIGGERS OF ASYMMETRIC AUTOCATALYSIS

There are many apparently achiral organic molecules that may become chiral by considering the randomly labeled isotopes in the enantiotopic moiety. However, it has been very difficult to recognize the chirality arising from the isotopic substitution because the chirality originates from the extremely small difference between the numbers of neutrons in the atomic nuclei. We have reported asymmetric autocatalysis triggered by the chiral isotopomers arising from carbon isotope ($^{12}C/^{13}C$) substitution (Figure 7.12) (Kawasaki et al. 2009a). The carbon isotope chirality triggers asymmetric autocatalysis between pyrimidine-5-carbaldehyde **2** and *i*-$Pr_2Zn$; therefore, enantiomerically amplified pyrimidyl alkanol **1** can be obtained with absolute configuration corresponding to that of the carbon isotope substitution. This stands as the first example of such a chiral effect, that is, the substitution of carbon isotopes ($^{12}C/^{13}C$) induces chirality in organic reaction.

Glycine is the simplest achiral α-amino acid and was detected in the coma of a comet (Sandford et al. 2006; Altwegg et al. 2016). It is also synthesized under prebiotic conditions (Miller 1953). Moreover, in addition to achiral glycine and α-methylalanine (Pizzarello and Huang 2005), L-enriched chiral α-methyl-substituted amino acids were identified in meteorites as deuterium-enriched forms (Engel and Macko 1997). We paid attention on the hydrogen isotope (H/D) chirality in the apparently achiral meteoritic amino acids. When one of the hydrogen atoms of the methylene group of glycine or one of the methyl groups of α-methylalanine is enantiomerically deuterated, these compounds become chiral because of the deuterium substitution. Thus, asymmetric autocatalysis triggered by these chiral isotopomers of meteoritic amino acids will become one of the approaches toward the extraterrestrial origin of biological homochirality.

We have synthesized the isotopomers of glycine-α-*d* and α-methyl-$d_3$-alanine. These were used as chiral triggers in asymmetric autocatalysis (Figure 7.12) (Kawasaki et al. 2009b). When aldehyde **2** and *i*-$Pr_2Zn$ were reacted in the presence of (*S*)-glycine-α-*d* as a chiral trigger, (*S*)-pyrimidyl alkanol **1** with high ee was formed. In contrast, the presence of (*R*)-glycine-α-*d* as a chiral trigger afforded (*R*)-**1** with high ee as the results of the asymmetric autocatalysis with amplification of ee. We also showed that α-methyl-$d_3$-alanine acts as a chiral trigger in asymmetric autocatalysis. We also have reported that the partially deuterated α-methylalanine derivatives could work as chiral inducers for asymmetric autocatalysis (Kawasaki et al. 2011b).

Nitrogen isotope ($^{15}N$) enrichment was also reported in the meteoritic organic compounds (Engel and Macko 1997). We have reported that the chiral nitrogen isotopomer of a *meso*-diamine is capable of acting as a chiral trigger in asymmetric autocatalysis. Highly enantioenriched pyrimidyl alkanol **1** with the corresponding absolute configurations to that of $N^2,N^2,N^3,N^3$-tetramethylbutane-2,3-diamine-containing nitrogen isotope ($^{14}N/^{15}N$) chirality is formed (Figure 7.12) (Matsumoto et al. 2016b). The addition of *i*-$Pr_2Zn$ to pyrimidine-5-carbaldehyde **2** in the presence of [$^{15}N$](*S*)-diamine afforded (*S*)-alkanol **1**. In sharp contrast, (*R*)-alkanol **1** was obtained from the reaction in the presence of [$^{15}N$](*R*)-diamine. The ee of alkanol **1** was amplified to >99.5% ee by a further asymmetric autocatalytic reaction using the obtained pyrimidyl alkanol **1**

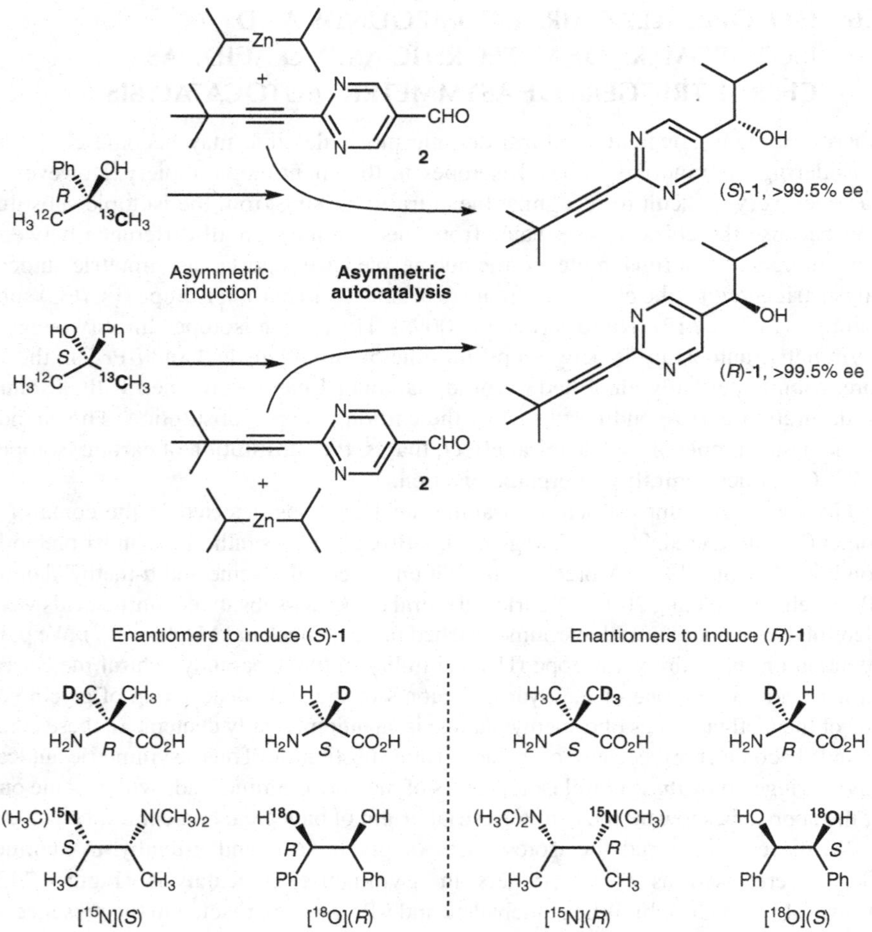

**FIGURE 7.12** Chiral isotopomers with hydrogen (D/H), nitrogen ($^{15}N/^{14}N$), carbon ($^{13}C/^{12}C$), and oxygen ($^{18}O/^{16}O$) trigger asymmetric autocatalysis.

as asymmetric autocatalyst. Moreover, asymmetric autocatalysis has enormous power to discriminate and amplify the chirality of the oxygen isotopomers with $^{18}O/^{16}O$ substitutions (Kawasaki et al. 2011c; Matsumoto et al. 2013).

By these results described earlier, the Soai reaction enhances the implications of chiral isotopomers as the origin of biological chirality.

## 7.7 CHIRAL CRYSTALS OF ACHIRAL NUCLEOBASES CYTOSINE AND ADENINE TRIGGER ASYMMETRIC AUTOCATALYSIS WITH THE AMPLIFICATION OF EE

Nucleobases are essential for life. They are achiral flat molecules and are considered to have existed on the prebiotic Earth before the emergence of the RNA world

(Gilbert 1986). Cytosine, for example, could be efficiently synthesized from interstellar molecules, i.e., hydrogen cyanide and cyanoacetylene (Robertson and Miller 1995; Shapiro 1999). The isolation of adenine from meteorites (Callahan et al. 2011) and prebiotic synthesis, i.e., pentamerization of hydrogen cyanide, under prebiotic conditions (Oro 1960; Schwartz, Joosten, and Voet 1982), has been reported. Therefore, the enantioselective synthesis induced by crystal chirality arising from an achiral nucleobase would be an important experiment for understanding the origins of both biological homochirality and life.

Crystallization of achiral cytosine from methanol affords a chiral form (space group: $P2_12_12_1$) that exhibits either plus or minus Cotton effects in the solid-state circular dichroism (CD) spectra at ca. 310 nm in Nujol. One of the enantiomorphs of cytosine is formed spontaneously using stirred crystallization (Figure 7.13) (Kawasaki et al. 2008a). Chiral crystals of cytosine have been used as chiral triggers for asymmetric autocatalysis. In the presence of [CD(–)$310_{\mathrm{Nujol}}$]-cytosine crystals, the reaction between pyrimidine-5-carbaldehyde **2** and *i*-$Pr_2Zn$ affords enantioenriched (*S*)-pyrimidyl alkanol **1** in conjunction with asymmetric autocatalysis (Figure 7.13). By contrast, [CD(+)$310_{\mathrm{Nujol}}$]-cytosine crystals trigger the formation of (*R*)-alkanol **1**. Meanwhile, the crystallization of cytosine from water gives achiral crystal of cytosine monohydrate. We have demonstrated the enantiospecific formation of chiral cytosine (anhydride) crystals by the dehydration of crystal water of achiral cytosine monohydrate under thermal or vacuum conditions (Kawasaki at al. 2010a; Mineki et al. 2013).

Crystallization of adenine from aqueous nitric acid affords adeninium dinitrate (adenine·$2HNO_3$) in enantiomorphic form (space group: $P2_12_12_1$). The chirality of adenine·$2HNO_3$ can be discriminated using solid-state CD analysis with Nujol (Mineki et al. 2012). Thus, one crystal exhibited a positive Cotton effect at 250 nm

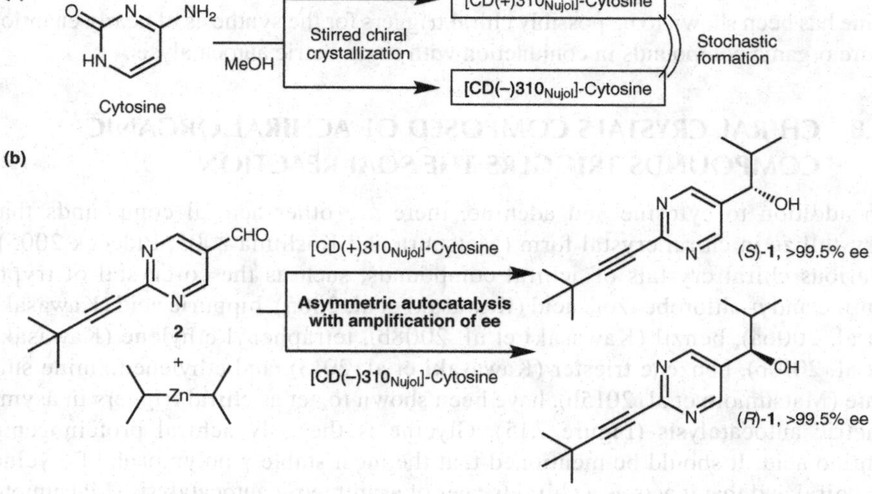

**FIGURE 7.13** (a) Crystallization of cytosine with stirring. (b) Asymmetric autocatalysis triggered by a spontaneously generated chiral crystal of achiral nucleobase cytosine.

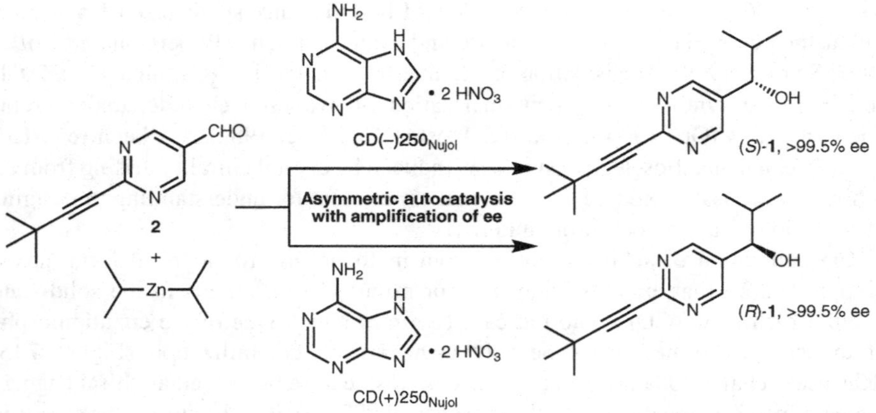

FIGURE 7.14 Chiral crystal of adeninium dinitrate triggers asymmetric autocatalysis.

(CD(+)250$_{Nujol}$), while the other enantiomorph had a negative Cotton effect at the same wavelength (CD(–)250$_{Nujol}$).

The enantiomorphs of adenine dinitrate trigger the enantioselective addition of *i*-$Pr_2Zn$ to pyrimidine-5-carbaldehyde **2** (Figure 7.14). The reaction of aldehyde **2** with *i*-$Pr_2Zn$ in the presence of finely powdered crystalline [CD(–)250$_{Nujol}$]-adeninium dinitrate afforded (*S*)-5-pyrimidyl alkanol **1** with >99.5% ee after the amplification of ee by asymmetric autocatalysis. On the contrary, the opposite (*R*)-alkanol **1** was obtained with >99.5% ee triggered by enantiomorphous [CD(+)250$_{Nujol}$]-adeninium dinitrate. Thus, the crystal chirality of the adenine salt in conjunction with asymmetric autocatalysis controlled the absolute configurations of the produced alkanol **1** with high ee.

As described, the crystal chirality of the prebiotic nucleobases cytosine and adenine has been shown to be possibly chiral triggers for the synthesis of nearly enantiopure organic compounds in conjunction with asymmetric autocatalysis.

## 7.8 CHIRAL CRYSTALS COMPOSED OF ACHIRAL ORGANIC COMPOUNDS TRIGGERS THE SOAI REACTION

In addition to cytosine and adenine, there are other achiral compounds that crystallize in chiral crystal form (Matsuura and Koshima 2005; Pidcock 2005). Various chiral crystals of achiral compounds, such as the co-crystal of tryptamine and *p*-chlorobenzoic acid (Kawasaki et al. 2005), hippuric acid (Kawasaki et al. 2006a), benzil (Kawasaki et al. 2008b), tetraphenyl ethylene (Kawasaki et al. 2010b), benzene triester (Kawasaki et al. 2013), and ethylenediamine sulfate (Matsumoto et al. 2015b), have been shown to act as chiral triggers in asymmetric autocatalysis (Figure 7.15). Glycine is the only achiral proteinogenic amino acid. It should be mentioned that the most stable γ-polymorph of glycine is chiral and that it acts as a chiral trigger of asymmetric autocatalysis (Matsumoto et al. 2019).

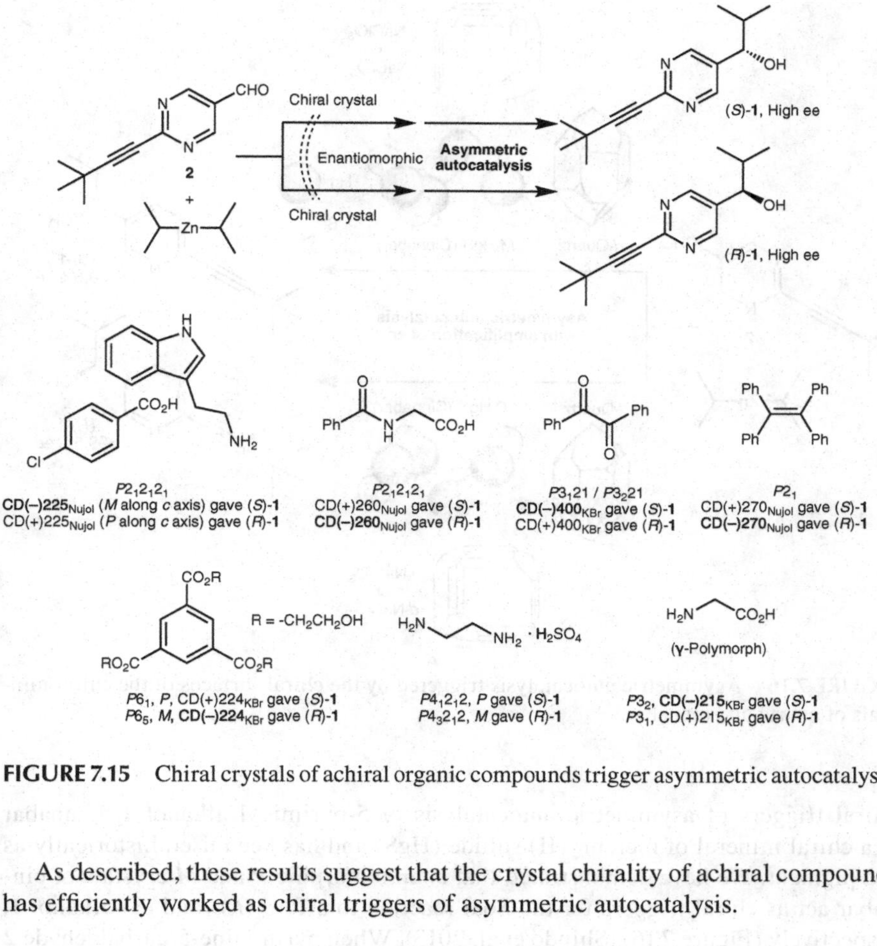

Chiral crystal

Enantiomorphic

Chiral crystal

Asymmetric autocatalysis

(*S*)-**1**, High ee

(*R*)-**1**, High ee

$P2_12_12_1$
CD(−)$225_{Nujol}$ (*M* along *c* axis) gave (*S*)-**1**
CD(+)$225_{Nujol}$ (*P* along *c* axis) gave (*R*)-**1**

$P2_12_12_1$
CD(+)$260_{Nujol}$ gave (*S*)-**1**
CD(−)$260_{Nujol}$ gave (*R*)-**1**

$P3_121$ / $P3_221$
CD(−)$400_{KBr}$ gave (*S*)-**1**
CD(+)$400_{KBr}$ gave (*R*)-**1**

$P2_1$
CD(+)$270_{Nujol}$ gave (*S*)-**1**
CD(−)$270_{Nujol}$ gave (*R*)-**1**

R = $-CH_2CH_2OH$

$P6_1$, *P*, CD(+)$224_{KBr}$ gave (*S*)-**1**
$P6_5$, *M*, CD(−)$224_{KBr}$ gave (*R*)-**1**

$\cdot H_2SO_4$

$P4_12_12$, *P* gave (*S*)-**1**
$P4_32_12$, *M* gave (*R*)-**1**

(γ-Polymorph)

$P3_2$, CD(−)$215_{KBr}$ gave (*S*)-**1**
$P3_1$, CD(+)$215_{KBr}$ gave (*R*)-**1**

**FIGURE 7.15** Chiral crystals of achiral organic compounds trigger asymmetric autocatalysis.

As described, these results suggest that the crystal chirality of achiral compounds has efficiently worked as chiral triggers of asymmetric autocatalysis.

## 7.9 ASYMMETRIC AUTOCATALYSIS TRIGGERED BY CHIRAL MINERALS AND CHIRAL SURFACES

The possible roles of chiral minerals and chiral surfaces in the origin of biological homochirality have been discussed previously (Hazen 2004). In the Earth's crust, there is a wide variety of chiral minerals that could serve as chiral surfaces in the prebiotic evolution of chiral organic molecules. However, only a very small degree of asymmetric induction has been observed in the enantiomer selective adsorption of a chiral compound on quartz (Bonner et al. 1974). We have demonstrated asymmetric autocatalysis triggered by powder of *d*- and *l*-quartz ($SiO_2$). When pyrimidine-5-carbaldehyde **2** was reacted with *i*-$Pr_2Zn$ in the presence of *d*- and *l*-quartz powder, respectively, (*S*)- and (*R*)-pyrimidyl alkanol **1** with high ee were formed in conjunction with asymmetric autocatalysis (Figure 7.16) (Soai et al. 1999). Chiral ionic crystals of sodium chlorate ($NaClO_3$) (Sato, Kadowaki, and Soai 2000) and of sodium bromate ($NaBrO_3$) (Sato et al. 2004) also act as

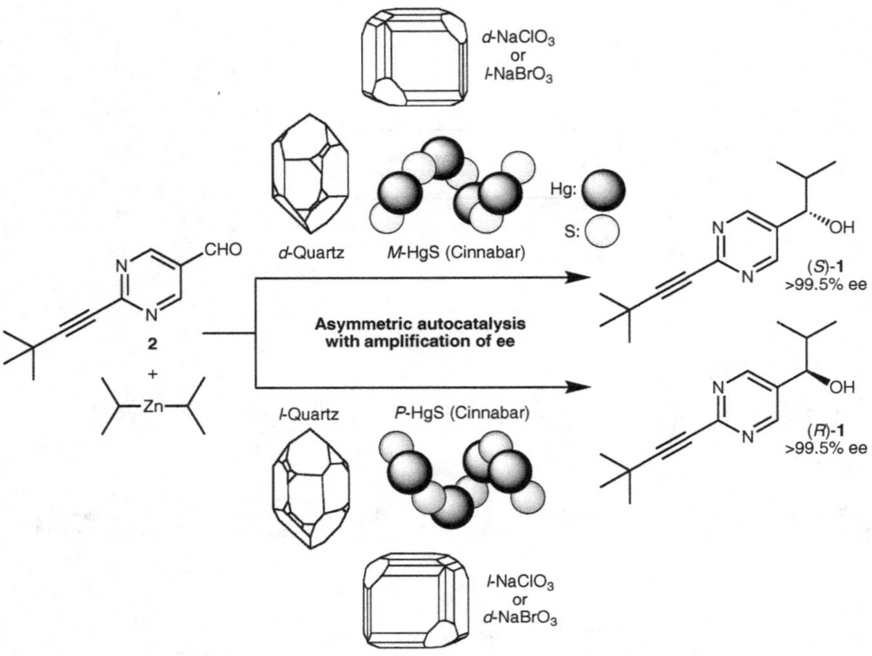

FIGURE 7.16 Asymmetric autocatalysis triggered by the chiral surfaces of the chiral minerals of quartz and cinnabar.

chiral triggers of asymmetric autocatalysis of 5-pyrimidyl alkanol **1**. Cinnabar is a chiral mineral of mercury(II) sulfide (HgS) and has been used historically as red pigment. We have demonstrated that enantiomorphic *P* and *M* crystals of cinnabar act as chiral triggers for the Soai reaction to afford (*R*)- and (*S*)-alkanol **1**, respectively (Figure 7.16) (Shindo et al. 2013). When pyrimidine-5-carbaldehyde **2** was treated with *i*-$Pr_2Zn$ in the presence of cinnabar with right-handed *P*-helicity, (*R*)-pyrimidyl alkanol **1** was formed in 88% ee and 87% yield. In contrast, (*M*)-HgS triggered the asymmetric autocatalysis to afford the opposite (*S*)-alkanol **1** with 92% ee in 91% yield. These ee values were further amplified to >99.5% ee by applying consecutive asymmetric autocatalysis.

The chiral enantiotopic surface of achiral minerals has also been considered as a possible origin of chirality. Recently, broad attention has been focused on two-dimensional surface chirality. Hazen reported the enantiomer-selective adsorption of racemic amino acids on the enantiotopic faces of the achiral mineral calcite ($CaCO_3$) (Hazen, Filley, and Goodfriend 2001). Certain metal surfaces, such as artificially prepared Cu (643), are enantiotopic. Gellman reported the enantiomer-selective desorption and decomposition on these enantiotopic metal surfaces (Gellman et al. 2013).

We have demonstrated that the enantiotopic (010) and (0–10) surfaces of achiral mineral gypsum ($CaSO_4{\cdot}2H_2O$) act as chiral environments to induce asymmetry in the asymmetric autocatalysis (Figure 7.17) (Matsumoto et al. 2017). On the

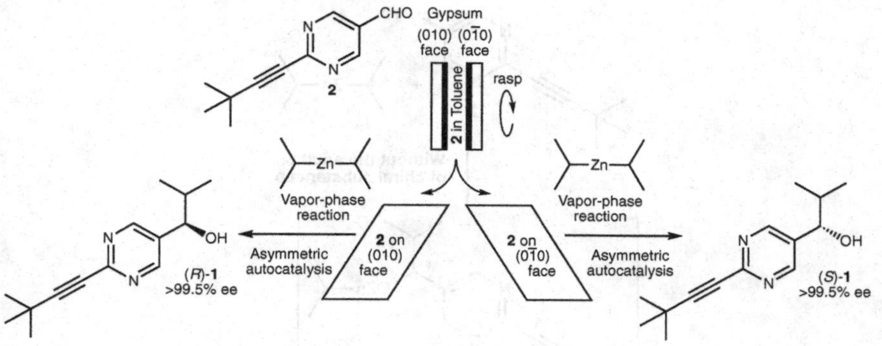

FIGURE 7.17 Asymmetric autocatalysis initiated on the enantiotopic face of achiral gypsum.

(010) surface of gypsum, the reaction of adsorbed pyrimidine-5-carbaldehyde **2** with the vapor of *i*-$Pr_2Zn$ afforded (*R*)-alkanol **2**. In contrast, (*S*)-alkanol **1** was formed through the asymmetric autocatalysis on the opposite (0–10) face. The ee values of pyrimidyl alkanol **1** were increased to >99.5% ee by the further asymmetric autocatalysis with amplification of ee. It is considered that the initial slightly imbalanced *Re* and *Si* face orientation of aldehyde **2** adsorbed on the enantiotopic (010) and (0–10) faces of gypsum is the origin of initial asymmetric induction.

## 7.10 ABSOLUTE ASYMMETRIC SYNTHESIS IN THE SOAI REACTION

The reaction of achiral substrates that produce enantioenriched products without the intervention of any chiral factor has been defined as spontaneous absolute asymmetric synthesis (Mislow 2003). Spontaneous absolute asymmetric synthesis has been proposed as one of the origins of homochirality. However, it has been widely accepted that common reactions without using any chiral factor always give equal (1:1) amounts of racemic mixture of two enantiomers. However, based on statistics, the numbers of the two enantiomers fluctuate. In addition, at the very initial stage of the reaction when 99 molecules of the product, i.e., isopropylzinc alkoxide of pyrimidyl alkanol, are formed, the lowest ee of the product is 50–49, that is, (50–49)/(50+49) = 1/99. This ee value is greater than 1% ee, which is apparently above the detection level of the Soai reaction (ca. 0.00005% ee). As described in the preceding section, the Soai reaction can amplify extremely low ee to >99.5% ee. We have reported asymmetric autocatalysis of pyrimidine-5-carbaldehyde **2** and *i*-$Pr_2Zn$ without adding any chiral substance (Figure 7.18) (Soai, Shibata, and Kowata 1996) in the mixed solvent of toluene and diethyl ether (Soai et al. 2003) and in the presence of achiral silica gel (Kawasaki et al. 2006b). Further investigations on spontaneous absolute asymmetric synthesis revealed that reactions in the presence of achiral amines spontaneously afforded pyrimidyl alkanol **1** with either (*S*)- or (*R*)-configurations with stochastic distributions (Suzuki et al. 2010). A stochastic distribution for the formation of either (*S*)- or (*R*)-5-pyrimidyl alkanol **1** was observed – for example, the formation of *S* 19

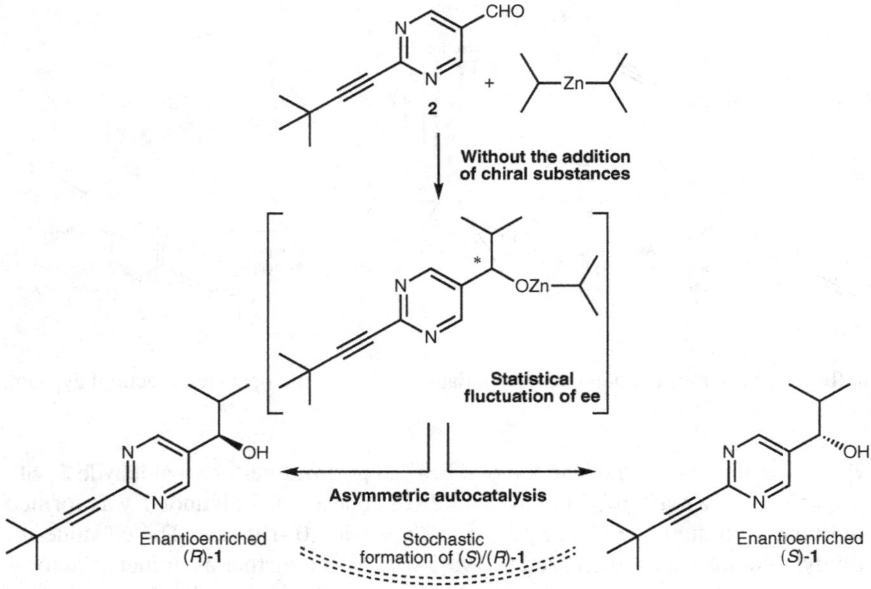

FIGURE 7.18 The first spontaneous absolute asymmetric synthesis: Amplification of the initial fluctuation of ee by asymmetric autocatalysis.

times and *R* 18 times for the 37 experiments (Soai et al. 2003). The fluctuation of ee produced spontaneously and stochastically in the initially forming alkanol **1** by the addition of *i*-$Pr_2Zn$ to achiral pyrimidine-5-carbaldehyde **2** can be amplified to a detectable value during the asymmetric autocatalysis. The stochastic behavior in the formation of (*S*)- and (*R*)-**1** fulfills one of the conditions necessary for spontaneous absolute asymmetric synthesis. Recently, we have also demonstrated that the reaction between vapor phase *i*-$Pr_2Zn$ and powder crystal of pyrimidine-5-carbaldehyde **2** achieve absolute asymmetric synthesis to afford the stochastic distribution of the (*S*) or (*R*)-product **1** (Kaimori et al. 2019).

Thus, the first spontaneous absolute asymmetric synthesis has been achieved by the Soai reaction.

## 7.11 DETECTION OF CHIRALITY OF METEORITIC SAMPLES BY ASYMMETRIC AUTOCATALYSIS

Asymmetric autocatalysis with amplification of ee, i.e., the Soai reaction, is a highly sensitive reaction to detect the chirality of a wide variety of materials, including chiral compounds such as isotopomers and chiral surfaces of minerals. Moreover, by absolute asymmetric synthesis, without the addition of any chiral materials, the Soai reaction affords (*S*)- or (*R*)-alkanol **1** with detectable ee in stochastic manner without the intervention of any chiral factor.

Thus, we have applied asymmetric autocatalysis to determine the chirality inside a meteorite, i.e., the Soai reaction was repeatedly performed in the presence of meteorite samples (Kawasaki et al. 2006c). As a result, we could judge that chirality remains in Murchison and Allende powders after the extraction with water and solvents in addition to the Murray kerogen-like insoluble organic materials (IOMs) after demineralization. On the other hand, the Murray IOM samples after hydrothermolytic treatment (IOM-H) gave both (*S*)- and (*R*)-pyrimidyl alkanol **1** in stochastic manner, which indicates the absence of chiral elements in the IOM-H sample. Chirality is not found in Murchison powders from which all organic components had been removed by oxygen ($O_2$) plasma at low temperature. Thus, the Soai reaction provides a powerful method to detect the extraterrestrial origin of chirality.

## 7.12 SUMMARY

5-Pyrimidyl alkanol **1** acts as a powerful asymmetric autocatalyst in the reaction between pyrimidine-5-carbaldehyde **2** and *i*-$Pr_2Zn$ with significant multiplication of the amount and with significant amplification of ee from extremely low (ca. 0.00005% ee) to near enantiopure (>99.5% ee) without the assistance of any other chiral auxiliary. Chiral compounds such as interstellar chiral propylene oxide and chiral isotopomers of meteoritic amino acids act as chiral triggers of asymmetric autocatalysis to afford enantioenriched alkanol **1** with absolute handedness corresponding to those of the chiral triggers. The spontaneous absolute asymmetric synthesis was realized in the Soai reaction. Furthermore, various origins of chirality, such as CPL; chiral minerals such as quartz; chiral organic crystals of achiral nucleobases; and enantiotopic surfaces of achiral mineral gypsum, have been directly correlated with nearly enantiopure pyrimidyl alkanol **1** by using asymmetric autocatalysis. The asymmetric autocatalysis of 5-pyrimidyl alkanol, i.e., the Soai reaction (Pályi, Zicchi, and Caglioti 2012), is a unique reaction leading to overwhelming one handedness, as observed in biological compounds such as L-amino acids and D-sugars. Carroll proposed the new definition of life that the Soai reaction is the simplest form of life – i.e., a self-amplifying autocatalytic reaction (Carroll 2009).

## REFERENCES

Altwegg, K., H. Balsiger, A. Bar-Nun, et al. 2016. Prebiotic chemicals-amino acid and phosphorus in the coma of comet 67P/Churyumov-Gerasimenko. *Sci. Adv.* 2: e1600285.

Ashkenasy, G., T.M. Hermans, S. Otto, et al. 2017. Systems chemistry. *Chem. Soc. Rev.* 46: 2543–2554.

Ashkenasy, G., R. Jagasia, M. Yadav, et al. 2004. Design of a directed molecular network. *Proc. Natl. Acad. Sci. USA*. 101: 10872–10877.

Athavale, S.V., A. Simon, K.N. Houk, et al. 2020. Demystifying the asymmetry-amplifying, autocatalytic behaviour of the Soai reaction through structural, mechanistic and computational studies. *Nat. Chem.* 12: 412–423.

Avalos, M., R. Babiano, P. Cintas, et al. 2000. Chiral autocatalysis: Where stereochemistry meets the origin of life. *Chem. Commun.* 887–892.

Bailey, J., A. Chrysostomou, J.H. Hough, et al. 1998. Circular polarization in star-formation regions: Implications for biomolecular homochirality. *Science* 281: 672–674.

Balavoine, G., A. Moradpour, and H.B. Kagan. 1974. Preparation of chiral compounds with high optical purity by irradiation with circularly polarized light, a model reaction for the prebiotic generation of optical activity. *J. Am. Chem. Soc.* 96: 5152–5158.

Becker, L., J.L. Bada, R.E. Winans, et al. 1994. Fullerenes in Allende meteorite. *Nature* 372: 507.

Bernstein, M.P., J.P. Dworkin, S.A. Sandford, et al. 2002. Racemic amino acids from the ultraviolet photolysis of interstellar ice analogues. *Nature* 416: 401–403.

Bissette, A.J., and S.P. Fletcher. 2013. Mechanisms of autocatalysis. *Angew. Chem. Int. Ed.* 52: 12800–12826.

Blackmond, D.G., C.R. McMillan, S. Ramdeehul, et al. 2001. Origins of asymmetric amplification in autocatalytic alkylzinc additions. *J. Am. Chem. Soc.* 123: 10103–10104.

Bolli, M., R. Micura, and A. Eschenmoser. 1997. Pyranosyl-RNA: chiroselective self-assembly of base sequences by ligative oligomerization of tetranucleotide-2 ,3 -cyclophosphates (with a commentary concerning the origin of biomolecular homochirality). *Chem. Biol.* 4: 309–320.

Bonner, W.A., P.R. Kavasmaneck, F.S. Martin, et al. 1974. Asymmetric adsorption of alanine by quartz. *Science* 186: 143–144.

Bonner, W.A., and E. Rubenstein. 1987. Supernovae, neutron stars and biomolecular chirality. *BioSystems* 20: 99–111.

Breslow, R., and Z.-L. Cheng. 2010. L-Amino acids catalyze the formation of an excess of D-glyceraldehyde, and thus of other D sugars, under credible prebiotic conditions. *Proc. Natl. Acad. Sci. USA* 107: 5723–5725.

Buono, F.G., and D.G. Blackmond. 2003. Kinetic evidence for a tetrameric transition state in the asymmetric autocatalytic alkylation of pyrimidyl aldehydes. *J. Am. Chem. Soc.* 125: 8978–8979.

Callahan, M.P., K.E. Smith, H.J. Cleaves II, et al. 2011. Carbonaceous meteorites contain a wide range of extraterrestrial nucleobases. *Proc. Natl. Acad. Sci. USA* 108: 13995–13998.

Cami, J., J.B.-Salas, E. Peeters, et al. 2010. Detection of C60 and C70 in a young planetary nebula. *Science* 329: 1180–1182.

Carroll, J.D. 2009. A new definition of life. *Chirality* 21: 354–358.

Cronin, J.R., and S. Pizzarello. 1997. Enantiomeric excesses in meteoritic amino acids. *Science* 275: 951–955.

Eliel, E.L., and S.H. Wilen. 1994. *Stereochemistry of organic compounds*. New York: John Wiley & Sons.

Engel, M.H., and S.A. Macko. 1997. Isotopic evidence for extraterrestrial non-racemic amino acids in the Murchison meteorite. *Nature* 389: 265–268.

Ernst, K.-H. 2012. Molecular chirality at surfaces. *Phys. Status Solid. B* 249: 2057–2088.

Feringa, B.L., and R.A. van Delden. 1999. Absolute asymmetric synthesis: The origin, control, and amplification of chirality. *Angew. Chem. Int. Ed.* 38: 3418–3438.

Frank, F.C. 1953. On spontaneous asymmetric synthesis. *Biochim. Biophys. Acta* 11: 459–463.

Gal, J. 2008. When Did Louis Pasteur present his memoir on the discovery of molecular chirality to the academie des sciences? Analysis of a discrepancy. *Chirality* 20: 1072–1084.

Gehring, T., M. Quaranta, B. Odell, et al. 2012. Observation of a transient intermediate in Soai's asymmetric autocatalysis: Insights from $^{1}$H NMR turnover in real time. *Angew. Chem. Int. Ed.* 51: 9539–9542.

Gellman, A.J., Y. Huang, X. Feng, et al. 2013. Superenantioselective chiral surface explosions. *J. Am. Chem. Soc.* 135: 19208–19214.

Gilbert, W. 1986. Origin of life: The RNA world. *Nature* 319: 618.

Girard, C., and H.B. Kagan. 1998. Nonlinear effects in asymmetric synthesis and stereoselective reactions: Ten years of investigation. *Angew. Chem. Int. Ed.* 37: 2922–2959.

Gonzalez-Campo, A., and D.B. Amabilino. 2013. Biomolecules at interfaces-chiral, naturally. *Top. Curr. Chem.* 333: 109–156.

Green, M.M., J.-W. Park, T. Sato, et al. 1999. The macromolecular route to chiral amplification. *Angew. Chem. Int. Ed.* 38: 3138–3154.

Gridnev, I.D., and J.M. Brown. 2004. Asymmetric autocatalysis: Novel structures, novel mechanism? *Proc. Natl. Acad. Sci. USA* 101: 5727–5731.

Gridnev, I.D., J.M. Serafimov, and J.M. Brown. 2004. Solution structure and reagent binding of the zinc alkoxide catalyst in the Soai asymmetric autocatalytic reaction. *Angew. Chem. Int. Ed.* 43: 4884–4887.

Gridnev, I.D., and A.K. Vorobiev. 2012. Quantification of sophisticated equilibria in the reaction pool and amplifying catalytic cycle of the Soai reaction. *ACS Catal.* 2: 2137–2149.

Hazen, R.M. 2004. Chiral crystal faces of common rock-forming minerals. In Pályi, G., and C. Zucchi, Eds. *Progress in biological chirality*. Oxford: Elsevier, chap. 11, p. 137–151.

Hazen, R.M., T.R. Filley, and G.A. Goodfriend, 2001. Selective adsorption of L- and D-amino acids on calcite: Implications for biochemical homochirality. *Proc. Natl. Acad. Sci. USA* 98: 5487–5490.

Hazen, R.M., and D.S. Sholl. 2003. Chiral selection on inorganic crystalline surfaces. *Nat. Mater.* 2: 367–374.

Hitosugi, S., A. Matsumoto, Y. Kaimori, et al. 2014. Asymmetric autocatalysis initiated by finite single-wall carbon nanotube molecules with helical chirality. *Org. Lett.* 16: 645–647.

Inoue, Y. 1992. Asymmetric photochemical reactions in solution. *Chem. Rev.* 92: 741–770.

Kagan, H., A. Moradpour, J.F. Nicoud, et al. 1971. Photochemistry with circularly polarized light. The synthesis of optically active hexahelicene. *J. Am. Chem. Soc.* 93: 2353–2354.

Kaimori, Y., Y. Hiyoshi, T. Kawasaki, et al. 2019. Formation of enantioenriched alkanol with stochastic distribution of enantiomers in the absolute asymmetric synthesis under heterogeneous solid – vapor phase conditions. *Chem. Commun.* 55: 5223–5226.

Kawasaki, T., Y. Hakoda, H. Mineki, et al. 2010a. Generation of absolute controlled crystal chirality by the removal of crystal water from achiral crystal of nucleobase cytosine. *J. Am. Chem. Soc.* 132: 2874–2875.

Kawasaki, T., Y. Harada, K. Suzuki, et al. 2008b. Enantioselective synthesis utilizing enantiomorphous organic crystal of achiral benzils as a source of chirality in asymmetric autocatalysis. *Org. Lett.* 10: 4085–4088.

Kawasaki, T., K. Hatase, Y. Fujii, et al. 2006c. The distribution of chiral asymmetry in meteorites: An investigation using asymmetric autocatalytic chiral sensors. *Geochim. Cosmochim. Acta* 70: 5395–5402.

Kawasaki, T., K. Jo, H. Igarashi, et al. 2005. Asymmetric amplification using chiral cocrystals formed from achiral organic molecules by asymmetric autocatalysis. *Angew. Chem. Int. Ed.* 44: 2774–2777.

Kawasaki, T., Y. Matsumura, T. Tsutsumi, et al. 2009a. Asymmetric autocatalysis triggered by carbon isotope (13C/12C) chirality. *Science* 324: 492–495.

Kawasaki, T., M. Nakaoda, N. Kaito, et al. 2010b. Asymmetric autocatalysis induced by chiral crystals of achiral tetraphenylethylenes. *Orig. Life Evol. Biosph.* 40: 65–78.

Kawasaki, T., Y. Okano, E. Suzuki, et al. 2011c. Asymmetric autocatalysis: Triggered by chiral isotopomer arising from oxygen isotope substitution. *Angew. Chem. Int. Ed.* 50: 8131–8133.

Kawasaki, T., H. Ozawa, M. Ito, et al. 2011b. Enantioselective synthesis induced by compounds with chirality arising from partially deuterated methyl groups in conjunction with asymmetric autocatalysis. *Chem. Lett.* 40: 320–321.

Kawasaki, T., T. Sasagawa, K. Shiozawa, et al. 2011a. Enantioselective synthesis induced by chiral crystal composed of DL-serine in conjunction with asymmetric autocatalysis. *Org. Lett.* 13: 2361–2363.

Kawasaki, T., M. Shimizu, D. Nishiyama, et al. 2009b. Asymmetric autocatalysis induced by meteoritic amino acids with hydrogen isotope chirality. *Chem. Commun.* 4396–4398.

Kawasaki, T., M. Shimizu, K. Suzuki, et al. 2004. Enantioselective synthesis induced by chiral epoxides in conjunction with asymmetric autocatalysis. *Tetrahedron: Asymmetry* 15: 3699–3701.

Kawasaki, T., K. Suzuki, Y. Hakoda, et al. 2008a. Achiral nucleobase cytosine acts as an origin of homochirality of biomolecules in conjunction with asymmetric autocatalysis. *Angew. Chem. Int. Ed.* 47: 496–499.

Kawasaki, T., K. Suzuki, K. Hatase, et al. 2006a. Enantioselective synthesis mediated by chiral crystal of achiral hippuric acid in conjunction with asymmetric autocatalysis. *Chem. Commun.* 1869–1871.

Kawasaki, T., K. Suzuki, M. Shimizu, et al. 2006b. Spontaneous absolute asymmetric synthesis in the presence of achiral silica gel in conjunction with asymmetric autocatalysis. *Chirality* 18: 479–482.

Kawasaki, T., M. Uchida, Y. Kaimori, et al. 2013. Enantioselective synthesis induced by the helical molecular arrangement in the chiral crystal of achiral tris(2-hydroxyethyl) 1,3,5-benzenetricarboxylate in conjunction with asymmetric autocatalysis. *Chem. Lett.* 42: 711–713.

Kitamura, M., S. Okada, S. Suga, et al. 1989. Enantioselective addition of dialkylzincs to aldehydes promoted by chiral amino alcohols. Mechanism and nonlinear effect. *J. Am. Chem. Soc.* 111: 4028–4036.

Kondepudi, D.K., R.J. Kaufman, and N. Singh. 1990. Chiral symmetry breaking in sodium chlorate crystallization. *Science* 250: 975–976.

Lee, D.H., J.R. Granja, J.A. Martinez, et al. 1996. A self-replicating peptide. *Nature* 382: 525–528.

Matsumoto, A., T. Abe, A. Hara, et al. 2015a. Crystal structure of isopropylzinc alkoxide of pyrimidyl alkanol: Mechanistic insights for asymmetric autocatalysis with amplification of enantiomeric excess. *Angew. Chem. Int. Ed.* 54: 15218–15221.

Matsumoto, A., S. Fujiwara, T. Abe, et al. 2016a. Elucidation of the structures of asymmetric autocatalyst based on x-ray crystallography. *Bull. Chem. Soc. Jpn.* 89: 1170–1177.

Matsumoto, A., T. Ide, Y. Kaimori, et al. 2015b. Asymmetric autocatalysis triggered by chiral crystal of achiral ethylenediamine sulfate. *Chem. Lett.* 44: 688–690.

Matsumoto, A., Y. Kaimori, M. Uchida, et al. 2017. Achiral inorganic gypsum acts as an origin of chirality through its enantiotopic surface in conjunction with asymmetric autocatalysis. *Angew. Chem. Int. Ed.* 56: 545–548.

Matsumoto, A., T. Kawasaki, and K. Soai. 2022. Structure analysis of asymmetric autocatalysis by X-ray crystallography and circular dichroism spectroscopy. In Soai, K., T. Kawasaki, and A. Matsumoto, Eds. *Asymmetric autocatalysis: The Soai reaction.* Cambridge: The Royal Society of Chemistry, chap. 12, pp. 273–288.

Matsumoto, A., S. Oji, S. Takano, et al. 2013. Asymmetric autocatalysis triggered by oxygen isotopically chiral glycerin. *Org. Biomol. Chem.* 11: 2928–2931.

Matsumoto, A., H. Ozaki, S. Harada, et al. 2016b. Asymmetric induction by nitrogen 14N/15N isotopomer in conjunction with asymmetric autocatalysis. *Angew. Chem. Int. Ed.* 55: 15246–15249.

Matsumoto, A., H. Ozaki, S. Tsuchiya, et al. 2019. Achiral amino acid glycine acts as an origin of homochirality in asymmetric autocatalysis. *Org. Biomol. Chem.* 17: 4200–4203.

Matsumoto, A., A. Tanaka, Y. Kaimori, et al. 2021. Circular dichroism spectroscopy of catalyst preequilibrium in asymmetric autocatalysis of pyrimidyl alkanol. *Chem. Comm.* 57: 11209–11212.

Matsuura, T., and H. Koshima. 2005. Introduction to chiral crystallization of achiral organic compounds spontaneous generation of chirality. *J. Photochem. Photobio. C* 6: 7–24.

McGuire, B., P.B. Carroll, R.A. Loomis, et al. 2016. Discovery of the interstellar chiral molecule propyleneoxide ($CH_3CHCH_2O$). *Science* 352: 1449–1452.

Meierhenrich, U.J., L. Nahon, C. Alcarez, et al. 2005. Asymmetric vacuum UV photolysis of the amino acids leucine in solid state. *Angew. Chem. Int. Ed.* 44: 5630–5636.

Micheau, J.-C., C. Coudret, J.-M. Cruz, et al. 2012. Amplification of enantiomeric excess, mirror-image symmetry breaking and kinetic proofreading in Soai reaction models with different oligomeric orders. *Phys. Chem. Chem. Phys.* 14: 13239–13248.

Miller, S.L. 1953. A production of amino acids under possible primitive earth conditions. *Science* 117: 528–529.

Miller, S.L., and H.C. Urey. 1959. Organic compound synthesis on the primitive earth. *Science* 130: 245–251.

Mineki, H., T. Hanasaki, A. Matsumoto, et al. 2012. Asymmetric autocatalysis initiated by achiral nucleic acid base adenine: implications on the origin of homochirality of biomolecules. *Chem. Commun.* 48: 10538–10540.

Mineki, H., Y. Kaimori, T. Kawasaki, et al. 2013. Enantiodivergent formation of a chiral cytosine crystal by removal of crystal water from an achiral monohydrate crystal under reduced pressure. *Tetrahedron: Asymmetry* 24: 1365–1367.

Mislow, K. 2003. Absolute asymmetric synthesis: A commentary. *Collect. Czech. Chem. Commun.* 68: 849–864.

Miyagawa, S., K. Yoshimura, Y. Yamazaki, et al. 2017. Asymmetric Strecker reaction arising from the molecular orientation of an achiral imine at the single-crystal face: Enantioenriched L- and D-amino acids. *Angew. Chem. Int. Ed.* 56: 1055–1058.

Munoz Caro, G.M., U.J. Meierhenrich, W.A. Schutte, et al. 2002. Amino acids from ultraviolet irradiation of interstellar ice analogues. *Nature* 416: 403–406.

Myrgorodska, I., C. Meinert, Z. Martins, et al. 2015. Molecular chirality in meteorites and interstellar ices, and the chirality experiment on board the ESA cometary Rosetta mission. *Angew. Chem. Int. Ed.* 54: 1402–1412.

Nanita, S.C., and R.G. Cooks. 2006. Serine octamers: Cluster formation, reactions, and implications for biomolecule homochirality. *Angew. Chem. Int. Ed.* 45: 554–569.

Olsson, S., P.M. Björemark, T. Kokoli, et al. 2015. Absolute asymmetric synthesis: Protected substrate oxidation. *Chem. Eur. J.* 21: 5211–5219.

Oro, J. 1960. Sÿnthesis of adenine from ammonium cyanide. *Biochem. Biophys. Res. Commun.* 2: 407–412.

Pályi, G., C. Zicchi, and L. Caglioti, eds. 2012. *The Soai reaction and related topic.* Modena: Academia Nationale di Scienze Lettere e Arti Modena.

Pidcock, E. 2005. Achiral molecules in non-centrosymmetric space groups. *Chem. Commun.* 3457–3459.

Pizzarello, S., and Y. Huang. 2005. The deuterium enrichment of individual amino acids in carbonaceous meteorites: A case for the presolar distribution of biomolecule precursors. *Geochim. Cosmochim. Acta.* 69: 599–605.

Pizzarello, S., and A.L. Weber. 2004. Prebiotic amino acids as asymmetric catalysts. *Science* 303: 1151.

Puchot, C., O. Samuel, E. Dunach, et al. 1986. Nonlinear effects in asymmetric synthesis. Examples in asymmetric oxidations and aldolization reactions. *J. Am. Chem. Soc.* 108: 2353–2357.

Quaranta, M., T. Gehring, B. Odell, et al. 2010. Unusual inverse temperature dependence on reaction rate in the asymmetric autocatalytic alkylation of pyrimidyl aldehydes. *J. Am. Chem. Soc.* 132: 15104–15107.

Raval, R. 2009. Chiral expression from molecular assemblies at metal surfaces: Insights from surface science techniques. *Chem. Soc. Rev.* 38: 707–721.

Ribó, J.M., J. Crusats, F. Sagués, et al. 2001. Chiral sign induction by vortices during the formation Of mesophases in stirred solutions. *Science* 292: 2063–2066.

Ring, D., Y. Wolman, N. Friedmann, et al. 1972. Prebiotic synthesis of hydrophobic and protein amino acids. *Proc. Natl. Acad. Sci. USA* 69: 765–768.

Robertson M.P., and S.L. Miller. 1995. An efficient prebiotic synthesis of cytosine and uracil. *Nature* 375: 772–774.

Saito, Y., and H. Hyuga. 2013. Homochirality: Symmetry breaking in systems driven far from equilibrium. *Rev. Mod. Phys.* 85: 603–621.

Sandford, S.A., J. Aléon, C.M.O'D. Alexander, et al. 2006. Organics captured from comet 81P/Wild 2 by the stardust spacecraft. *Science* 314: 1720–1724.

Sato, I., K. Kadowaki, Y. Ohgo, et al. 2004. Highly enantioselective asymmetric autocatalysis induced by chiral ionic crystals of sodium chlorate and sodium bromate. *J. Mol. Cat. A: Chemical.* 216: 209–214.

Sato, I., K. Kadowaki, and K. Soai. 2000. Asymmetric synthesis of an organic compound with high enantiomeric excess induced by inorganic ionic sodium chlorate. *Angew. Chem. Int. Ed.* 39: 1510–1512.

Sato, I., Y. Ohgo, H. Igarashi, et al. 2007. Determination of absolute configurations of amino acids by asymmetric autocatalysis of 2-alkynylpyrimidyl alkanol as a chiral sensor. *J. Organomet. Chem.* 692: 1783–1787.

Sato, I., D. Omiya, H. Igarashi, et al. 2003b. Relationship between the time, yield, and enantiomeric excess of asymmetric autocatalysis of chiral 2-alkynyl-5-pyrimidyl alkanol with amplification of enantiomeric excess. *Tetrahedron: Asymmetry* 14: 975–979.

Sato, I., D. Omiya, K. Tsukiyama, et al. 2001a. Evidence of asymmetric autocatalysis in the enantioselective addition of diisopropylzinc to pyrimidine-5-carbaldehyde using chiral pyrimidyl alkanol. *Tetrahedron: Asymmetry* 12: 1965–1969.

Sato, I., S. Osanai, K. Kadowaki, et al. 2002. Asymmetric autocatalysis of pyrimidyl alkanol induced by optically active 1,1¢-binaphthyl, an atropisomeric hydrocarbon, generated from spontaneous resolution on crystallization. *Chem. Lett.* 31: 168–169.

Sato, I., H. Urabe, S. Ishiguro, et al. 2003a. Amplification of chirality from extremely low to greater than 99.5% ee by asymmetric autocatalysis. *Angew. Chem. Int. Ed.* 42: 315–317; *Angew. Chem.* 115: 329–331.

Sato, I., R. Yamashima, K. Kadowaki, et al. 2001b. Asymmetric induction by helical hydrocarbons. *Angew. Chem., Int. Ed.* 40: 1096–1098.

Schiaffino, L., and G. Ercolani. 2008. Unraveling the mechanism of the Soai asymmetric autocatalytic reaction by first-principles calculations: Induction and amplification of chirality by self-assembly of hexamolecular complexes. *Angew. Chem. Int. Ed.* 47: 6832–6835.

Schwartz, A.W., H. Joosten, and A.B. Voet. 1982. Prebiotic adenine synthesis via HCN oligomerization in ice. *Biosystems* 15: 191–193.

Shapiro, R. 1999. Prebiotic cytosine synthesis: A critical analysis and implications for the origin of life. *Proc. Natl. Acad. Sci. USA* 96: 4396–4401.

Shibata, T., J. Yamamoto, N. Matsumoto, et al. 1998. Amplification of a slight enantiomeric imbalance in molecules based on asymmetric autocatalysis. The first correlation between high enantiomeric enrichment in a chiral molecule and circularly polarized light. *J. Am. Chem. Soc.* 120: 12157–12158.

Shibata, T., S. Yonekubo, and K. Soai. 1999. Practically perfect asymmetric autocatalysis with (2-alkynyl-5-pyrimidyl)alkanols. *Angew. Chem. Int. Ed.* 38: 659–661; *Angew. Chem.* 111: 749–751.

Shindo, H., Y. Shirota, K. Niki, et al. 2013. Asymmetric autocatalysis induced by cinnabar: observation of the enantioselective adsorption of a 5-pyrimidyl alkanol on the crystal surface. *Angew. Chem. Int. Ed.* 52: 9135–9138.

Siegel, J.S. 1998. Homochiral imperative of molecular evolution. *Chirality* 10: 24–27.

Sievers, D., and G. von Kiedrowski. 1994. Self-replication of complementary nucleotide-based oligomers. *Nature* 369: 221.

Soai, K. 2019. Asymmetric autocatalysis. Chiral symmetry breaking and the origins of homochirality of organic molecules. *Proc. Jpn. Acad.*, Ser. B 95: 89–109.

Soai, K. 2022. The Soai reaction and its implications with the life's characteristic features of self-replication and homochirality. *Tetrahedron* 124: 133017.

Soai, K., and T. Kawasaki. 2008. Asymmetric autocatalysis with amplification of chirality. *Top. Curr. Chem.* 284: 1–33.

Soai, K., T. Kawasaki, and A. Matsumoto. 2014a. Asymmetric autocatalysis of pyrimidyl alkanol and its application to the study on the origin of homochirality. *Acc. Chem. Res.* 47: 3643–3654.

Soai, K., T. Kawasaki, and A. Matsumoto. 2014b. The origins of homochirality examined by using asymmetric autocatalysis. *Chem. Rec.* 14: 70–83.

Soai, K., T. Kawasaki, and A. Matsumoto. 2019. Role of asymmetric autocatalysis in the elucidation of origins of homochirality of organic compounds. *Symmetry* 11: 694.

Soai, K., T. Kawasaki, and A. Matsumoto. 2022. Asymmetric autocatalysis: The Soai reaction. An overview. In Soai, K., T. Kawasaki, and A. Matsumoto, Eds. *Asymmetric autocatalysis: The Soai reaction*. Cambridge: The Royal Society of Chemistry, chap. 1, pp. 1–32.

Soai, K., A. Matsumoto, and T. Kawasaki. 2017. Asymmetric autocatalysis and the origins of homochirality of organic compounds. An overview. In Pályi, G., R. Kurdi, and C. Zucchi, Eds. *Advances in asymmetric autocatalysis and related topic*. Cambridge: Elsevier, chap. 1, pp. 1–30.

Soai, K., S. Niwa, and H. Hori. 1990. Asymmetric self-catalytic reaction. Self-production of chiral 1-(3-Pyridyl)alkanols as chiral self-catalysts in the enantioselective addition of dialkylzinc reagents to pyridine-3-carbaldehyde. *J. Chem. Soc. Che. Commun*. 982–983.

Soai, K., S. Osanai, K. Kadowaki, et al. 1999. *d*- and *l*-Quartz-promoted highly enantioselective synthesis of a chiral compound. *J. Am. Chem. Soc.* 121: 11235–11236.

Soai, K., I. Sato, T. Shibata, et al. 2003. Asymmetric synthesis of pyrimidyl alkanol without adding chiral substances by the addition of diisopropylzinc to pyrimidine-5-carbaldehyde in conjunction with asymmetric autocatalysis. *Tetrahedron: Asymmetry* 14: 185–188.

Soai, K., T. Shibata, and Y. Kowata, 1996. Production of optically active pyrimidylalkyl alcohol by spontaneous asymmetric synthesis. *JPN.* Kokai Tokkyo Koho JP 1996012114. An abstract is readily available as JPH09268179 from the European Patent Office (http://worldwide.espacenet.com).

Soai, K., T. Shibata, H. Morioka, et al. 1995. Asymmetric autocatalysis and amplification of enantiomeric excess of a chiral molecule. *Nature* 378: 767–768.

Soai, K., T. Shibata, and I. Sato. 2000. Enantioselective automultiplication of chiral molecules by asymmetric autocatalysis. *Acc. Chem. Res.* 33: 382–390.

Soloshonok, V.A., C. Roussel, O. Kitagawa, et al. 2012. Self-disproportionation of enantiomers via achiral chromatography: A warning and an extra dimension in optical purifications. *Chem. Soc. Rev.* 41: 4180–4188.

Suarez, M., and G.B. Schuster. 1995. Photoresolution of an axially chiral bicyclo[3.3.0]octan-3-one: Phototriggers for a liquid-crystal-based optical switch. *J. Am. Chem. Soc.* 117: 6732–6738.

Suzuki, K., K. Hatase, D. Nishiyama, et al. 2010. Spontaneous absolute asymmetric synthesis promoted by achiral amines in conjunction with asymmetric autocatalysis. *J. Systems Chem.* 1: 5.

Tielens, A.G.G.M. 2005. *The physics and chemistry of the interstellar medium*. Cambridge: Cambridge University Press.

Todd, M.H. 2002. Asymmetric autocatalysis: Product recruitment for the increase in the chiral environment (PRICE). *Chem. Soc. Rev.* 31: 211–222.

Trapp, O., S. Lamour, F. Maier, et al. 2020. In Situ mass spectrometric and kinetic investigations of Soai's asymmetric autocatalysis. *Chem. – A Eur. J.* 15871–15880.

Viedma, C. 2005. Chiral symmetry breaking during crystallization: Complete chiral purity induced by nonlinear autocatalysis and recycling. *Phys. Rev. Lett.* 94: 065504.

Weissbuch, I., L. Addadi, Z. Berkovitch-Yellin, et al. 1984. Spontaneous generation and amplification of optical activity in α-amino acids by enantioselective occlusion into centrosymmetric crystals of glycine. *Nature* 310: 161–164.

Weissbuch, I., and M. Lahav. 2011. Crystalline architectures as templates of relevance to the origins of homochirality. *Chem. Rev.* 111: 3236–3267.

Wintner, E.A., M.M. Conn, and J.J. Rebek. 1994. Studies in molecular replication. *Acc. Chem. Res.* 27: 198–203.

Zepik, H., E. Shavit, M. Tang, et al. 2002. Chiral amplification of oligopeptides in two-dimensional crystalline self-assemblies on water. *Science* 295: 1266–1269.

# 8 Abiotic-to-Biotic Transition of Organic Systems Starting with Thermodynamic Inversion

*Vladimir N. Kompanichenko*

## 8.1 INTRODUCTION

At present time, there exist a large number of concepts of prebiotic chemical evolution, the latest stage of which was the origin of life on Earth. Based on them, various types of prebiotic microsystems that preceded living cells have been proposed: Coacervate droplets (Oparin 1953), protein world (Fox et al. 1996; Ikehara 2015), the world of RNA (Joyce 2002; Budin and Szostak 2010), liposomes (Luisi 2000; Deamer 2011), the world of aromatic hydrocarbons (Ehrenfreund et al. 2006), formamides (Saladino et al. 2012), as well as various combined models. Some of them are now of historical interest. Many chemical models are formed during spontaneous self-assembly (e.g., liposomes), but for others, different mechanisms of formation are envisaged – in particular, synthesis of organic material on mineral surfaces: Some clay, apatite, etc. (Wachtershauser 1988; Ferris 2002). The next stage of chemical evolution consists in the formation of more complex structures that are usually called prebiotic microsystems, or "protocells", with the following complication up to (artificial) cells (Kurihara et al. 2011; Sugawara et al. 2012; Damer and Deamer 2020).

Some researchers express the opinion that the next stage of evolution (following protocells and "artificial" cells) were progenotes. Progenote is the conventional name of a group of now extinct life forms that had a more primitive organization than the simplest modern cells. They were theoretically reconstructed by Woese (1987, 1998) on the basis of the analysis of the phylogenetic tree of microorganisms. Progenotes were alive cells. In contrast, any types of artificial cells (models) synthesized from abiogenic material are not alive. They do not manifest two key properties of living cells populations – the ability for expansion in the environment and the ability to purposeful behavior. This means that the actual moment of life origin (i.e., the key transformation that determines the appearance of the living state in prebiotic microsystems) is "hidden" within the evolutionary gap between non-living model cells and living progenote cells.

 DOI: 10.1201/9781003294276-8

So, the key question is this: Why we cannot transform prebiotic systems of any chemical composition into primary life forms capable of purposeful behavior and expansion in the environment? To better understand this, one must first consider the fundamental differences between the two types of natural systems: Non-living chemical and living biological.

## 8.2 COMPARISON OF NON-LIVING AND LIVING SYSTEMS IN TERMS OF CHEMISTRY AND THERMODYNAMICS

Let us consider the boundary between non-living and living systems, shown in Figure 8.1 from the point of view of chemistry and thermodynamics.

The central part of Figure 8.1 shows the main types of prebiotic microsystems – the RNA world, protein world, lipids' world, and at the top – the root region of the phylogenetic tree of life on Earth, represented by the simplest prokaryotes (archaea and bacteria). In the right part of the Figure 8.1, the main trends are indicated, reflecting the direction of chemical transformations in the course of prebiotic (by arrows)

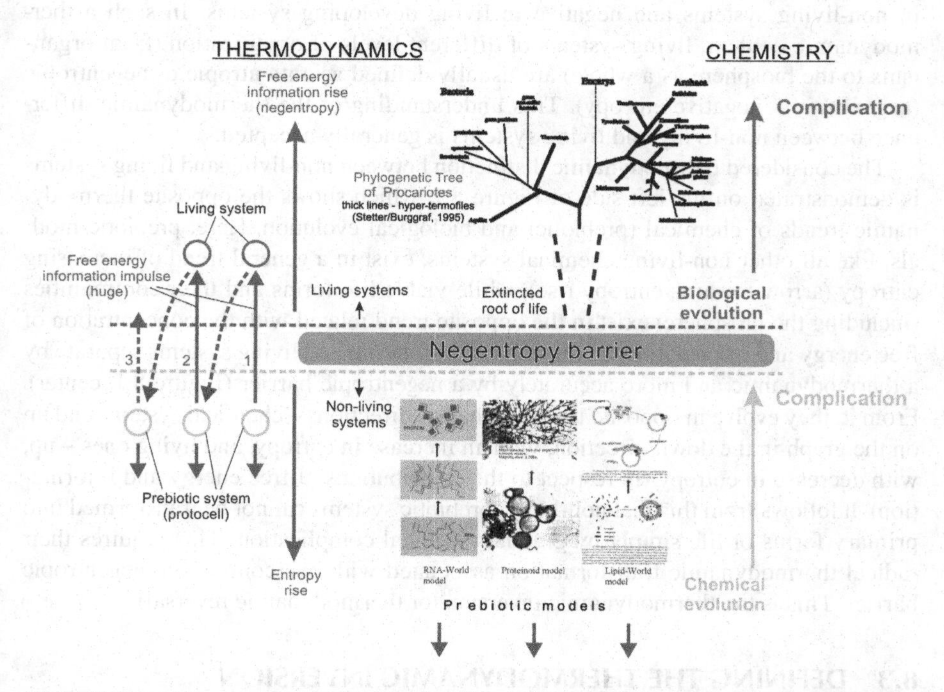


**FIGURE 8.1** Negentropic barrier between non-living and living systems and the scheme of its possible overcoming. The left side shows a jump-like transition of a non-living organic microsystem through the negentropic barrier, initiated by a powerful impulse of free energy (vertical arrow 1). The repetition of such impulses (arrows 2, 3, . . .) leads to the emergence in the microsystem of an intermediate state between nonlife and life, which exists in an oscillatory mode and is associated with periodic overcoming of the barrier and returning back.

*Source:* Vladimir N. Kompanichenko

and biological (arrow on the top right) evolution. In general, these trends look similar, since in both cases, evolution proceeds with the complexification of organic molecules, molecular structures, and an increase in their chemical diversity. Accordingly, biological evolution looks like a natural continuation of chemical evolution.

Consideration of non-living and living systems in the context of thermodynamics, which determines the macrostate of the system, leads to a different result. It is known that non-living natural systems eventually evolve in the direction of increasing entropy, which follows from the second law of thermodynamics. It is also known that biological evolution paradoxically proceeds in the thermodynamically opposite (negentropic) direction – with an increase in free energy and information in the system due to a relative decrease in entropy. This difference expresses through the general thermodynamic balance of a natural system: "The total contribution of entropy ($S_c$)/the total contribution of free energy ($F_c$)" (Kompanichenko 2017). According to the book by Feistel and Ebeling (2011) and other researchers, a small part of entropy relates with informational processes. In this regard, the aforementioned ratio supplements by one more balance: "The total contribution of informational entropy ($Si_c$)/ the total contribution of information ($I_c$)". Accordingly, both balances are positive in non-living systems and negative in living developing systems. In such a thermodynamic context, living systems of different levels of organization (from organisms to the biosphere as a whole) are usually defined as antientropic or negentropic (negentropy – negative entropy). This understanding of the thermodynamic difference between non-living and living systems is generally accepted.

The considered thermodynamic distinction between non-living and living systems is demonstrated on the left side of Figure 8.1, which shows the opposite thermodynamic trends of chemical (prebiotic) and biological evolution. Here, prebiotic models, like all other non-living chemical systems, exist in a general trend of increasing entropy (arrow towards entropy rise), while viable life forms and their communities (including the biosphere) exist in the opposite trend related with the concentration of free energy and information. In this regard, non-living and living systems separate by a thermodynamic and more accurately by a negentropic barrier (Figure 8.1, center). From it, they evolve in opposite thermodynamic directions – chemical systems end up on the graph in the down direction, with an increase in entropy, and living ones – up, with decrease in entropy (in respect to the contributions of free energy and information). It follows from this that non-living prebiotic systems cannot be transformed into primary forms of life simply by gradual chemical complication. This requires their radical thermodynamic transformation associated with overcoming the negentropic barrier. This is the thermodynamic inversion (or thermodynamic reversal).

## 8.3 DEFINING THE THERMODYNAMIC INVERSION

Thermodynamic inversion is a decisive transformation of a chemical system in the process of acquiring a living state. Its essence can be substantiated in the following way. Based on the opposite thermodynamic trends of chemical and biological evolution (Figure 8.1), we can conclude that during the origin of life on the early Earth, there was a transition from a chemical system with a dominant contribution of entropy to a biotic system in which free energy and information prevailed. This transition is called thermodynamic inversion (Kompanichenko 2017). It's like if we observe a car moving first in one direction and then in the opposite direction, then we understand that it has turned around somewhere (even if we did not see it). Usually, the need for thermodynamic inversion during the transformation of organic systems into primary microorganisms is not taken into account by researchers dealing with issues related to the origin of life.

Starting from the moment of thermodynamic inversion, the transformed organic system can no longer exist as an ordinary chemical system. It is forced to build own internal chemical (as well as informational) processes in such a way as to generate more free energy than can be compensated by entropy. This way of organizing the system is referred to as "negentropic". It is laid down at the very beginning of life appearance on Earth or other planet. It means that chemical evolution by itself (as a process of increasing complexity of organic matter) cannot lead to the origin of life, since it does not imply the transition of the system to a qualitatively different (negentropic) thermodynamic state (Kompanichenko 2017). Such a transition requires a prebiotic chemical "revolution", which includes the obligatory presence of both evolving organic systems of a suitable composition and the described thermodynamic transformation.

The carried-out analysis of Figure 8.1 leads to the conclusion about the only permissible way to transform chemical systems into primary life forms, which does not contradict the second law of thermodynamics. First of all, this transition can only be carried out in an oscillatory mode, passing through an intermediate state between nonlife and life. To start this process, the chemical system must receive such a strong negentropy impulse (a large contribution of free energy and information) that it greatly exceeds the current entropy production in the system. Such an impulse can be initiated by rapid and high-amplitude changes in parameters in the external environment with the appearance of conditions that are far from equilibrium. Such an impulse should lock the trend of the prebiotic system toward an increase in entropy and create a large excess of free energy (and information) in the system, which entropy cannot immediately devalue. For some time, such a system exists in a peculiar and thermodynamically extremely nonequilibrium (unstable) state of entropy deficit or, the same, negentropy excess. The subsequent increase in entropy, mandatory according to the second law, will return the organic system to its initial state (as shown by the dotted arrow on the left in Figure 8.1).

With the repetition of such strong negentropic impulses due to external fluctuations in physicochemical parameters (arrows 2–3), the prebiotic system begins to oscillate above and below the negentropic barrier and passes into an intermediate state between nonlife and life. Being in a position above the negentropic barrier, it has the potential for evolution to life through directed reorganization using the accumulated reserves of supra-entropic free energy and information. This mechanism of thermodynamic inversion considers in more detail in a number of publications (Kompanichenko 2008, 2012, 2017, 2019). Another interpretation of the appearance of living state in organic systems, based on the transition to life through a gradual thermodynamically un-directional process, will be in conflict with the second law.

Possible trends in the further evolution of a prebiotic system, which is in the intermediate oscillatory state between nonlife and life, are shown in Figure 8.2. In its left part I, the evolution of a system is shown in which the ratio "total contribution of entropy ($S_c$)/total contribution of free energy ($F_c$)" fluctuates all the time, sometimes spontaneously "jumping" over the negentropic barrier (horizontal axis). During such periods, the system passes into the intermediate state, either rising above the barrier or falling below it. If it is able to provide an effective (intensified and purposeful) response to the impact of external stress factors, then with each such fluctuation, it accumulates more and more supra-entropic free energy, developing along an upward trend toward life (Figure 8.2, bottom part II). Such a process is called *stepwise activation* (Kompanichenko and El-Registan 2022). Conversely, the inability of the system to effectively respond to periodic external influences initiates an irreversible increase in entropy, accompanied by degradation and elimination from the process of germinal biological evolution (Figure 8.2, right part III). In general, a role of effective stress response in life emergence is considered in (Kompanichenko and Kotsyurbenko 2022). It should be added that, in accordance with the provisions of the theory of stress, external influences can be both beneficial for the system ("stress$^+$") and unfavorable ("stress$^-$") (Selye 1974).

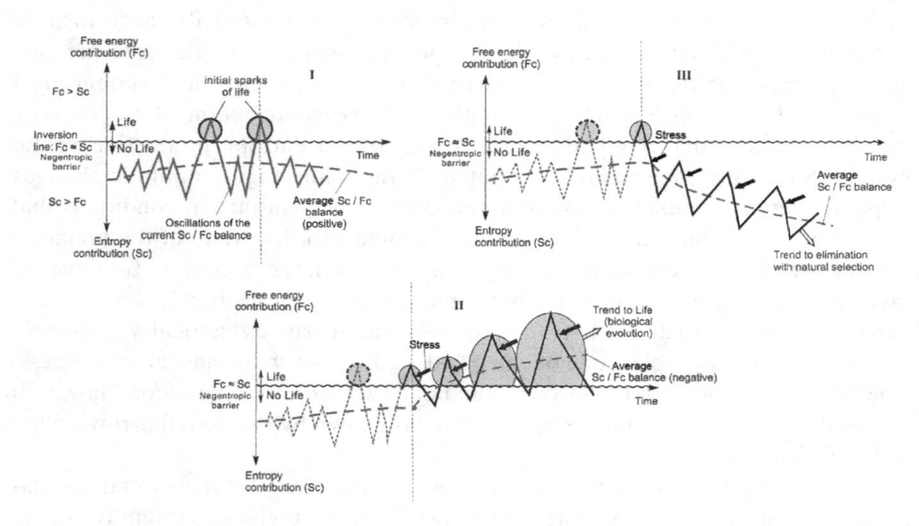


**FIGURE 8.2** Trends in the evolution of a prebiotic system from the intermediate state between nonlife and life, determined by the thermodynamic relation "total contribution of entropy ($S_c$)/total contribution of free energy ($F_c$)". I – general scheme of $S_c/F_c$ ratio fluctuations with periodic random transitions ("jumps") of the prebiotic microsystem above the line of approximate equality of the contributions ($S_c \approx F_c$, horizontal axis); II – stepwise activation of a nascent living subcell with a progressive prevalence of the total contribution of free energy over the total contribution of entropy ($S_c < F_c$); and III – return of the microsystem to the prebiotic state (its degradation as a potentially possible life form) ($S_c > F_c$).

*Source:* Vladimir N. Kompanichenko

## 8.4 MEDIUM OF THE ORIGIN OF LIFE

It follows from the previous sections that physicochemical parameters (pressure, temperature, pH, Eh, concentrations of components, electrical potential, etc.) in the environment for the origin of life should fluctuate in various modes with changes in amplitudes, frequencies, periods, etc. In a framework of the TI concept, such fluctuations consider as the fourth required condition for the emergence of life (in addition to the three generally accepted ones: The presence of organic matter, water, and an energy source). The combination of fluctuations of the listed parameters of different scales is a characteristic feature of hydrothermal systems. But they are not peculiar to the ocean or the ice sheet, where physicochemical conditions are more or less stable.

In this regard, hydrothermal systems considered by the author are the most suitable environment for the origin of life, which coincides with the opinion of many experts in the field of astrobiology (Corliss et al. 1981; Shock et al. 1998; Holm and Andersson 2005; Martin and Russell 2007; Cleaves et al. 2009; Barge et al. 2017; Deamer et al. 2019; Kolb 2019; Damer and Deamer 2020). In recent years, the most active discussion has been between supporters of the origin of life in hot springs on the continents (Deamer 2021) and in hot vents at a bottom of ocean (Russell 2021). Note that in both cases, hot solutions come to the surface from deeper regions of hydrothermal systems. Within the framework of the TI model, the subsurface environment considers the initial generator of the processes that led to the emergence of populations of primary life forms (subcells). Some scientists have also considered the subsurface regions of hydrothermal systems as the most suitable environment for the emergence of life (Washington 2000). In this area, sharp changes in pressure, temperature, and other parameters periodically occur (especially during volcanic eruptions and earthquakes). According to the TI concept, just these conditions are necessary for the transition of clusters of organic microsystems to a far from equilibrium state (maintained by bifurcations). Such highly fluctuating conditions in hydrothermal systems are interrupted from time to time by periods of relative calm, when low-amplitude fluctuations appear (often referred to as "trembling of the earth's crust"). Such types of oscillations are widely described in the literature on the dynamics of hydrothermal systems (Kralj and Kralj 2000; Kiryukhin et al. 2002; Kompanichenko and Shlufman 2013). According to the TI model, abrupt changes in the environment contribute to the transition of clusters of organic microsystems to a nonequilibrium bifurcation state.

One more peculiarity of the hydrothermal fluid in subsurface areas is that it migrates irreversibly to the surface against the backdrop of a decrease in temperature and pressure (Kompanichenko and El-Registan 2022). This is important for the diversification of kinetic traps, the role of which in the course of life emergence is considered in detail by Ross and Deamer (2016). Their meaning is that the polycondensation of organic molecules (e.g., amino acids) in an aqueous medium inevitably ends with their hydrolysis, that is, decomposition. Such a decomposition must inevitably complete the process of formation of kinetic traps for macromolecules, which usually proceeds under reversible conditions (however, sometimes reversibility can decrease due to volcanic eruptions). On the contrary, the parameters of a hydrothermal fluid migrating from a depth of about 1 km to the surface change irreversibly. For instance, in the hydrothermal systems of Kamchatka, the temperature in such a vertical range decreases from values of the order of 200°C to 80–100°C, and the pressure decreases from 40–80 bar to 1–2 bar (Kiryukhin et al. 2002; Kompanichenko et al. 2015). Such a directed change in conditions during the vertical migration of a hot solution, taking into account fluctuations of different ranks of parameters, provides unlimited possibilities for the expansion and irreversible modification of kinetic traps for chemical reactions, when there is predominance of the synthesis and polymerization of macromolecules over their destruction. So, according to the TI model being developed, the hydrothermal system was the most appropriate cradle of life on Earth.

## 8.5 CORRELATION OF THE SEQUENCE OF EXIT OF A BACTERIAL CELL FROM ANABIOSIS WITH THE STAGES OF THE ORIGIN OF LIFE

The considered general thermodynamics of the transition to life cannot give a definite answer to the question of chemical transformations in the system during the transition period and the stages of the emergence of metabolism. The sequence of metabolism formation during the origin of life has not yet been reliably revealed by researchers. There are many conflicting views on this. However, in microbiology, within the framework of the theory of anabiosis, a resting (dormant) bacterial cell occupies a similar intermediate position between non-living and living: on the one hand, it is no longer able to counteract the growth of entropy, and, on the other hand, it retains structural memory of the previous living state. At the same time, the sequence of changes in metabolic processes in the simplest bacterial cell, entering the state of anabiosis and leaving it, has been well studied (Frenkel-Krispin et al. 2001; Mazur 2004; El-Registan et al. 2006; Lewis 2010; Mika et al. 2010; Parry et al. 2014; Fonesca et al. 2016; Loiko et al. 2017). The restoration of metabolic processes in a resting (dormant) cell in the course of its exit from anabiosis includes three general steps. They are listed here and described in detail in Kompanichenko and El-Registan (2022).

- Step of activation or resuscitation: Stepwise increasing the level of stored free energy; the appearance of weak energy-giving process of respiration; an increase in membrane fluidity; an acceleration of the movement of lipids and membrane proteins.
- Step of initiation: A sharp increase in respiratory activity and devitrification of the cytoplasm; the beginning of entry into the cell of $K^+$ and $Na^+$ ions; the cytoplasm watering and the loss of thermal stability; the restoration of the energy function of the cytoplasmic membrane and the structure of the protein-synthesizing apparatus of the cell.
- Step of growth: Replication of nucleotide DNA beginning with parallel syntheses of informational and transfer RNA, as well as the synthesis of new ribosomes; the beginning of the growth cell cycle; the formation of the first mature vegetative cell.

Note that the both processes – the transition of prebiotic microsystems to life (in accordance with the TI concept) and the restoration of metabolic processes in resting bacterial cells – begin with the intermediate state between nonlife and life. Moreover, in the both types of processes, the further development of metabolism from the initial intermediate state proceeded through an increase in the level of (supra-entropic) free energy in the system (through "stepwise activation") that provides the tendency of the synthesis prevalence over destruction (or polymerization over hydrolysis). In both cases, external stressors and the ability of systems to respond effectively to them are required for the process: Exit of a resting bacterial cell from anabiosis begins with optimal changes in the environment (impact of stress factors), while the need for an effective response to stress during the emergence of life is substantiated in Section 8.3. Some details in this way are given in Kompanichenko and El-Registan (2022); Kompanichenko and Kotsyurbenko (2022).

From the foregoing follows the idea of the potential correlation of these processes – prebiotic and bacterial. The following thesis is put forward for discussion: Since the transition of a bacterial cell from a passive (anabiotic) state to an active state always occurs through the same sequence of changes in metabolic processes, it can be assumed that this sequence itself arose in the earliest period of the emergence of primary life forms. And since then, this sequence has been repeated with variations. This means that the stages of transformation of prebiotic microsystems into primary life forms had been recorded in the sequence of the exit of a bacterial cell from anabiosis. Therefore, the latter can be used to reconstruct the processes that took place during the period of the emergence of the simplest microorganisms and their populations on Earth. This reconstruction is given in the next section.

## 8.6 STAGES OF THE ORIGIN OF LIFE

Following the comparison of the TI concept with the theory of anabiosis (suspended animation), the four stages of metabolism emergence in a cluster of prebiotic microsystems have been distinguished (Figure 8.3). This process occurred in a pulsating hydrothermal fluid moving to the surface in the background of a decrease in temperature and pressure and corresponding changes in other physicochemical parameters.

*First stage:* the beginning of the prebiotic process, accompanied by the complexification of organic matter. Organic matter of various genesis could be involved in the water cycle between surface and subsurface reservoirs of the planet. Subsequent self-assembly of three-dimensional prebiotic microsystems of predominantly lipid-protein composition occurred in local areas of ascending hydrothermal fluid. The fluid temperature range in this stage was $100°C \ll T < 200°C$. This is consistent with data from modern hydrothermal fields in which hot solutions rise from a depth of about 1 km to depths of a few hundred meters from the surface. Suitable types of chemical reactions under such conditions ("in superheated water" and "on-water") are investigated by many researchers (Breslow, 2006; Liotta et al., 2007; Avola et al., 2013; Braakman, 2013; Manna and Kumar, 2013). In the context of prebiotic chemistry, these reactions are generalized by Kolb (2016, 2019). The flow of chemical reactions in hydrothermal environments was stimulated by conditions far from

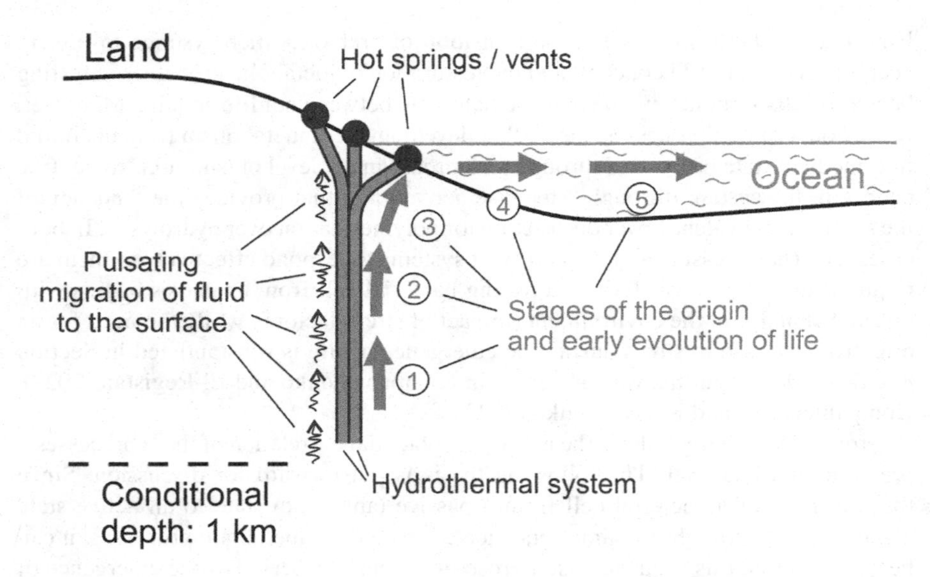


**FIGURE 8.3** Scheme of the stages of the origin of life and early biological evolution (1 – self-assembly of three-dimensional prebiotic microsystems and their chemical evolution; 2 – formation of clusters of protocells that are in an intermediate state between nonlife and life; 3 – formation of primary populations of living subcells; 4 – formation of populations of the simplest living cells; 5 – early biological evolution of initial prokaryotes to the oldest species of archaea and bacteria).

*Source:* Vladimir N. Kompanichenko

equilibrium, associated with high-frequency oscillations of physicochemical parameters and periodic bifurcation transitions of prebiotic systems.

*Second stage*: the formation of the clusters of protocells that are in an intermediate state between non-life and life, maintained in an oscillatory mode. The transition of prebiotic microsystems into protocells was carried out through thermodynamic inversion by means of their active (enhanced and purposeful) response to physicochemical fluctuations in the environment, i.e., to periodic stress (Kompanichenko and Kotsyurbenko, 2022). In protocells, there appeared a weak energy-giving process of respiration due to redox reactions and local watering of the membrane, accompanied by some acceleration of the movement of molecules.

*Third stage*: formation of the primary populations of living subcells in which the tendency to the concentration of free energy and information at the expense of entropy decrease is maintained (in accordance with the thermodynamic balances: $S_c < F_c$ and $Si_c < I_c$). During this stage, ions of $K^+$ and $Na^+$ entered into the evolving subcells; watering of the cytoplasm and loss of thermal stability enhanced; the nonenzymatic antioxidant system and protein-synthesizing apparatus began to form.

*Fourth stage*: populations of the simplest living cells appeared. The replication of nucleotide DNA along with parallel syntheses of informational and transfer RNA arose in them, as well as the synthesis of new ribosomes. The growth cell cycle and the formation of the initial mature vegetative cells started. These cells can correlate with the extinct progenotes reconstructed by Woese (1987, 1998).

*Fifth stage*: early biological evolution of prokaryotic microorganisms in oceans with the emergence of the oldest species of archaea and bacteria.

In the framework of this reconstruction, the stages of life emergence correspond to the steps of bacterial cells' exit from anabiosis: the stage 2 corresponds with the step of activation, the stage 3 – with the step of initiation, and the stage 4 – with the step of growth (see Section 8.5). The stages of this reconstruction are considered in more detail in another work (Kompanichenko and El-Registan, 2022).

## 8.7 CONCLUSION

Some of the key points of this chapter should be briefly emphasized. The transition of prebiotic (micro)systems to life could not occur only due to chemical evolution directed to complexification of organic matter. It also required thermodynamic inversion. Since this crucial transformation, the organic systems began to concentrate supra-entropic free energy and information connected with relative decrease in entropy. This transformation occurred through an effective (enhanced and purposeful) response of prebiotic systems to periodic stress initiated by high-frequency fluctuations in physicochemical parameters (temperature, pressure, concentration of components, pH, Eh, electric potentials, etc.) in hydrothermal environment. It was the start of biological evolution.

The direction of chemical processes in the period of origin and early evolution of life was revealed on a basis of the established sequence of metabolic transformations during the recovery of a bacterial cell from anabiosis. According to the TI approach, it is this sequence that was laid at the basis of life processes. Since then, the sequence has been repeated with some variations every time a bacterial cell exits from anabiosis. Based on this thesis, one can also explain the Haeckel–Muller biogenetic law, according to which the stages of phylogenesis in general terms repeat in ontogenesis.

The emergence of living state in prebiotic systems (organic microsystems and their clusters) began with a weak energy-giving process of respiration associated with the transfer of electrons in conditions of a hydrothermal medium far from equilibrium. At the same time, there was a local watering of the membrane and some acceleration of the movement of molecules (stage 2 of the process). The RNA world molecules appeared in stage 3 when the hydrothermal fluid moved closer to the surface, where the temperature was lower. At the end of the process (stage 4), primary vegetative cells were formed, which already had the simplest genetic apparatus.

## REFERENCES

Avola S, Guillot M, da Silva-Perez D, Pellet-Rostaing S, Kunz W, Goettmann F (2013) Organic chemistry under hydrothermal conditions. *Pure. Appl. Chem.* 85: 89–103.

Barge LM, Branscomb E, Brucato JR, et al. (2017) Thermodynamics, disequilibrium, evolution: Far-from-equilibrium geological and chemical considerations for origin-of-life research. *Orig. Life Evol. Biosph.* 47: 39–56.

Braakman R (2013) Mapping metabolism onto the prebiotic organic chemistry of hydrothermal vents. *Proc. Natl. Acad. Sci. U.S.A.* 110: 13236–13237.

Breslow R (2006) The hydrophobic effect in reaction mechanism studies and in catalysis by artificial enzymes. *J. Phys. Org. Chem.* 19: 813–822.

Budin I, Szostak JW (2010) Expanding roles for diverse physical phenomena during the origin of life. *Annu. Rev. Biophys.* 39: 245–263.

Cleaves HJ, Aubrey AD, Bada JL (2009) An evaluation of critical parameters for abiotic peptide synthesis in submarine hydrothermal systems. *Orig. Life Evol. B.* 39: 109–126.

Corliss JB, Baross JA, Hoffman SE (1981) An hypothesis concerning the relationship between submarine hot springs and the origin of life on the Earth. *Oceanol. Acta SP.* 4: 59–69.

Damer B, Deamer D (2020) The hot spring hypothesis for an origin of life. *Astrobiology* 20: 429–452.

Deamer DW (2011) *First Life*. Berkeley, CA: University of California Press.

Deamer DW (2021) Where did life begin? Testing ideas in prebiotic analogue conditions. *Life* 11: 134.

Deamer DW, Damer B, Kompanichenko V (2019) Hydrothermal chemistry and the origin of cellular life. *Astrobiology* 19: 1523–1537.

Ehrenfreund P, Rasmussen S, Cleaves J, Chen L (2006) Experimentally tracing the key steps in the origin of life: The aromatic world. *Astrobiology* 6: 490–520.

El-Registan GI, Mylyukin AL, Nikolaev Yu A, Suzina NE, Galcheko VF, Duda VI (2006) Adaptogenic functions of extracellular autoregulators of microorganisms. *Microbiology* 75(4)

Feistel R, Ebeling W (2011) *Physics of Self-organization and Evolution*. Berlin: VCH; Wiley.

Ferris JP (2002) Montmorillonite catalysis of 30–50 mer oligonucleotides: Laboratory demonstration of potential steps in the origin of the RNA world. *Orig. Life Evol. Biosph.* 32: 311–332.

Fonesca F, Meneghel J, Cenard S, Passot S, Morris GJ (2016) Determination of intracellular vitrification temperatures for unicellular microorganisms under condition relevant for cryopreservation. *PLOS ONE*. http://doi.org/10.1371/jornal.pone.0152939.

Fox S, Balin P, Pappelis A, Yu B (1996) Experimental retracement of terrestrial origin of an excitable cell: Was it predictable? In *Chemical Evolution: Physics of the Origin and Evolution of Life*. Chela-Flores J, Raulin F, Eds. Dordrecht: Kluwer, pp. 21–32.

Frenkel-Krispin D, Levin-Zaidman S, Shimoni E (2001) Regulated phase transitions of bacterial chromatin: A non-enzymatic pathway for generic DNA protection. *EMBO J.* 20: 1184–2292.

Holm NG, Andersson E (2005) Hydrothermal simulation experiments as a tool for studies for the origin of life on Earth and other terrestrial planets: A review. *Astrobiology* 5(4): 444–460.

Ikehara K (2015) [GADV]-protein world hypothesis on the origin of life. *Orig. Life Evol. Biosph.* http://doi.org/10.1007/s11084-014-9383-4.

Joyce GF (2002) The antiquity of RNA-world evolution. *Nature* 418: 214–221.

Kiryukhin AV, Lesnyikh MD, Polyakov AY (2002) Natural hydrodynamic mode of the Mutnovsky geothermal reservoir and its connection with seismic activity. *Volc. Seis.* 1: 51–60.

Kolb VM (2016) *Green Organic Chemistry and Its Interdisciplinary Applications*. Boca Raton, FL: CRC Press/Taylor & Francis Group.

Kolb VM (2019) Chapter 5.5: Prebiotic reactions in water, "on water," in superheated water, solventless, and in the solid state. In *Handbook of Astrobiology*. Kolb, VM, Ed. Dordrecht: CRC Press, pp. 331–340.

Kompanichenko VN (2008) Three stages of the origin-of-life process: Bifurcation, stabilization and inversion. *Int. J. Astrobiol.* 7(1): 27–46.

Kompanichenko VN (2012) Inversion concept of the origin of life. *Orig. Life Evolut. Biosph.* 42(2–3): 153–178.

Kompanichenko VN (2017) *Thermodynamic Inversion: Origin of Living Systems.* Cham, Switzerland: Springer International Publishing.

Kompanichenko VN (2019) The rise of a habitable planet: Four required conditions for the origin of life in the universe. *Geosciences* 9: 92.

Kompanichenko VN, El-Registan G (2022) Advancement of the TI concept: Defining the origin-of-life stages based on the succession of a bacterial cell exit from anabiosis. *AIMS Geosci.* 8(3): 398–437.

Kompanichenko VN, Kotsyurbenko O (2022) Role of stress in the origin of life. *Life* 12(11): #1930.

Kompanichenko VN, Poturay VA, Shlufman KV (2015) Hydrothermal systems of Kamchatka as the model for prebiotic environment. *Orig. Life Evolut. Biosph.* 45(1–2): 93–103.

Kompanichenko VN, Shlyufman KV (2013) The amplitude-frequency function of pressure variations: steam-water mixture in the Verkhne-Mutnovskii hydrothermal system. *J. Volcanol. Seismol.* 7(5): 338–344.

Kralj P, Kralj P (2000) Thermal and mineral waters in north-eastern Slovenia. *Environ. Geol.* 39(5): 488–498.

Kurihara K, Tamura M, Shohda K, Toyota T, Suzuki K, Sugawara T (2011) Self-reproduction of supramolecular giant vesicles combined with the amplification of encapsulated DNA. *Nat. Chem.* 3: 775–781.

Lewis K (2010) Persister cells. *Annu. Rev. Microbiol.* 64: 357–372.

Liotta CL, Hallett JP, Pollet P, Eckert CA (2007) Reactions in near critical water. In *Organic Reactions in Water, Principles, Strategies and Applications.* Lindström UM, Ed. Oxford: Blackwell, pp. 256–300.

Loiko NG, Suzina NE, Soina VS, Smirnova TA, Zubasheva MV, Azizbekyan RR, et al. (2017) Biocrysteline structure in the nucleoids of the stationary and dormant prokaryotic cells. *Microbiology.* 86(6): 703–719.

Luisi PL (2000) The relevance of supramolecular chemistry for the origin of life. *Adv. Supramol. Chem.* 6: 287–307.

Manna A, Kumar A (2013) Why does water accelerate organic reactions under heterogeneous condition? *J. Phys. Chem.* 117: 2446–2454.

Martin W, Russell JM (2007) On the origin of biochemistry at an alkaline hydrothermal vent. *Philos. Trans. R Soc. B* 362: 1887–1925.

Mazur P (2004) Principles of cryobiology. In *Life in the Frozen State.* Fuller BJ, Lane N, Benson EE, Eds. Boca Raton, FL: CRC Press, pp. 3–65.

Mika JT, van der Bogaart G, Veenhoff L, et al. (2010) Molecular sieving properties of the cytoplasm of Escherichia coli and consequences of osmotic stress: Molecule diffusion and barriers in the cytoplasm. *Mol. Microbiol.* 77(1).

Oparin AI (1953) *The Origin of Life*, 2nd ed. New York: Dover Publishing.

Parry BR, Surovtsev IV, Cabeen MT, O'Hem CS, Dufresne ER, Jacobs-Wagner C (2014) Bacterial lytoplasm has glass-like properties and fluidized by metabolic activity. *Cell.* 156(1–2): 183–194.

Ross DS, Deamer D (2016) Dry/Wet cycling and the thermodynamics and kinetics of prebiotic polymer synthesis. *Life* 6: x.

Russell MJ (2021) The "water problem" (sic), the illusory pond and life's submarine emergence – A review. *Life* 11: 429.

Saladino R, Crestini C, Pino S, et al. (2012) Formamide and the origin of life. *Phys Life Rev* 9: 84–104.

Selye H (1974) *Stress Without Distress*. Philadelfia and New York: JB Lippincott Company.

Shock EL, McCollom TM, Schulte MD (1998) The emergence of metabolism from within hydrothermal systems. In *Thermophiles: The Keys to Molecular Evolution and the Origin of Life*. Wiegel J, Adams MWW, Eds. Washington, DC: Taylor & Francis Group, pp. 59–76.

Sugawara T, Kurihara K, Suzuki K (2012) Constructive approach toward protocells. In *Engineering of chemical complexity*. Mikhailov A, Ed. World Scientific Review, pp. 1–17.

Wächtershäuser G (1988) Before enzymes and templates: Theory of surface metabolism. *Microbiol Rev* 52: 452–484.

Washington J (2000) The possible role of volcanic aquifers in prebiotic genesis of organic compounds and RNA. *Orig. Life Evol. Biosph.* 30: 53–79.

Woese CR (1987) Microbial evolution. *Microbiol. Rev.* 51: 221–270.

Woese CR (1998) The universal ancestor. *Proc. Nat. Acad. Sci.* 95: 6854–6859. https://doi.org/10.1073/pnas.95.12.6854

# 9 Systems Biology

*Iman Tavassoly*

## 9.1 HISTORICAL PERSPECTIVES ON SYSTEMS BIOLOGY

In the field of medicine, holistic approaches to health and disease have been foundational since the early days, even before the understanding of modern biology and biological mechanisms. The initial attempts to explain physiology and pathophysiology exhibited holistic insights (Bakhtiar, Gruner et al. 2014). In these ancient traditions, the focus was on perceiving each patient as a whole being and taking into account numerous factors and their intricate interactions that contribute to overall health and well-being.

Through the emergence of modern physiology, the systems approach continued to play a vital role in defining and understanding health and disease. Organ systems such as the cardiovascular and nervous systems were investigated to comprehend their functions. However, a paradigm shift occurred with the advent of molecular biology, which focused on unraveling the building blocks of these organ systems, such as tissues, cells, and intercellular and genetic components. While the reductionist approach yielded a wealth of information about the constituents of living systems, a comprehensive understanding of biological systems necessitated the study of their functional aspects, which arise from the temporal and spatial interactions of all components. Systems biology was the solution that bridged the gap between vast amounts of data on the building blocks of biological systems and their emergent phenotypes in time and space.

Although the concept of systems-level methods applicable in biology was initially introduced by Claude Bernard in 1865, but the term "systems biology" was first coined by Mihajlo D. Mesarović in 1968, paving the way for further development of the field by contemporary scientists (Mesarović 1968; Kitano 2002; Hood 2003; Noble 2008). With the application of advanced computational capabilities in the late 20th and 21st centuries, systems biology evolved into a mainstream interdisciplinary scientific area, rejuvenating the holistic medicine viewpoint under the new names of "systems medicine" and "precision medicine" (Tavassoly, Goldfarb et al. 2018). Recent years have witnessed the removal of barriers to solving complex computational and mathematical problems in systems biology, thanks to high performance and cloud computing. Nonetheless, the field remains in a state of constant evolution, with the development of novel computational technologies, including quantum computing and Artificial Intelligence (AI), pushing its boundaries further.

## 9.2 COMPLEX BIOLOGICAL SYSTEMS

The cell functions as a complex computational unit, responding to environmental signals in various ways (Bray 1995). It possesses a sophisticated network of

DOI: 10.1201/9781003294276-9

molecular interactions that act as a control system. The behaviors and responses of cells are nonlinear and dynamic, emerging from interconnected networks comprising genes, RNAs, proteins, and organelles. Adaptation and robustness are two critical characteristics of cellular functions, which contribute to the overall cell phenotype. Moreover, the presence of diverse feedback loops within cellular regulatory networks plays a vital role in generating emergent properties observed in cellular phenotypes (Alon 2019).

This complexity and nonlinearity extend to larger scales, including tissues, organs, and the human body. Interactions occur between cells, within cells, among tissues, within tissues, among organs, within organs, and between the human body and its environment. Understanding these intricate dynamics necessitates a nonlinear and systems-level approach.

To comprehend the interactions and dynamics of these systems, a holistic perspective is required, recognizing the interconnectedness of components, the emergence of properties, and the sensitivity to initial conditions. This systems-level approach accounts for the complexity and nonlinearity present at various scales, providing a comprehensive understanding of biological phenomena from the cellular level to the entire human organism.

## 9.3 BIOLOGICAL SYSTEMS IN TIME AND SPACE

At the cellular level, the conversion of genomic data into the phenome (encompassing all possible and existing phenotypes) is a dynamic process. The interaction networks within cells undergo rewiring and disturbances, triggered by environmental factors and genomic aberrations, leading to a diseased state. To restore cells to a healthy state (physiological condition), therapeutic interventions are necessary. It is important to consider that disease is a process rather than a static state, and, therefore, the timing of these interventions is crucial as the disease process may progress to an irreversible state (Figure 9.1).

The understanding of health and disease requires a comprehensive systems-level perspective, encompassing the complex interactions between various regulatory systems at different scales. These systems, represented by complex interaction networks, can be perturbed by environmental factors and/or genomic aberrations, leading to a transition from a physiological state to a pathological one, commonly referred to as "disease". In this context, all therapeutic interventions aim to restore the system to a state of "health" by influencing the network structures or dynamics. By targeting the nodes or pathways within these networks, therapeutic strategies aim to reverse or counteract the perturbations that drive the disease state. Through interventions that modulate the network dynamics, such as altering signaling cascades or restoring cellular homeostasis, the goal is to promote new network dynamics that align with a healthy state. By adopting a systems-level approach, researchers and clinicians gain a deeper understanding of the complex interplay among environmental factors, genomic determinants, and phenotypes. This perspective facilitates the development of precise and tailored therapeutic approaches that address the underlying causes of diseases. Ultimately, by focusing on restoring the system's equilibrium and promoting a return to the state of "health", we can make significant progresses in precision medicine.

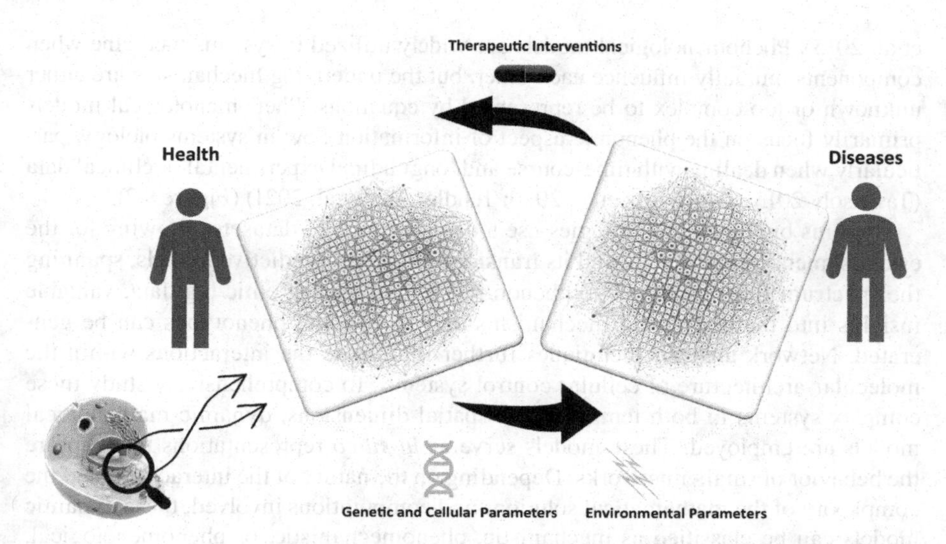


**FIGURE 9.1** A systems-level perspective on health and disease and understanding transitions between physiological and pathological states.

## 9.4 SYSTEMS BIOLOGY METHODOLOGIES

Systems biology is an interdisciplinary field that employs computational sciences, mathematics, cell and molecular biology, and physics to tackle the complexities of biological systems across various scales. Omics studies, such as genomics, transcriptomics, metabolomics, epigenomics, and lipidomics, generate vast datasets that serve as a valuable resource for knowledge discovery. Methodologies like differential expression analysis and pathway analysis enable the understanding of molecular changes on a large scale, identifying crucial components that contribute to different phenotypes and transitions between them (Tavassoly, Goldfarb et al. 2018).

This information forms the foundation for constructing interaction networks, such as protein–protein interaction networks (PPIs), which map the connectivity of these elements. Static analysis of these networks provides insights, but to comprehend their temporal dynamics, they need to be translated into mathematical equations and dynamic models, often involving ordinary differential equations (ODEs). However, due to the computational complexity and cost of translating large networks into mathematical formalisms, smaller networks are often extracted for *in silico* study through mathematical modeling. Different modalities can be utilized depending on the knowledge of interaction mechanisms within these networks (Alon 2019; Dorvash, Farahmandnia et al. 2020).

The first modality involves mechanistic models, where the mechanisms are well-known, allowing for a straightforward translation of interactions and elements into mathematical equations (Dorvash, Farahmandnia et al. 2019). In cases where some mechanisms are unclear, a hybrid approach can be employed to construct phenomechanistic models. These models incorporate elements with known activation or inhibition relationships, while the nature of these activities remains unknown (Tavassoly, Parmar et al. 2015). Phenomenological models are widely utilized in systems medicine when components mutually influence each other, but the underlying mechanisms are either unknown or too complex to be represented by equations. Phenomenological models primarily focus on the phenomic aspect of information flow in systems biology, particularly when dealing with time-course and longitudinal experimental or clinical data (Tavassoly 2015; Karin, Raz et al. 2020; Tendler, Bar et al. 2021) (Figure 9.2).

Systems biology methodologies use a diverse range of datasets, allowing for the establishment of knowledge and its transformation into predictive models, spanning the spectrum from genomics to phenomics. By harnessing omic big data, valuable insights into the underlying mechanisms driving specific phenotypes can be generated. Network analysis techniques further illuminate the interactions within the molecular architecture of cellular control systems. To comprehensively study these complex systems in both temporal and spatial dimensions, dynamic mathematical models are employed. These models serve as *in silico* representations that capture the behavior of smaller networks. Depending on the nature of the interactions and the complexity of the mathematical solutions and computations involved, these dynamic models can be classified as mechanistic, phenomechanistic, or phenomenological. As we delve deeper into the realm of phenomics, the datasets become more phenomenological, primarily comprising time-course and temporal data. However, systems biology methods allow for the conversion of these phenotypic time-course data into phenomenological models. Moreover, these models can be expanded to incorporate other types of models using diverse mathematical techniques, such as parameter sensitivity analysis and network analysis. This integration of methodologies provides a deeper understanding of the underlying molecular signatures associated with specific phenotypes. By employing systems biology approaches, researchers can bridge the gap between genomics and phenomics, capturing the dynamic nature of biological systems and revealing the intricate relationships that drive phenotypic outcomes.

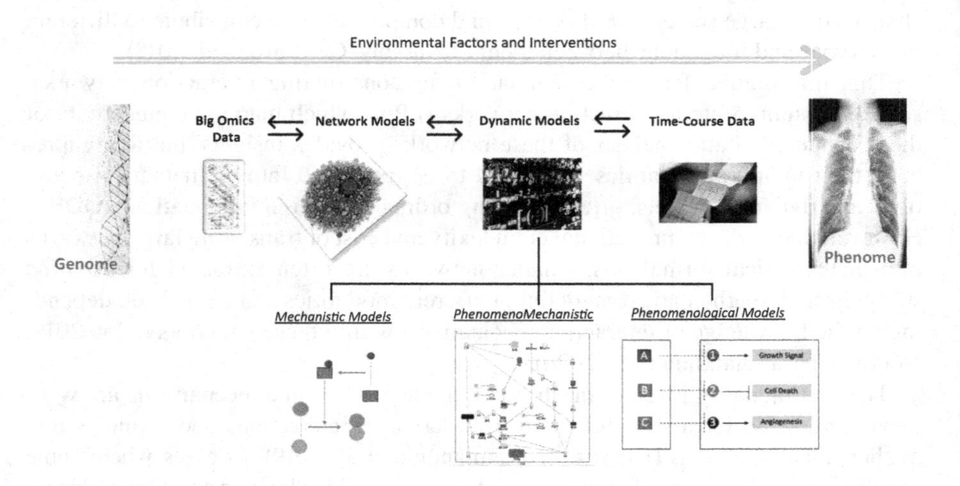


**FIGURE 9.2** Leveraging systems biology methodologies and integrating data types for knowledge extraction and predictive modeling.

## 9.5 SYSTEMS BIOLOGY APPLICATIONS

Systems biology offers a powerful framework for creating precise and accurate models that drive the design and definition of novel diagnostic and therapeutic approaches, with the ultimate goal of transitioning from a diseased state to a state of health. This encompasses various applications, including quantitative patient monitoring, early disease detection, drug discovery, drug repurposing, and personalized treatment selection (Hamburg and Collins 2010; Crowley, Di Nicolantonio et al. 2013; Tavassoly, Hu et al. 2019; Tavassoly, Barbieri et al. 2020; Mehrpooya, Saberi-Movahed et al. 2022). The insights gained from systems biology studies play a crucial role in transforming medicine from a qualitative discipline to a quantitative one.

The advancements in systems biology have laid the foundation for the field of "personalized and precision medicine". This emerging field aims to leverage molecular and clinical data to administer the right treatment at the right time, with the right dosage, specifically tailored to each individual patient. By harnessing the principles of systems biology, personalized and precision medicine seeks to optimize patient care and improve outcomes by delivering tailored interventions based on a deep understanding of each patient's unique characteristics (Hood 2013).

Overall, systems biology serves as a key driver in revolutionizing medical practices, enabling the shift toward a more quantitative and individualized approach to healthcare. It empowers clinicians and researchers to leverage comprehensive molecular and clinical data, paving the way for personalized and precise interventions that hold the promise of enhancing patient well-being (Figure 9.3).

As shown in this schematic figure as an illustrative example in the field of immuno-oncology, to address immuno-oncology challenges effectively, a comprehensive

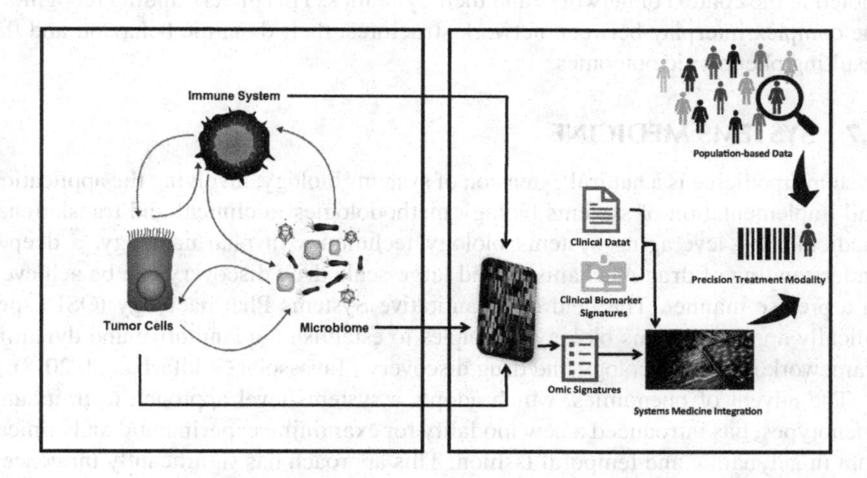


**FIGURE 9.3** Systems medicine modality and data integration: Systems medicine leverages the methods and findings of systems biology to develop diagnostic, monitoring, and therapeutic platforms for precision medicine in a personalized manner.

collection of molecular datasets from the tumor, immune system, and microbiome is necessary. These datasets are then combined with clinical biomarkers such as blood biomarkers and relevant clinical data. By employing AI-based tools and integrating the principles of systems biology, data from clinical trials and diverse patient populations can be amalgamated to extract precise and personalized treatment strategies tailored to the unique needs of individual patients. The core objective of this approach is to administer the right treatment to the right patient at the right time and in the correct dosage.

## 9.6 FROM MOLECULES AND NETWORKS TO PHYSIOLOGICAL AND PATHOLOGICAL PHENOTYPES

A disease is a dynamic process characterized by its trajectory and evolving nature. Within different organ systems, the transition from health to disease can exhibit multi-stable characteristics, indicating that at specific times and with appropriate treatments, it is possible to recover health and halt disease progression (Mackey and Milton 1987).

The functioning of cells relies on the dynamics of their interconnected networks, which can be studied using mathematical models. Alterations in the structure of these networks, such as the addition or removal of nodes and edges, can surpass the robustness threshold of the networks, leading to a transition into a disease state. Simultaneously, the nonlinear ordinary differential equations (ODEs) that describe the dynamics of these networks are highly complex, with an expansive parameter space. Only specific sets of parameters within this space maintains the system's robustness. Any intrinsic or extrinsic signal that modifies this set of parameters can also trigger a disease state (Tyson, Chen et al. 2001; Tavassoly, Parmar et al. 2015).

In the field of systems biology, the definition of phenotype as health and disease is rooted in the context of networks and their dynamics. This understanding recognizes the complex interplay between network structures, their dynamic behavior, and the resulting phenotypic outcomes.

## 9.7 SYSTEMS MEDICINE

Systems medicine is a natural extension of systems biology, involving the application and implementation of systems biology methodologies in clinical and translational medicine. By leveraging systems biology techniques in pharmacology, a deeper understanding of drug mechanisms and large-scale drug discovery can be achieved in a precise manner. The field of Quantitative Systems Pharmacology (QSP) specifically applies systems biology principles to establish a quantitative and dynamic framework in pharmacology and drug discovery (Tavassoly, Goldfarb et al. 2018).

The advent of phenomics, which adopts a systems-level approach to traits and phenotypes, has introduced a new modality for examining experimental and clinical data in a dynamic and temporal fashion. This approach has significantly influenced both translational and clinical medicine (Fang, Chen et al. 2023). Viewing diseases as dynamic processes has paved the way for precision and personalized medicine, revolutionizing medical research and practice. In this paradigm, each patient is regarded as a unique individual progressing through time, and real-time monitoring is employed to track disease progression at different stages. Tailored therapeutic interventions specific to each patient's condition are implemented at the appropriate time to address various disease states.

Systems biology, alongside its counterpart systems medicine, represents crucial advancements toward the utilization of real-time, quantitative, and accurate diagnostic and therapeutic tools in clinical practice. These interdisciplinary approaches are essential for navigating the new era of healthcare, where personalized and data-driven strategies play a fundamental role in improving patient outcomes.

## 9.8 FUTURE DIRECTIONS IN SYSTEMS BIOLOGY

One of the fundamental challenges in systems biology lies in constructing a virtual cell system capable of emulating the behavior of a real cell, albeit in a computer simulation (*in silico*). When it comes to astrobiology, the development of such a platform helps to understand living systems that can advance the astrobiology by providing a system to analyze human cell biology and physiology in the space (Kolb, Chapter 12 in this book). This endeavor necessitates the development of a mathematical model that integrates all components of a cell into a cohesive framework. Such a model entails a vast parameter space and a complex multidimensional parameter landscape, demanding more powerful computational resources. Recent advancements in quantum computing offer a promising avenue to explore for enhanced computational capabilities in this pursuit (Emani, Warrell et al. 2021; Davids, Lidströmer et al. 2022). Additionally, selecting the appropriate mathematical formalism to accurately represent and simulate the *in silico* model is a crucial concern that warrants careful consideration.

Furthermore, the ability to measure these parameters or design quantitative experiments for extracting the necessary data to inform the model poses a significant challenge. It is a complex problem that requires advancements in quantitative experimental biology to be overcome. Addressing this challenge will be instrumental in refining and calibrating the mathematical model, ensuring its accuracy and relevance.

Overall, the development of a comprehensive virtual cell system and the associated mathematical model represents a frontier in systems biology, necessitating the convergence of computational advancements, mathematical expertise, and innovative experimental techniques.

## ACKNOWLEDGMENT

Iman Tavassoly contributed to this chapter at the Cellular Energetics Program of the Kavli Institute for Theoretical Physics (KITP), Santa Barbara, United States, supported in part by the National Science Foundation Grant NSF PHY-1748958, NIH Grant R25GM067110, and the Gordon and Betty Moore Foundation Grant 2919.02. He currently works as a physician-scientist in Manhattan, New York, United States.

## REFERENCES

Alon, U. (2019). *An Introduction to Systems Biology: Design Principles of Biological Circuits*, CRC Press.

Auffray, C., et al. (2009). "Systems medicine: The future of medical genomics and healthcare." *Genome Medicine* **1**: 1–11.

Bakhtiar, L., et al. (2014). *The Canon of Medicine (al-Qānūn Fī'l-ṭibb)*, Great Books of the Islamic World.

Bray, D. (1995). "Protein molecules as computational elements in living cells." *Nature* **376**(6538): 307–312.

Crowley, E., et al. (2013). "Liquid biopsy: Monitoring cancer-genetics in the blood." *Nature Reviews Clinical Oncology* **10**(8): 472–484.

Davids, J., et al. (2022). "Artificial intelligence in medicine using quantum computing in the future of healthcare." *Artificial Intelligence in Medicine*, Springer: 423–446.

Dorvash, M., et al. (2019). "Dynamic modeling of signal transduction by mTOR complexes in cancer." *Journal of Theoretical Biology* **483**: 109992.

Dorvash, M., et al. (2020). "A systems biology roadmap to decode mTOR control system in cancer." *Interdisciplinary Sciences: Computational Life Sciences* **12**: 1–11.

Emani, P. S., et al. (2021). "Quantum computing at the frontiers of biological sciences." *Nature Methods* **18**(7): 701–709.

Fang, H., et al. (2023). "Translational phenomics and its applications in immunotherapy." *Frontiers Media SA* **14**: 1211704.

Hamburg, M. A. and F. S. Collins (2010). "The path to personalized medicine." *New England Journal of Medicine* **363**(4): 301–304.

Hood, L. (2003). "Systems biology: Integrating technology, biology, and computation." *Mechanisms of Ageing and Development* **124**(1): 9–16.

Hood, L. (2013). "Systems biology and p4 medicine: Past, present, and future." *Rambam Maimonides Medical Journal* **4**(2).

Ideker, T., et al. (2001). "A new approach to decoding life: Systems biology." *Annual Review of Genomics and Human Genetics* **2**(1): 343–372.

Karin, O., et al. (2020). "A new model for the HPA axis explains dysregulation of stress hormones on the timescale of weeks." *Molecular Systems Biology* **16**(7): e9510.

Kitano, H. (2002a). "Computational systems biology." *Nature* **420**(6912): 206–210.

Kitano, H. (2002b). "Systems biology: A brief overview." *Science* **295**(5560): 1662–1664.

Kolb, V.M. (2024 in press). "Knowledge Gaps in Astrobiology: Can Systems Approach Help Elucidate Them?" Chapter 12 in the Guidebook for Systems Applications in Astrobiology, Kolb, V M., Editor, CRC Press, Boca Raton, Florida, pp. 223–238.

Mackey, M. C. and J. G. Milton (1987). "Dynamical diseases." *Neuro – Ophthalmology* **21**: 24.

Mehrpooya, A., et al. (2022). "High dimensionality reduction by matrix factorization for systems pharmacology." *Briefings in Bioinformatics* **23**(1): bbab410.

Mesarović, M. D. (1968). *Systems Theory and Biology – View of a Theoretician*. Systems Theory and Biology: Proceedings of the III Systems Symposium at Case Institute of Technology, Springer.

Noble, D. (2008). "Claude Bernard, the first systems biologist, and the future of physiology." *Experimental Physiology* **93**(1): 16–26.

Tavassoly, I. (2015). *Dynamics of Cell Fate Decision Mediated by the Interplay of Autophagy and Apoptosis in Cancer Cells: Mathematical Modeling and Experimental Observations*, Springer.

Tavassoly, I., et al. (2015). "Dynamic modeling of the interaction between autophagy and apoptosis in mammalian cells." *CPT: Pharmacometrics & Systems Pharmacology* **4**(4): 263–272.

Autophagy is a conserved biological stress response in mammalian cells that is responsible for clearing damaged proteins and organelles from the cytoplasm and recycling their contents via the lysosomal pathway. In cases of mild stress, autophagy acts as a survival mechanism, while in cases of severe stress cells may switch to programmed cell death. Understanding the decision process that moves a cell from autophagy to apoptosis is important since abnormal regulation of autophagy occurs in many diseases, including cancer. To integrate existing knowledge about this decision process into a rigorous, analytical framework, we built a mathematical model of cell fate decisions mediated by autophagy. Our dynamical model is consistent with existing quantitative measurements of autophagy and apoptosis in rat kidney proximal tubular cells responding to cisplatin-induced stress.

Tavassoly, I., et al. (2018). "Systems biology primer: The basic methods and approaches." *Essays in Biochemistry* **62**(4): 487–500.

Tavassoly, I., et al. (2019). "Genomic signatures defining responsiveness to allopurinol and combination therapy for lung cancer identified by systems therapeutics analyses." *Molecular Oncology* **13**(8): 1725–1743.

Tavassoly, I., et al. (2020). "A tissue-and organ-based cell biological atlas of obesity-related human genes and cellular pathways." *bioRxiv*. http://doi.org/10.1101/2020.03.16.993824.

Tendler, A., et al. (2021). "Hormone seasonality in medical records suggests circannual endocrine circuits." *Proceedings of the National Academy of Sciences* **118**(7): e2003926118.

Tyson, J. J., et al. (2001). "Network dynamics and cell physiology." *Nature Reviews Molecular Cell Biology* **2**(12): 908–916.

Tyson, J. J., et al. (2011). "Dynamic modelling of oestrogen signalling and cell fate in breast cancer cells." *Nature Reviews Cancer* **11**(7): 523–532.

Cancers of the breast and other tissues arise from aberrant decision-making by cells regarding their survival or death, proliferation or quiescence, damage repair or bypass. These decisions are made by molecular signalling networks that process information from outside and from within the breast cancer cell and initiate responses that determine the cell's survival and reproduction. Because the molecular logic of these circuits is difficult to comprehend by intuitive reasoning alone, we present some preliminary mathematical models of the basic decision circuits in breast cancer cells that may aid our understanding of their susceptibility or resistance to endocrine therapy.

# 10 Systems Approach to Microbial Communities and Its Application to Astrobiology

*Oleg R. Kotsyurbenko, D.A. Skladnev and S. Jheeta*

## 10.1 BACKGROUND

We first provide the basic principles of systems approach toward study of biological systems, including the historical background in Section 10.2. Then, we address the systems approach to microbiology in Section 10.3. This will provide a background to the functioning of biological systems of different level and further analysis of issues in the framework of astrobiology. The modern stage of the development of science in general and biology in particular, is characterized by a systematic approach to the evaluation of various phenomena. In the concept of hierarchical holism, which dominates the systems approach, various biological systems form a hierarchical structure in which an element of one system is an independent system of a lower level. In any individual system, the key points are the interaction of its components and the structure that determines the stability of the system.

The summary of the chapter is as follows:

The highest biological system on Earth is its biosphere in which microorganisms play a key role. They are responsible for the implementation of global biogeochemical cycles of various elements. The biosphere includes various ecosystems in which microorganisms are involved in various processes for the transformation of matter. One such ecosystem is the wetlands of cold regions, where methane, an important greenhouse gas in the concept of global climate change and atmospheric composition, is microbiologically produced. Methane is produced by an anaerobic microbial community, which is a complex system with close trophic interactions. The key terminal microbial groups are the methanogenic archaea, which are one of the most ancient microbial groups, and the acetyl-CoA pathway they use is considered to be oldest metabolic process. The system hierarchy in biology can be effectively used for astrobiological tasks in the context of searching for various kinds of biosignatures, where each biosystem has its own bioindicators and subsequent detection methods.

 DOI: 10.1201/9781003294276-10

## 10.2 BASIC PRINCIPLES OF SYSTEMS APPROACH TOWARD THE STUDY OF BIOLOGICAL SYSTEMS

The idea to consider the object under study as a whole system is mentioned in ancient philosophy and science, for example in Anaxagoras' naturphilosophic theory of the indestructible elements, stating that all things emerge due to the combination of the qualities of these elements (c. 500 BC–428 BC), or in Aristotle's systemic worldview of space and biological expediency (384 BC–322 BC). Subsequently, in the Middle Ages, the idea of a system organization of knowledge was suggested and then most developed in German classical philosophy and, in particular, in the ideas relating to the system interrelationship of animate and inanimate nature of I. Kant (1724–1804) and the necessity for subordination and coordination in the concept of I.G. Lambert (1728–1777).

In biology, the first ideas of system organization were associated with the views of Claudius Galen (ca. 130–ca. 217), who postulated an alternative systemic function of arterial and venous blood in the human body. Much later, attempts to systematize nature as a whole and to study the influence of environmental factors on organisms were related to the works of C. Linnaeus (1707–1778), J.-B. Lamarck (1744–1829), C. Darwin (1809–1882), and I.V. Vernadsky (1863–1945).

In the early 1920s, an Austrian biologist, namely Ludwig von Bertalanffy (1901–1972) began to study living organisms as systems and summarized his findings in the book entitled: "*Modern Theories of Development*" (1929). It is Bertalanffy who is considered to be the developer, by this token the founder, of a systems approach to the study of biological organisms, although he had several noted predecessors, in particular, the Russian scientist A.A. Bogdanov (1873–1928), who was developing the theory of organization.

In the 1930s, Bertalanffy created a general theory of systems – a scientific and methodological concept for the study of objects, including living organisms, which were considered as systems that constantly exchange matter and energy with the environment. In the framework of this theory, a system is a set of interacting elements, which in turn can be a system of a lower sub-level. The structure of the system is a set of stable relationships between elements. The properties of an object as an integral system are determined by the special system-forming emergent properties of its structure, rather than simple summation of the properties of its individual elements (Glagolev and Fastovets, 2012).

Further development of systems theory within the framework of biology led to the emergence of *systems biology*. Currently, this scientific direction has gained a powerful experimental base with the development of various omix technologies in biology and the accumulation of a huge amount of data that requires processing and systematization.

In general, the systems approach is focused on studying the object as an integrated system, on identifying its elements and types of interactions between them and eventually on summarizing the results in a single theoretical concept. The laws structural similarities established in various fields of science and practice allows identifying system-wide patterns that are the basis of a systems approach.

## 10.3 SYSTEMS APPROACH TO MICROBIOLOGY, INCLUDING HISTORICAL BACKGROUND

Microbiology is a discipline that is at the core of all life sciences as microbiology is the study of microscopic organisms (i.e., microorganisms); further, it is the active interaction of non-living matter which is considered by other related disciplines, such as ecology and biogeochemistry to be important.

In relation to microbial ecology and biogeochemistry, it is necessary to note the outstanding works of the Russian academician G.A. Zavarzin (1933–2011), who proactively used the basic principles of systems approach in the study of microbial communities and their role in global ecology (Zavarzin, 2011). Microorganisms associated in communities based on trophic interactions are complex biological systems involved in various biogeochemical cycles of elements on the Earth.

The application of systems approach to the study of biological systems is associated with the development of the concept of biological organization levels. For the most comprehensive analysis of the system, it is first necessary to determine its hierarchical level and its relationships with other systems. It is generally accepted to assign the whole organism to the "organism level" as well as subsystems (e.g., cells) to the suborganismal level, and the population-species organization to supraorganismal levels. However, when considering microbial systems, the researchers mostly deal with a single unicellular organism. In this case a cell, for example, a bacterial cell as a unit of living matter is considered at the "organism level". Then systems of higher and lower (e.g., cellular structures) levels belong to supraorganismic and suborganismic levels, respectively. Finally, the interaction of various systems characterizing the biological organization levels, as well as their relationship with the environment is very close. This means that a comprehensive study of any systemic phenomenon requires determining its position in the hierarchical rank of subordinate systems and determining the main intersystem and intrasystem interactions.

The most accurate theoretical concept of this approach is hierarchical holism (Zavarzin, 1995a), which has been created in modern natural science and, as applied to biology, claims that biological phenomena or processes in nature are combined into systems, each of which can in turn be an element of another system of a higher rank, and the precision in describing the studied object is dependent primarily on the correct identification of the system in which this object operates. An incorrect characterization of the system under study leads to false conclusions.

The variety of levels of the structural biological organization specifies the differentiation of biology in which each such level is studied in more detail by the corresponding biological science. Additionally, a systems approach promotes the integration of individual biological disciplines, offering a more general view of various phenomena that can only be understood by their coordinated investigation in various biological disciplines.

### 10.3.1 Global Biogeochemical Cycles and the Role of Microorganisms

The system of the highest level of organization associated with the vital activity of the living organisms of the planet is its biosphere.

On Earth, microorganisms are involved in a variety of elemental cycles and are directly or indirectly responsible for the formation of various rocks on the planet's surface (Figure 10.1).

The close connection of biological and geological processes makes it possible to refer them as a geospheric–biospheric system in which cyclicality is the most important condition for its stability.

The organic carbon cycle is the leading cycle, at the first stage of which the photoautotrophic assimilation of carbon dioxide ($CO_2 + H_2O = [CH_2O] + O_2$) and conjugation with the oxygen and carbon dioxide cycles occur. At the second stage, $O_2$ is removed as a result of oxygen respiration ($[CH_2O] + O_2 = CO_2 + H_2O$). However, the destruction of the substance turns out to be incomplete, which leads to the disruption of the cycle and the burial of part of the organic carbon in the form of kerogen, the main reservoir of reduced carbon on Earth. At the same time, the imbalance of the carbon cycle leads to an excess accumulation of oxygen with its subsequent sink into oxidized forms of iron and sulfates of the world ocean. Accordingly, a certain part of the carbon dioxide generated during the microbial fermentation removed as deposits of carbonates – limestones and dolomites, together with clays as a result of chemical weathering (Zavarzin, 2002).

Such non-closure of cycles leads to biogeochemical succession, which is the main driving force behind the evolution of the geospheric–biospheric system with the key process of sedimentogenesis.

As a result of the biogeochemical succession of 2.5 billion years, the composition of the modern atmosphere was formed, which is closely related to the biogeochemical cycles of elements (Zavarzin, 2001). All major components of the atmosphere ($O_2$, $CO_2$, $CH_4$, CO, $H_2$, $N_2$, NOx, $NH_3$, and sulfur species) are intermediate products of various metabolic reactions carried out by microorganisms.

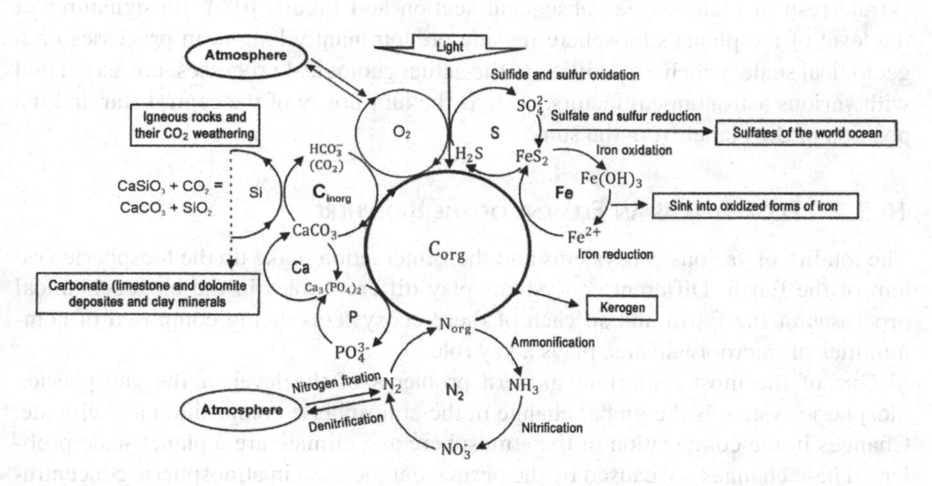


**FIGURE 10.1** Biogeochemical Corg-centered cycles and various groups of microorganisms involved in them. The imbalance in cycles leads to biogeochemical successions with sedimentogenesis.

*Source:* Modified from Zavarzin (2004a)

The cycle of organic carbon can be completely carried out by the cyanobacterial community, as it was in the conditions of ancient Earth, when eukaryotic organisms did not exist. Cyanobacterial communities created a fully-fledged biosphere of the modern geochemical type, which functioned for quite a long time – two to three billion years, which served as the basis for the sustainable biological evolution and development of various life forms on the planet.

Under modern conditions in the biosphere, phototrophic algae and protists play the role of primary producers both in aquatic environments and vascular plants on land. The role of destructors in the sea is executed by invertebrates, and on land by fungi. However, the cycles of nitrogen, sulfur, and iron are completely under the control of prokaryotes, namely bacteria and archaea. Thus, it is important to note that the biogeochemical system of the planet was undertaken by prokaryotes. Their functional diversity is quite sufficient to catalyze the biogeochemical "machineries" of the planet.

Biogeochemical cycles are the main system for converting solar energy into reduced organic compounds needed by living entities for their survival strategies. The relationship of biogeochemical cycles makes the survival strategies of microorganisms in the biosphere the most fundamental (Zavarzin, 2008).

The active involvement of living organisms in biogeochemical control, which determines the geology of the planet, was due to the work of an outstanding scientist of the beginning of the 20th century, namely V.I. Vernadsky (Marov, 2013). Manifestations of such activities on the global scale—for example, in the form of specific geological features of the surface of presumably biogenic origin, pigmentation of its individual areas, or violation of thermodynamic equilibrium in the atmosphere, necessary for the implementation of redox reactions by organisms to obtain energy, can be considered as the most important biosignatures in the search for life on other extraterrestrial planets (see subsequent section and Figure 10.6). Biosignatures at the level of the planet's biosphere include predominantly long-term processes on a geological scale, which, in addition to the actual geological processes, are associated with various astronomical factors, such as the luminosity of the central star and the position of the planet from the sun.

### 10.3.2 Ecosystem as an Element of the Biosphere

The totality of various ecosystems and their interaction make up the biospheric system of the Earth. Different ecosystems play different roles in the biogeochemical processes of the Earth and so each of these ecosystems, being composed of communities of microorganisms, plays a key role.

One of the most important modern problems of the level of the geospheric–biospheric system is the global change in the atmospheric composition and climate. Changes in the composition of the atmosphere and climate are a planet-scale problem. These changes are caused by the permanent increase in atmospheric concentration of the main greenhouse gases including methane and $CO_2$ (IPCC, 2013). Along with an increase in the concentration of such gases, the rise of the average global temperature on Earth is observed at a rate of about 0.15°C per decade. Such a temperature rise can subsequently have a strong impact on various regions of Earth.

These impacts can be the melting of permafrost and polar glaciers, an increase in the frequency of extreme weather events, and global changes in climatic zones.

The most important greenhouse gas, carbon dioxide, is the result of predominantly anthropogenic activity.

Methane is the second-most important greenhouse gas after $CO_2$. Despite the fact that the concentration of methane in the atmosphere is much lower than that of carbon dioxide, the $CH_4$ molecule has a global warming potential (GWP) that is 28 times stronger when calculated over the past 100 years, and 84 times stronger for a 20 years perspective than the GWP of the $CO_2$ molecule (IPCC, 2014, p. 87).

Another important feature of methane is its primarily biological origin. In total, the microbial communities of various ecosystems generate 85–90% of methane passing to the atmosphere (Ehhalt and Schmidt, 1978). Methane is released in the atmosphere unequally. Its basic amount was determined to be emitted by the ecosystems of northern latitudes and more precisely by wetland ecosystems located in the boreal climate zone (Matthews and Fung, 1987; Cicerone and Oremland, 1988; IPCC, 2013, pp. 505–510; Kotsyurbenko et al., 2019). Boreal wetlands are defined as natural ecosystems being influenced by rather low average annual temperatures with their strong fluctuations on the surface in summer and winter seasons and with more stable temperatures in the deep layers (Whalen and Reeburgh, 2000).

Methane is formed in such ecosystems due to the activity of microorganisms. Its release into the atmosphere is the result of the incompleteness of the microbial cycle of methane, which is also called the Söhngen cycle (Zavarzin, 1995b) (Figure 10.2).

The methane cycle is driven by the interaction of methanogenic and methanotrophic microbial communities (Zavarzin, 1995b; Kallistova et al., 2017), which are clearly spatially separated depending on the levels of oxygen present in the environment. Microorganisms involved in the formation of methane are anaerobes. Methanogenesis cannot happen in the presence of oxygen. In contrast, microbial oxidation of methane occurs with the participation of oxygen. In the water layer, the concentration of dissolved oxygen decreases and anaerobic conditions are created at a certain depth under which the methanogenic community can operate. Methane is formed as a result of anaerobic degradation of organic matter and then enters the upper layers, where it is oxidized by aerobic methanotrophs in the presence of oxygen. Methanotrophic microorganisms are the main component of the so-called bacterial filter, a community of aerobic bacteria that trap out the products of activity of anaerobic microorganisms coming from the lower layers. If all the methane formed in the anaerobic zone is trapped by the bacterial filter, then the methane cycle is closed. However, in many types of wetland ecosystems, part of the methane passes through a bacterial filter and releases into the atmosphere, where it can then participate in various photochemical reactions and contribute to the greenhouse effect.

Photochemical oxidation depends on solar radiation and the contribution of photochemically active gases, primarily nitrogen oxides, and volatile organic compounds such as terpenes. This reaction system produces tropospheric ozone and carbon monoxide.

Thus, the microbial methane cycle is a complex system in which trophic interactions between its two main components (methanogenic and methanotrophic microbial communities) occur (Zavarzin, 1995c; Conrad, 2007). The activity of these

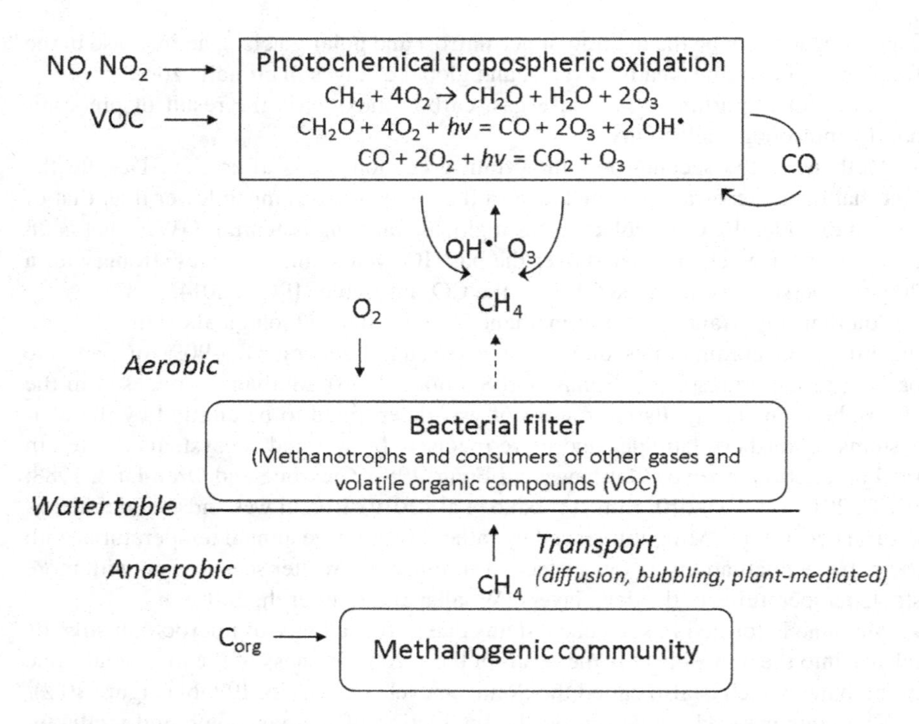


**FIGURE 10.2** Open cycle of methane in the bog ecosystem and its subsequent photooxidation in the atmosphere.
*Source:* Modified from Zavarzin (1997)

communities is regulated by various environmental factors such as temperature, pH, Eh, oxygen and mineral elements' concentrations, and availability of organic substrates. Since these communities are spatially separated, their regulation by the aforementioned factors can occur to a large extent independently. Specifically, temperature conditions directly impacting the rate of metabolic processes can be substantially different on the surface and in the deep peat layers, and these differences can vary depending on the season. Accordingly, the result of the methane cycle operation should be also be variable.

In extensive wetlands areas, the environmental conditions are favorable to anaerobic methane formation, and cold conditions impede the activity of bacteria in the aerobic filter that facilitates the emission of methane into the atmosphere.

Due to the spatial separation of two microbial communities, the process of delivery of the substrate (methane) from one community to another has a certain specificity. Methane enters the upper layers in three main ways: (1) by diffusion, (2) with gas bubbles, or (3) due to transport mechanism associated with plants (Lansdown et al., 1992; Kutzbach et al., 2004).

Thus, the methane cycle is a complex system of interaction between microbial communities functioning in different spatial layers and is regulated by a large number of factors (Conrad, 1996; Limpens et al., 2008). In addition to methane, a number of other greenhouse gases enter the atmosphere, as well as a variety of volatile organic compounds that are involved in various photooxidation reactions. Taken together, the aforementioned processes constitute a complex system of methane transformations in which its main source is the anaerobic methanogenic community of microorganisms.

The biological cycle of methane is carried out only by prokaryotic organisms. The predominantly biological origin of atmospheric methane makes it an important biomarker in astrobiological research. Since methane is actively involved in photochemical transformations in the atmosphere and can be completely oxidized, its stable presence there can be associated with its continuous supply from the surface of the planet, where it can be formed as a result of biological processes.

### 10.3.3 Microbial Community

The most important element of every ecosystem is the microbial community.

The anaerobic microbial community producing methane is an excellent example of a complex biological system, where all its components are closely related to each other by trophic interactions. Complex organic substances decompose in such a community at stages that the products of one microbial group are substrates of another one. As a result, the total energy of degraded complex compounds (e.g., polymeric molecules) are shared between different microbial groups in the community (Figure 10.3).

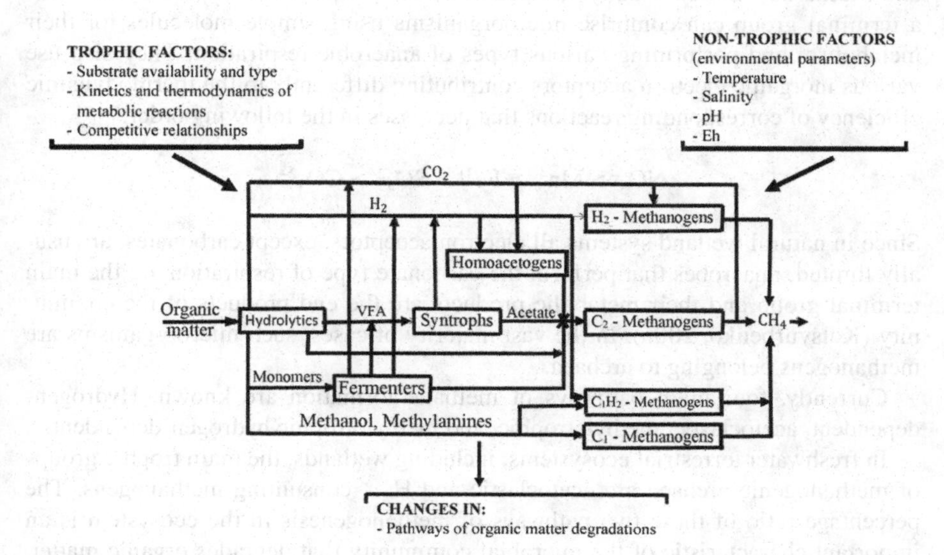


**FIGURE 10.3** Methanogenic microbial community as a system: Trophic relationships, main regulating factors, and sustainability strategies.

*Source:* Modified from Kotsyurbenko et al. (2020)

The feasibility of sequential metabolic reactions in the microbial community is dependent on their thermodynamics. Gibbs energy value of a certain reaction must be negative enough in order to allow the microbial group gaining sufficient energy to support life.

Anaerobic processes have energy efficiencies much lower compared to those of aerobic processes; therefore, anaerobic organisms are forced to form close trophic cooperative relationships (McInerney and Beaty, 1988; Schink, 1997; McInerney et al., 2010).

Depending on the type of organic matter entering the anaerobic zone, the trophic structure of the methanogenic community can vary greatly. If the main primary component is plant debris, then the complete trophic chain is formed. This is the most general situation happening in natural ecosystems.

Cellulose (the main component of plant debris) is broken down by exoenzymes of hydrolytic microorganisms with the release of glucose molecules, which are then consumed by the same hydrolytics, as well as a large group of microorganisms that perform various types of fermentation. The main products of fermentation of glucose and other saccharides, as well as compounds of a different biochemical nature, for example amino acids, which can also be present in the system and decomposed parallel to the main metabolic pathway, are volatile fatty acids (VFAs), as well as hydrogen and carbon dioxide.

The decomposition of VFA is a result of syntrophic relationships. They are the closest microbial interactions in the anaerobic community due to thermodynamics of such processes that become energy-yielding for microorganisms only if the products of these metabolic reactions (acetate and hydrogen) are actively removed from the system as a result of consumption by another microbial group (Schink, 1997; Jackson and McInerney, 2002), usually situated at the very end of the trophic chain. Such a terminal group can comprise microorganisms using simple molecules for their metabolism and performing various types of anaerobic respiration. They also use various inorganic electron acceptors contributing differently to the thermodynamic efficiency of corresponding reactions that decreases in the following order:

$$NO_3^- > Mn^{4+} > Fe^{3+} > SO_4^{2-} > CO_3^{2-}$$

Since in natural wetland systems all electron acceptors, except carbonates, are usually limited, anaerobes that perform the carbonate type of respiration are the main terminal group and their metabolic products are the end products of the community (Kotsyurbenko, 2005). In the vast majority of cases, such microorganisms are methanogens belonging to archaea.

Currently, four main pathways of methane formation are known: Hydrogen-dependent, acetoclastic, methylotrophic, and methylotrophic hydrogen-dependent.

In freshwater terrestrial ecosystems, including wetlands, the main trophic groups of methanogenic archaea are acetoclastic and $H_2$ – consuming methanogens. The percentage ratio of these two pathways of methanogenesis in the ecosystem is an important characteristic of the microbial community that degrades organic matter. Usually, the ratio of acetoclastic and hydrogen-dependent methanogenesis is 2/1 in a balanced system at moderate temperatures, neutral pH, and the absence of electron acceptors other than $CO_2$. This is due to the fact that the decomposition of organic matter in natural terrestrial ecosystems normally occurs with the formation of approximately 1 mole of acetate and 2 moles of hydrogen, when methanogens are the terminal group. Then, the above ratio (2/1) of these methanogenesis pathways is composed on the basis of the stoichiometry of the reactions of methane formation from acetate as well as hydrogen and $CO_2$,

$$CH_3COOH = CH_4 + CO_2 \quad \text{Eq. (1)}$$
$$4H_2 + CO_2 = CH_4 + H_2O \quad \text{Eq. (2)}$$

A shift in this ratio can be due to a change in set of products of organic matter degradation. The formation of more hydrogen equivalents or consumption of acetate as a result of another metabolic process shifts the ratio toward $H_2$-methanogenesis, whereas an increase in the competition of methanogens with other microbial groups for hydrogen is favorable to acetoclastic methanogenesis.

In freshwater terrestrial systems, the main competitors of methanogens for hydrogen are homoacetogenic bacteria (Kotsyurbenko et al., 2001), which also perform the autotrophic consumption of hydrogen and carbon dioxide, but with the formation of acetate instead of methane:

$$4H_2 + 2CO_2 = CH_3COOH + 2H_2O \quad (3)$$

The methanogenic microbial system is regulated by various factors, which can be divided into two main groups – trophic and non-trophic (Figure 10.3). Trophic microbial interactions are the structural basis for the functioning of the methanogenic community. When balanced functioning, the community as a system has a closed trophic structure, and methane is the final product of its work. This is facilitated by stable environmental conditions and regular intake of necessary nutrients into the system. Under these conditions, trophic interactions are formed in the community in such a way that the concentration of intermediate metabolites is at a stable low level. If a linear increase in methane concentration is observed in the system, then the methanogenic system is balanced. Such a state, for example, can be achieved in laboratory incubation experiments in which methanogenic degradation of organic matter occurs under controlled conditions.

In nature, this state can be achieved transiently. Seasonal changes in environmental conditions, and particularly temperature as a non-trophic factor, affect the stability of the methanogenic community, which tends to adapt to the changes through a rearrangement in its trophic structure. As a result, the main pathways of degradation of organic compounds can also change, which leads to redirection of organic matter flows in the microbial community.

The main changes in the trophic structure of the methanogenic community are related to changes in metabolic pathways driven by the same microbial group to produce, for example, more reduced compounds, or to the replacement of key microbial groups in the trophic chain. The former changes are most characteristic of microbial groups located at the beginning of the trophic chain and capable of fermenting complex organic compounds with the formation of several products. For example, if the consumption of hydrogen, the main intermediate molecule in the microbial system involved in various regulatory interactions, is impeded, then the fermentation shifts from acetate as the most oxidized product to ethanol or more reduced forms of VFA in order to compensate for the excess of reducing equivalents in system. The latter change relates more to the terminal microorganisms using simple molecules and synthesizing one or two end products.

In general, the strategy of the methanogenic microbial community is aimed at maintaining its basic system structure under changing environmental conditions.

It is also important to note that during biological processes, fractionation of molecules occurs in which their constituent elements can be in the form of various isotopes (Whiticar, 1999). Living organisms primarily prefer to use substrates with lighter isotopes of elements, due to which their percentage in products increases. Thus, an increase in the $^{13}CH_4/^{14}CH_4$ ratio is an indicator of the biological pathway of methane formation, and different values of this isotope ratio provide additional information on the metabolic pathways of the product formation. Thus, the measurement of the isotopic ratio of elements is an important indicator of the biological origin of the measured compound.

### 10.3.4 System Properties of Microbial Communities

The basic principle of hierarchical holism is the subordination of various systems. Microbial communities are key systems in environmental microbiology. Elements of such systems are microorganisms as units of living matter. They are trophically connected and actively interacting with the environment. The main systemic properties of the microbial community were examined in detail in the works of G.A. Zavarzin (2015).

The formation of the microbial community occurs according to the systemic principle of expediency. The determining factor in the formation of the microbial community is trophic relationships. The inclusion of an organism in its composition does not depend on how high the potential this organism has in relation to a certain metabolic process, but rather on how beneficial is its inclusion for the community in relation to increasing the environmental potential of the community as a whole and increasing its sustainability as a system.

In addition, microorganisms included in the community enter into cooperative mutually beneficial relations. This is an indication of the fact that the microbial community is formed on the basis of non-Darwinian laws, and the main driving force in its formation is cooperation rather than competition. The latter only plays an important regulatory role (Zavarzin, 1995a).

The proper functioning of a community depends on the coordinated work of all its elements, and its structural stability is determined by Le Chatelier's principle, formulated initially for physical and chemical systems: If a system in equilibrium is subjected to any deviating influence from an equilibrium state, then the system is adjusted such that the deviation is counteracted. In this regard, the community as a system resists the inclusion of an alien element in its structure that can reduce or interfere with its stability. The artificial introduction of an organism into the natural system will lead to its elimination in case of a conflict with the systemic principle of expediency for the microbial community.

Cyclicity is a necessary condition for the long-term existence of any trophic system. Non-cyclic systems with an imbalance of production and destruction can exist temporarily, for example, in certain climatic seasons or locally introduced matter will be removed by precipitation, perhaps. The autonomous cyclic system is maintained.

In addition to being cyclical, the stability of the microbial community as a system is determined by its ability to maintain the basic structure under the influence of various environmental factors. The stability of such a biological system depends on keeping its functional system intact and of the highest order. The methanogenic microbial community has various mechanisms of adaptation to changing environmental conditions. As already mentioned, the system reacts to the impacts of various environmental factors by restructuring its trophic structure and changing the composition of key microorganisms (Figure 10.3).

The more diverse the metabolic pathways and microbial groups performing them, the higher is the ability of the community to adaptation. An important systemic feature of the microbial community is the preponderance of the functional constituent over the taxonomic one. For example, it was found that in different types of wetlands (eutrophic, mesotrophic, and oligotrophic), methanogenic microbial communities are functionally similar but can be different to a great extent taxonomically (Hunger et al., 2015). The key microbial groups, characteristic of the trophic structure of each of these communities, can be represented by phylogenetically different microorganisms (Stoeva et al., 2014).

The more extreme the external conditions, the less microbial diversity, and the more difficult is for the microbial community to maintain the functionality when the external conditions change and, hence, the lower is its adaptive potential. Nevertheless, under quite constant conditions, a microbial system consisting of extremophilic microorganisms is formed. Its sustainability is primarily determined by specific conditions of the ecological niche in which unique non-competitive adaptive survival mechanisms of extremophiles and their proper functionalities are in demand. That is, under unfavorable conditions, the stability of the community is associated with its minimization, while under favorable conditions, opportunities are created to increase the metabolic and taxonomic diversity. Modern extreme ecosystems are analogues of the ecosystems that prevailed on the ancient Earth. They are inhabited only by prokaryotes and are inaccessible to eukaryotes. As conditions approached modern ones and their extremeness decreased, the diversity of living organisms increased. Accordingly, the corridor of preferred habitats shrinks as the complexity of organisms increases (Zavarzin, 2003a, 2003b). The basis of life for higher forms is limited to "normal" conditions. The prokaryotic system is self-sufficient and able to maintain its existence for an indefinitely long time. It is such a system that is primarily considered as inhabiting hypothetical extreme extraterrestrial systems.

However, organisms emerging in the process of phylogenetic complication had to adapt to an already existing system. Thus, evolution took place not as a result of the displacement of the old by the new but by superimposing one on the other, that is, additively. Moreover, the preservation of the "old" is a necessary condition for the further evolution of the system (Zavarzin, 2006). The system of higher organisms is superimposed on the original system formed by prokaryotes, which is the basis for maintaining life on the planet (Zavarzin, 2008). Moreover, a higher-level system determines the possibilities for the evolution of a lower-level system (Zavarzin, 2003a).

The trend toward increased specialization in the environment correlates with a decrease in the genome of microorganisms, while universalism, on the other hand, leads to a more complex genome (Suslov et al., 2013). It is interesting that the same tendency to increase the genome exists in microorganisms in their relation to the ambient temperature of the environment. It should be noted that both thermophiles and hyperthermophiles, being the most ancient and specialized organisms, have a minimal genome (Suslov et al., 2013).

Thus, the functionality is the most important characteristic of the microbial community and determines the feasibility of its existence in a particular ecosystem. Specialization increases the potential of an organism to perform a function, while versatility increases its adaptive capabilities. Obviously, nature chooses one or another strategy depending on ecosystem requirements. If the adaptation capabilities of the community are not enough to withstand external impacting factors, in other words, if the impact on the system is too strong, then Le Chatelier's law is no longer fulfilled, and the community is to be destroyed or changes its structure, which means a transition to a new stationary state – a new biological system.

### 10.3.5 Methanogens as a Key Microbial Group of Methane-Producing Microbial Community

A microorganism is a biological system that is a unit of life. Life is an emergent attribute of a system whose components assemble an organism (Zavarzin, 2004b).

Individual components of a microbial cell are no longer alive.

The purpose of a microorganism in a microbial community is to make it more stable and efficient.

The methanogens belong to the domain of archaea, to the kingdom of Euryarchaeotes. Methanogens are placed in 34 genera, representing 14 families and 7 orders (Luy and Liu, 2019).

Methanogens are common in moderate as well as extreme habitats. Thus, growth temperatures of methanogens span 100°C, from psychrophilic to hyperthermophilic. Optimal salinities for growth vary from freshwater to saturated brine. Lastly, methanogens are found from being moderately alkaline to having neutral to acidic pH values.

Representatives of methanogens differ morphologically, including such forms as rods, cocci, sarcinas, and spirilla. They are all united by a similar cell dimension and methanogenesis as a metabolic pathway for obtaining energy.

Because modern methanogens are monophyletic, it also is likely that methanogenesis evolved only once and that all modern methanogens share a single ancestor. The wide diversity within the group suggests that methanogenesis is an ancient lifestyle.

The prokaryotic cell contains in its composition various components, the combined functioning of which allows it to carry out vital activity (Figure 10.4). The life activity of the cell is based on cyclic metabolic processes occurring in the cell cytoplasm, as a result of which the substance (substrate) entering the cell is metabolized in a series of successive reactions to obtain either molecules for cellular structures, or to obtain

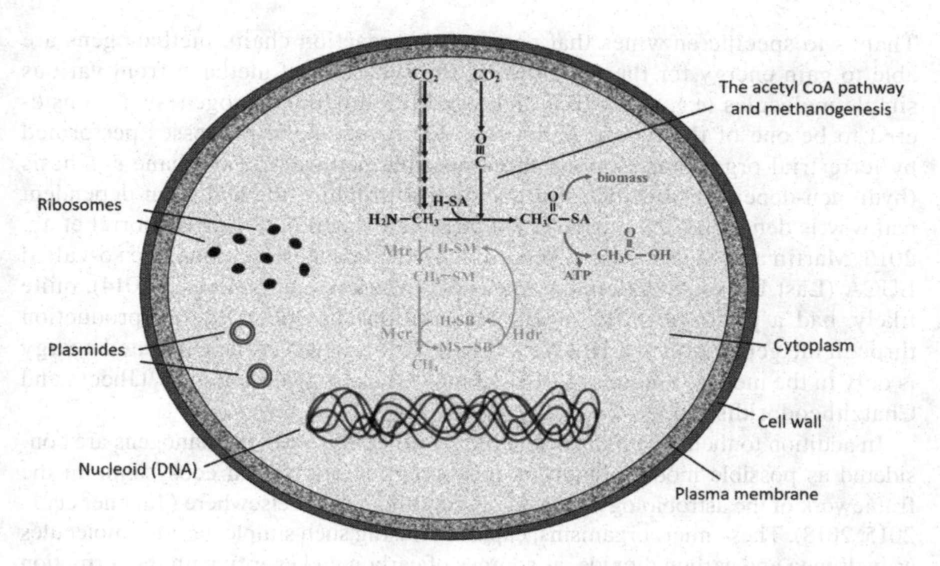


**FIGURE 10.4** The acetyl-CoA pathway and methanogenesis (modified from Oren (2014). Whether H4F or H4MPT is the pterin cofactor (-N) in the methyl branch of the acetyl-CoA pathway (black) is not indicated; activated acetyl group is either used anabolically (Part of the figure on the right just above CH3(CO)SA) or catabolically (Part of the figure on the right just below CH3(CO)SA); methanogenesis (Top part of the figure, arrows on the far left and the figure just below), reduction of methyl-N to methane involves Mtr (N5-methyl-H4MPT:HS-CoM methyltransferase), Mcr (methyl-S-CoM reductase), and Hdr (heterodisulfide reductase); for simplicity, thiol-containing cofactors coenzyme A, B, and M are abbreviated, and the input of electrons is not shown.

energy. In methanogens, energy is obtained as a result of the action of the oldest metabolic process of chemosynthesis along the acetyl-CoA pathway with the formation of methane as an end product. The storage of energy in the form of ATP occurs in the electron transport system when a transmembrane electrochemical potential is created.

Ribosomes are involved in the process of synthesis of structural components, and genetic information is contained in the nucleoid, which consists of a circular DNA molecule and plasmids. Each cell must maintain its structure and at the same time actively exchange components with the environment, which is facilitated by the cell wall and cytoplasmic membrane. In most archaea, to which methanogens belong, the cell wall is formed by protein molecules. Inside the cell, a near-neutral pH is maintained, despite the fact that the pH in the environment can change.

In general, all components of the system act in concert and ensure the effective functioning of the microbial cell as the simplest biological system.

### 10.3.6 Biomacromolecules and Metabolic Processes

Methanogenesis as a metabolic process is one of the subsystems that function in the cells of methanogenic archaea. A feature of this metabolic pathway is a unique enzyme system, which is found only in methanogens (Thauer, 1998).

Thanks to specific enzymes that perform this reaction chain, methanogens are able to gain energy for their life during the synthesis of methane from various simple molecules (e.g., from $CO_2$ and $H_2$); Microbial methanogenesis is considered to be one of the oldest, if not the oldest, metabolic processes performed by terrestrial organisms. Among three possible pathways of methane synthesis (hydrogen-dependent, acetoclastic, and methylotrophic), the hydrogen-dependent pathway is defined as the most ancient one (Battistuzzi et al., 2004; Borrel et al., 2016; Martin and Sousa, 2016). According to the latest estimations, the so-called LUCA (Last Universal Common Ancestor) (Markov and Naimark, 2014), quite likely, had a protosystem of metabolic reactions based on energy production through the generation of $CH_4$ (Weiss et al., 2016), however the generated energy is only in the meagre amount of $-131$ kJ $mol^{-1}$ during oxidation of $H_2$ (Jheeta and Chatzitheodoridis, 2023).

In addition to their living in ancient terrestrial ecosystems, methanogens are considered as possible model organisms inhabiting extraterrestrial ecosystems in the framework of the astrobiology concept of searching for life elsewhere (Taubner et al., 2015, 2018). These microorganisms, capable of using such simple gaseous molecules as hydrogen and carbon dioxide, as sources of carbon and energy with the formation of methane molecule, can exist in almost any ecosystems poor in organics, where the aforementioned gases are available in sufficient quantity. In this respect, the microbial community of the deep aquafer ecosystem is noteworthy, where the trophic chain begins just with the aforementioned gases, and where lithotrophic methanogenic archaea and homoacetogenic bacteria were found to be highly predominant microbial groups (Kotelnikova and Pedersen, 1997; Chapelle et al., 2002). Methane is, in particular, suggested as an important biosignatures of possible microbial activity on other cosmic bodies (Seager et al., 2016).

The uniqueness of methanogenesis enzymes was the basis for the development of a method of specific markers of methanogenic archaea for their qualitative and quantitative identification in the studied microbial community. This method is based on the specific sequence of a gene encoding one of the key enzymes of methanogenesis, namely methyl coenzyme M reductase (MCR), and allows detecting the presence of various methanogenic archaea in ecosystems (Friedrich, 2005). This method of identifying methanogens by functional genes is also widely used to study the diversity of methanogens in wetlands. It has been established that in wetlands located in various regions, including Western Siberia, methanogens of all trophic groups are present, and their diversity strongly depends on the type of the wetland system (Kotsyurbenko et al., 2004, 2007; Yavitt et al., 2006, 2012; Andersen et al., 2013). The main predominant families of methanogens are as follows: *Methanomicrobiaceae*, *Methanobacteriaceae*, *Methancoccaceae*, and, the relatively recently discovered acid-tolerant methanogens of the *Methanoregulaceae* family, which perform the hydrogen-dependent methanogenesis pathway, as well as acetoclastic representatives of the *Methanosaetaceae* and the *Methanosarcinoaceae* (Basiliko et al., 2003; Kotsyurbenko et al., 2004; Galand et al., 2005; Narrowe et al., 2017).

The active site of MCR contains the hydroporphinoid nickel complex coenzyme $F_{430}$. The nickel center of coenzyme $F_{430}$ catalyzes the unusual reaction (Figure 10.5).

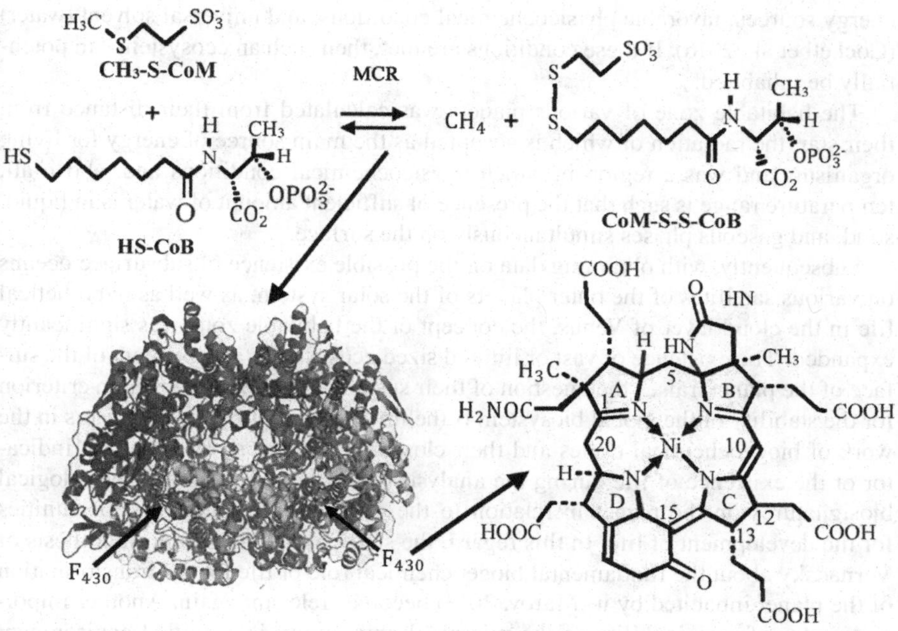

**FIGURE 10.5** Reaction catalyzed by methyl-coenzyme M reductase (MCR).

*ABBREVIATIONS:* $CH_3-S-CoM$, methyl-coenzyme M (2-(methylthio)ethanesulfonate); HS – CoB, coenzyme B (N-7-mercaptoheptanoyl-O-phospho-L-threonine); CoM – S – S – CoB, heterodisulfide of coenzyme M (2-mercaptoethanesulfonate) and coenzyme B; $F_{430}$, coenzyme (factor) $F_{430}$ in the catalytically active Ni(I) valence form.

The typical uniqueness of the enzymatic system of methanogenesis as the most ancient metabolic system allows us to consider its key components, such as factor 430, along with the end product methane, as the most important biomarkers.

In addition, methanogenesis as one of the types of chemosynthesis allows methanogens to exist in an ecosystem with a minimum amount of organic matter, since for their full life they can use only gaseous $H_2$ and $CO_2$ molecules, which can serve as both energy sources and carbon sources for constructive exchange.

## 10.4 THE CONTRIBUTION OF SYSTEMS ANALYSIS OF TERRESTRIAL BIOSYSTEMS TO ASTROBIOLOGY

The Earth's biosphere is the only known biosystem on a planetary scale that can be considered as a model in the search for extraterrestrial life. Accordingly, all the systemic interactions and properties discussed earlier must be taken into account when discussing possible habitable extraterrestrial ecosystems.

The dominant concept of habitability includes the mandatory presence of four necessary factors for the existence of life: The main biofriendly elements CHNOPS, energy sources, favorable physicochemical conditions, and universal solvent (water) (Cockell et al., 2016). If these conditions are met, then such an ecosystem can potentially be inhabited.

The habitable zone of various planets was calculated from their distance from their star, the radiation of which is accepted as the main source of energy for living organisms and was a region in which physicochemical conditions and, above all, temperature range is such that the presence of sufficient amount of water is in liquid, solid, and gaseous phases simultaneously on the surface.

Subsequently, with obtaining data on the possible existence of subsurface oceans on various satellites of the outer planets of the solar system, as well as hypothetical life in the cloud layer of Venus, the concept of the habitable zone was significantly expanded. The existence of vast or limited sized ecosystems, independent of the surface of the planet, raises the question of their sustainability. Since the main criterion for the stability of the global biosystem is the involvement of living organisms in the work of biogeochemical cycles and their closed nature, the most important indicator of the existence of life during the analysis of the geochemical and hydrological biosignatures on the planet in relation to the existence of potential opportunities for the development of life. In this regard, the study of the most important thesis of Vernadsky about the fundamental biogeochemical role of life for the transformation of the planet inhabited by it (Marov, 2013) becomes relevant again. Another important factor of sustainability of the microbial community is a unified environment where close trophic interactions between microorganisms are feasible (Skladnev et al., 2021).

If we suggest the existence of a relatively closed biological system, which is limited by favorable physicochemical conditions, such as the Venusian cloud system, which is restricted from above by open space, and from below by ultra-high temperatures on the surface of the planet (Kotsyurbenko et al., 2021), or some other system where favorable conditions are stable only within a certain econiche, then we must assume its openness geochemically, so that the cyclical processes of transformation of matter and energy extend to the entire or most of the planet. This will increase the stability of the entire system, including its biological component. In other words, the more global the cyclic process in which living organisms are involved, the more stable the biosystem of that planet.

Since at the moment there is no evidence of the existence of extraterrestrial life, practical astrobiology is aimed at developing a methodology for its search in which the main aspect considered is relevant biosignatures. In this regard, systems analysis of biological systems on Earth can make an important contribution to determining the type of such biosignatures and their relationship with manifestations of hypothetical life activity at different levels (Figure 10.6).

At various system levels, their biosignatures are determined. It would be ideal to fix them at several levels, which would give more evidence of the presence of biological activity on the planet. Nevertheless, due to the limited technical capabilities of an individual space mission, researchers, as a rule, develop a methodology for detecting only a certain number of possible biosignatures. Obviously, the type of detected biosignature can give direct (biomacromolecules or cells) or only indirect (gas composition, geological processes) indication of the presence of extraterrestrial

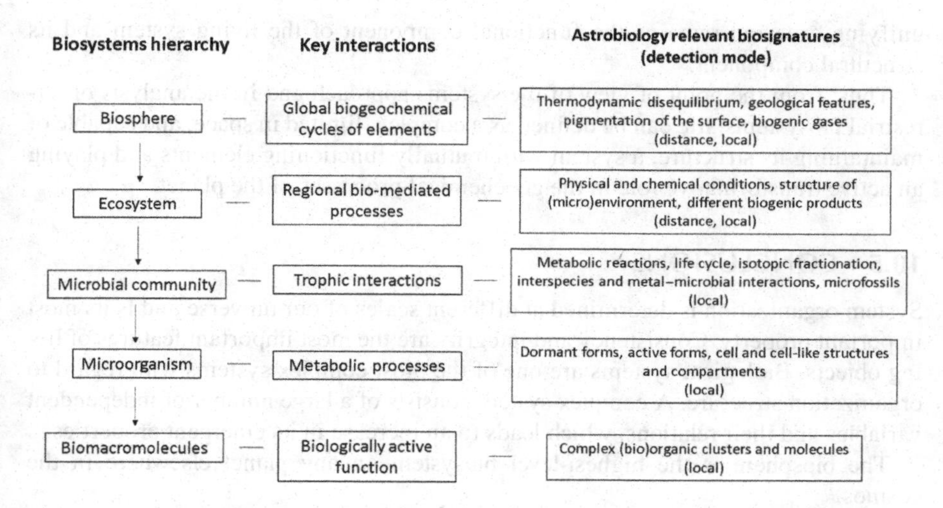


**FIGURE 10.6** Types of biosignatures are quite diverse. They include chemical, morphological, sedimentary, or isotopic processes or structures that are biofriendly and could be detected to infer the past or present life activities.

life on celestial bodies. However, when searching for direct evidence, it must be considered that extraterrestrial life forms may differ significantly from the terrestrial life and, therefore, will not necessarily be detected by methods developed for terrestrial organisms.

In this respect, it is important to search for the most general principles and features of living organisms that could be applied not to the terrestrial form of life alone. Accordingly, the four main criteria for habitability must be considered in a broader sense.

So, despite the fact that the chemical elements are universal, their ratio and role in a living organism may differ from those found in the earthly system. Similarly, energy sources such as photo- and chemosynthetic processes can potentially be represented by alternative biochemical reactions. Moreover, hypothetically, water as a solvent could be replaced by another solvent (e.g., ammonia or formamide), which may be much more common on another planet (Bains et al., 2021; Norman and Fortes, 2011). In this respect, it makes sense to study any signs of thermodynamic imbalance on the planet (Seager et al., 2022), a red-ox pair of compounds in planetary ecosystems that can potentially be considered as energy-producing for an extraterrestrial life form in the physicochemical conditions of another planet.

Finally, it is necessary to bear in mind a possible different basis of living organisms and a different metabolism in conditions very different from terrestrial conditions on another planet (Ksanfomality et al., 2018).

The unifying principle, in this case, can be considering the systemic concept of life, as a certain structure limited in space (if we do not consider the concept of a field form of life, the functioning of which is difficult to describe in our ideas), which includes various interacting elements, in our understanding of cell structure. Thus, even if the interacting elements of the system are very different from each other, the unifying criterion remains the functional component of the living system and its structural component.

Thus, from the point of view of the systems approach and in the analysis of terrestrial biosystems, life can be defined as a complex, limited in space, and capable of maintaining its structure, a system with mutually functioning elements and playing an active transformative role in the geochemical processes of the planet.

## 10.5 CONCLUSIONS

System organization is determined at different scales of our universe and is its most important property. Consistency and integrity are the most important features of living objects. Biological systems are one of the most complex systems with regard to organization structure. A complex system consists of a large number of independent variables and their relations, which leads to an increase in its emergent properties.

The biosphere is the highest-level biosystem on any planet elsewhere in the cosmos.

Microorganisms, especially bacteria and archaea (collectively called: prokaryotes), represent the first stable system within the planet's "sphere" (technically called biosphere) on which all other living entities were successively superimposed and included. The stability of such a system depends on the determining role of prokaryotes in the cycles of biofriendly elements (in the cyclicity of these processes) and on their mediating role in the transformation of other elements.

Microbial systems (i.e., communities of microbes) are formed on the basis of trophic interactions, and their role is determined by the requirement of a higher-level system. The functional diversity and abilities of microorganisms are the main factor in the formation of the trophic structure of the community.

The activity of microbial communities is closely related to geochemical processes on the planet. The openness of element cycles led to biogeochemical succession, the most noticeable result of which was the saturation of the atmosphere with oxygen about 2.4 billion years ago and its subsequent use by organisms in the process of aerobic respiration. This, in turn, has led to a significant increase in the complexity of organisms and biodiversity on the planet.

However, excessive biodiversity and the complication of organisms in and of themselves do not contribute to the stability of biosystems. The work of the main biogeochemical mechanism of the planet is provided by prokaryotes, which have the necessary and sufficient functional diversity for this (Zavarzin, 2003a).

The expediency of the existence of a biological system is determined by its compliance with the requirements of a system of a higher rank. In this sense, the evolution of a biosystem follows the path of maintaining its correspondence to the system of a higher rank, which in turn changes in time with changes in the factors of the environment in which it functions (Zavarzin, 2006).

Accordingly, from the point of view of a systems approach, the origin of life should also be considered as a systemic phenomenon. Zavarzin postulated that the concept of the origin of life should be inexorably linked with the origin of the microbial community, since only such a biological system can be stable in nature. An individual organism can only survive as an "element" in a community.

Developing this hypothesis, we can assume that the first microorganisms should have been included in the already existing cycles of elements, replacing the non-living catalysts of such a system. Thus, the geochemical cyclic system gradually evolved into a biogeochemical one with the gradual replacement of abiogenic process catalysts with biogenic agents within the framework of a functioning cycle of elements or any coordinated processes within any system. The first protocellular systems could act as bioagents. Subsequently, after the emergence of the first microbial community, evolution proceeded as the evolution of microbial communities.

Thus, the emergence of life is the evolution of a system with the gradual replacement of its key elements in favor of biological agents. Of course, this is only a general conceptual approach that needs further careful theoretical and practical study.

It is important to note that there are already theories in which prebiological processes are considered from the point of view of a systematic approach. The principles of the processes of gradual self-assembly and complication of precellular systems are described within the framework of systemic prebiology (Lancet et al., 2018).

In relation to astrobiology, the study of systemic patterns in terrestrial biosystems provides an understanding of which phenomena and biosignatures should be attended to when looking for manifestations of extraterrestrial life. Despite the fact that extraterrestrial ecosystems and the biosphere as a whole may differ significantly from the terrestrial ones, the basic principles of interaction in such systems should be common.

Microbial systems as basic biosystems are a key object of search within the framework of the concept of extraterrestrial life.

One of the most interesting systems in this regard is the methanogenic microbial community, which plays a crucial role as a source of the greenhouse gas methane into the Earth's atmosphere. Under the conditions of the Earth's biosphere, methanogens are included in a complex trophically connected community, which decomposes organic matter in stages under anaerobic conditions.

In extraterrestrial ecosystems, methanogens can potentially function as part of a simplified biosystem. They are able to grow exclusively on inorganic compounds, synthesizing all the substances necessary for the cell in the processes of constructive exchange using carbon dioxide as the only source of carbon.

The complication of organisms occurs with the loss of functional properties and, accordingly, their ability to survive in extreme habitats.

Obviously, the reverse process is possible in which microbial systems are capable of evolution toward simplification under gradually worsening environmental conditions. Examples would be hypothetical Venusian cloud layer organisms under highly arid and low acidic conditions and Martian microorganisms, if they exist subsurface, since changing conditions on the surface of the above planets, and especially on Venus, have made the existence of life almost impossible. Such harsh conditions will contribute to the emergence of polyextremophilia in microorganisms.

In general, the systems approach is an excellent methodological tool for studying the manifestations of life at different levels. A systematic study of microbial processes requires the involvement of specialists from various disciplines from atmospheric physicists to molecular biologists. The same trend of scientific integration exists in astrobiology for solving the problems in which astrophysicists, geochemists, and microbiologists etc., are involved, which will make it possible to obtain new knowledge that would otherwise be impossible to obtain within the framework of only one discipline.

## REFERENCES

Andersen R., Chapman S.J., Artz R.R.E. (2013) Microbial communities in natural and disturbed peatlands: a review. *Soil Biol. Biochem.* V. 57. P. 979–994.

Bains W., Petkowski J.J., Zhan Z., Seager S. (2021) Evaluating alternatives to water as solvents for life: the example of sulfuric acid. *Life.* V. 11. P. 400. https://doi.org/10.3390/life11050400.

Basiliko N., Yavitt J.B., Dees P.M., Merkel S.M. (2003) Methane biogeochemistry and methanogen communities in two northern peatland ecosystems, New York State. *Geomicrobiol. J.* V. 20. P. 563–577.

Battistuzzi F.U., Feijao A., Hedges S.B. (2004) A genomic timescale of prokaryote evolution: insights into the origin of methanogenesis, phototrophy, and the colonization of land. *BMC Evol. Biol.* V. 4. P. 44. http://doi.org/10.1186/1471-2148-4-44.

Borrel G., Adam P.S., Gribaldo S. (2016) Methanogenesis and the wood – Ljungdahl pathway: an ancient, versatile, and fragile association. *Genome. Biol. Evol.* V. 8. No. 6. P. 1706–1711. http://doi.org/10.1093/gbe/evw114.

Chapelle F.H., O'Neill K., Bradley P.M., Methe Â.B.A., Ciufo S.A., Knobel L.L., Lovley D.R. (2002) A hydrogen-based subsurface microbial community dominated by methanogens. *Nature.* V. 415. P. 312–315.

Cicerone R.J., Oremland R.S. (1988) Biogeochemical aspects of atmospheric methane. *Global Biogeochem. Cycles.* V. 2. P. 299–327.

Cockell C.S., Bush T., Bryce C.S., Direito S., Fox-Powell M., Harrison P., Lammer H., Landenmark H., Martin-Torres J., Nicholson N., Noack L., O'Malley-James J., Payler S.J., Rushby A., Samuels T., Schwendner P., Wadsworth J., Zorzano M.P. (2016) Habitability: a review. *Astrobiology.* V. 16, P. 89–117.

Conrad R. (1996) Soil microorganisms as controllers of atmospheric trace gases ($H_2$, CO, $CH_4$, OCS, $N_2O$, NO). *Microbiol. Rev.* V. 60. P. 609–640.

Conrad R. (2007) Microbial ecology of methanogens and methanotrophs. *Advan. Agron.* V. 96. P. 1–63. http://doi.org/10.1016/s0065-2113(07)96005-8.

Ehhalt D.H., Schmidt U. (1978) Sources and sinks of atmospheric methane. *Pure Appl. Geophys.* V. 116. P. 452–464.

Friedrich M.W. (2005) Methyl-Coenzyme M reductase genes: unique functional markers for methanogenic and anaerobic methane-oxidizing archaea. *Methods Enzymol.* V. 397. P. 428–442. http://doi.org/10.1016/s0076-6879(05)97026-2

Galand P.E., Fritze H., Conrad R., Yrjälä K. (2005) Pathways for methanogenesis and diversity of methanogenic archaea in three boreal peatland ecosystems. *Appl. Environ. Microbiol.* V. 71. P. 2195–2198.

Glagolev M.V., Fastovets I.A. (2012) Apology of reductionism (reductionism as the worldview of mathematical modeling). *Environm. Dynam. Glob. Clim. Change.* V. 3. No. 2(6). P. 1–24 (In Russian).

Hunger S., Gößner A.S., Drake H.L. (2015) Anaerobic trophic interactions of contrasting methane-emitting mire soils: processes versus taxa. *FEMS Microbiol. Ecol.* V. 91. No. 5. http://doi.org/10.1093/femsec/fiv045.

IPCC. (2013) *Carbon and other biogeochemical cycles. Chapter 6: Climate change. The physical science basis. Global methane budget.* Cambridge University Press, Cambridge, UK and New York, NY. P. 505–510.

IPCC. (2014) Climate change 2014: Synthesis report. *Contribution of working groups I, II and III to the fifth assessment report of the intergovernmental panel on climate change* [Core Writing Team, R.K. Pachauri and L.A. Meyer (eds.)]. IPCC, Geneva, Switzerland, 151 p. www.ipcc. ch/site/assets/uploads/2018/02/SYR_AR5_FINAL_full.pdf

Jackson B.E., McInerney M.J. (2002) Anaerobic microbial metabolism can proceed close to thermodynamic limits. *Nature*. V. 415. P. 454–456.

Jheeta S., Chatzitheodoridis E. (2023) Origin of Life: Conflicting Models for the Origin of Life. In Origin of Life. Smoukov Stoyan. K., Seckbach J., and Gordon R. (Eds). Wiley Online Library. (page 1 – 31).

Kallistova A. Yu., Merkel A. Yu., Tarnovetskii I. Yu., Pimenov N.V. (2017) Methane formation and oxidation by prokaryotes. *Microbiology*. V. 86. No. 6. P. 671–691. http://doi.org/10.1134/S0026261717060091.

Kotelnikova S., Pedersen K. (1997) Evidence for methanogenic Archaea and homoacetogenic Bacteria in deep granitic rock aquifers. *FEMS Microbiol. Rev.* V. 20. P. 339–349.

Kotsyurbenko O.R. (2005) Trophic interactions in the methanogenic microbial community of low-temperature terrestrial ecosystems. *Mini-Rev. FEMS Microbiol. Ecol.* V. 53. P. 3–13.

Kotsyurbenko O.R., Chin K.-J., Glagolev M.V., Stubner S., Simankova M.V., Nozhevnikova A.N., Conrad R. (2004) Acetoclastic and hydrogenotrophic methane production and methanogenic populations in an acidic West-Siberian peat bog. *Environ. Microbiol.* V. 6. No. 11. P. 1159–1173.

Kotsyurbenko O.R., Cordova Jr. J.A., Belov A.A., Cheptsov V.S., Khrunyk J., Kölbl D., Kryuchkova M.O., Milojevic T., Sasaki S., Słowik G.P., Snytnikov V., Vorobyova E.A. (2021) Exobiology of Venus clouds: new insights into habitability through terrestrial models and methods of detection. *Astrobiology*. V. 21, No. 9. P. 1186–1205. http://doi.org/10.1089/ast.2020.2296.

Kotsyurbenko O.R., Friedrich M.W., Simankova M.V., Nozhevnikova A.N., Golyshin P., Timmis K., Conrad R. (2007) Shift from acetoclastic to $H_2$-dependent methanogenesis in a West Siberian peat bog at low pH. *Appl. Environ. Microbiol.* V. 73. No. 7. P. 2344–2348.

Kotsyurbenko O.R., Glagolev M.V., Merkel A.Y., Sabrekov A.F., Terentieva I.E. (2019) Methanogenesis in soils, wetlands and peat. In *Handbook of hydrocarbon and lipid microbiology Series. Biogenesis of hydrocarbons*. A.J.M. Stams and Diana Z. Sousa (Eds.). Springer-Verlag, Berlin and Heidelberg. http://doi.org/10.1007/978-3-319-53114-4_9-1.

Kotsyurbenko O.R., Glagolev M.V., Nozhevnikova A.N., Conrad R. (2001) Competition between homoacetogenic bacteria and methanogenic archaea for hydrogen at low temperature. *FEMS Microbiol. Ecol.* V. 38. P. 153–159.

Kotsyurbenko O.R., Glagolev M.V., Sabrekov A.F., Terenieva I.E. (2020) Systems approach to the study of microbial methanogenesis in West-Siberian wetlands. *Environ. Dynam. Glob. Clim. Chang.* V. 11, No. 1, P. 54–68. https://doi.org/10.17816/edgcc15809.

Ksanfomality L.V., Selivanov A.S., Gektin Yu.M. (2018) Signs of hypothetical flora and fauna of the planet Venus: Returning to archive of the old TV-experiments. *Int. J. Opt. Photonic Eng.* V. 3, P. 007.

Kutzbach L., Wagner D., Pfeiffer E.M. (2004) Effect of microrelief and vegetation on methane emission from wet polygonal tundra, Lena Delta, Northern Siberia. *Biogeochemistry*. V. 69. P. 341–362. http://doi.org/10.1023/B:BIOG.0000031053.81520.db.

Lancet D., Zidovetzki R., Markovitch O. (2018) Systems protobiology: origin of life in lipid catalytic networks. *J. Royal Soc. Interface*. V. 15. No. 144, P. 20180159. http://doi.org/10.1098/rsif.2018.0159.

Lansdown J.M., Quay P.D., King S.L. (1992) $CH_4$ production via $CO_2$ reduction in a temperate bog: a source of $^{13}C$-depleted $CH_4$. *Geochim. Cosmochim. Acta.* V. 56. P. 3493–3503.

Limpens J., Berendse F., Blodau C., Canadell J.G., Freeman C., Holden J., Roulet N., Rydin H., Schaepman-Strub G. (2008) Peatlands and the carbon cycle: from local processes to global implications – a synthesis. *Biogeosciences.* V. 5. P. 1475–1491.

Lyu, Z., Liu, Y. (2019). Diversity and Taxonomy of Methanogens. In: Stams, A., Sousa, D. (eds) *Biogenesis of Hydrocarbons. Handbook of Hydrocarbon and Lipid Microbiology.* Springer, Cham. https://doi.org/10.1007/978-3-319-78108-2_5.

Markov A., Naimark E. (2014) *Evolyutsiya. Klassicheskie Idei v Svete Novykh Otkrytiy.* M.: AST: CORPUS. P. 57 (In Russian).

Marov M.Ya. (2013) Vladimir Ivanovich Vernadsky: studies of the biosphere and astrobiology. *Noosfera.* No. 3, P. 111–131.

Martin W.F., Sousa F.L. (2016) Early microbial evolution: the age of anaerobes. *Cold Spring Harb. Perspect. Biol.* V. 8. P. a018127. http://doi.org/10.1101/cshperspect.a018127.

Matthews E., Fung I. (1987) Methane emission from natural wetlands: global distribution, area and environmental characteristics of sources. *Global Biogeochem. Cycles.* V. 1. P. 61–86.

McInerney M.J., Beaty P.S. (1988) Anaerobic community structure from a nonequilibrium thermodynamic perspective. *Can. J. Microbiol.* V. 34. P. 487–493.

McInerney M.J., Hoehler T., Gunsalus R.P., Schink B. (2010) Introduction to microbial hydrocarbon production: bioenergetics. In *Handbook of hydrocarbon and lipid microbiology.* K.N. Timmis (Ed.). Springer, Berlin and Heidelberg.

Narrowe A.B., Angle J.C., Daly R.A, Stefanik K.C., Wrighton K.C., Miller C.S. (2017) High-resolution sequencing reveals unexplored archaeal diversity in freshwater wetland soils. *Environ. Microbiol.* V. 19. P. 2192–2209.

Norman L.H., Fortes A.D. (2011) Is there life on . . . Titan? *Astron. Geophys.* V. 52, No. 1, P. 1.39–1.42. https://doi.org/10.1111/j.1468-4004.2011.52139.x

Oren, A. (2014) The family methanosarcinaceae. Prokaryotes. P. 259–281. https://doi.org/10.1007/978-3-642-38954-2_408.

Schink B. (1997) Energetics of syntrophic cooperation in methanogenic degradation. *Microbiol. Mol. Biol. Rev.* V. 61. No. 2. P. 262–280.

Seager S., Bains W., Petkowski J.J. (2016) Toward a list of molecules as potential biosignature gases for the search for life on exoplanets and applications to terrestrial biochemistry. *Astrobiology.* V. 16. No. 6. https://doi.org/10.1089/ast.2015.1404.

Seager S., Petkowski J.J., Carr C.E., Grinspoon D.H., Ehlmann B.L., Saikia S.J., Agrawal R., Buchanan W.P., Weber M.U., French R., Klupar P., Worden S.P., Baumgardner D. on behalf of the Venus Life Finder Mission Team. (2022) Venus life finder missions motivation and summary. *Aerospace.* V. 9. No. 7. P. 385. https://doi.org/10.3390/aerospace9070385.

Skladnev D.A., Karlov S.P., Khrunyk Y.Y., Kotsyurbenko O.R. (2021) Water – sulfuric acid foam as a possible habitat for hypothetical microbial community in the cloud layer of Venus. *Life.* V. 11. P. 1034. https://doi.org/10.3390/life11101034.

Stoeva M.K., Aris-Brosou S., Chételat J., Hintelmann H., Pelletier P., Poulain A.J. (2014) Microbial community structure in lake and wetland sediments from a high arctic polar desert revealed by targeted transcriptomics. *PLOS ONE.* V. 9. No. 3. P. e89531.

Suslov V.V., Afonnikov D.A., Podkolodny N.L., Orlov Yu.L. (2013) Genome features and GC content in prokaryotic genomes in connection with environmental evolution. *Paleontol. J.* V. 47. P. 9. P. 1056–1060. http://doi.org/10.1134/S0031030113090220.

Taubner R.-S., Pappenreiter P., Zwicker J., Smrzka D., Pruckner C., Kolar P., Bernacchi S., Seifert A.H., Krajete A., Bach W., Peckmann J., Paulik C., Firneis M.G., Schleper C., Rittmann S.K.-M.R. (2018) Biological methane production under putative Enceladus-like conditions. *Nat. Com.* V. 9. P. 748.

Taubner R.-S., Schleper C., Firneis M.G., Simon Rittmann S.K.-M.R. (2015) Assessing the ecophysiology of methanogens in the context of recent astrobiological and planetological studies. *Life (Basel).* V. 5. No. 4. P. 1652–1686.

Thauer R.K. (1998) Biochemistry of methanogenesis: a tribute to Marjory Stephenson. *Microbiology.* V. 144. P. 2377–2406.

Weiss M.C., Sousa F.L., Mrnjavac N., Neukirchen S., Roettger M., Nelson-Sathi S., Martin W.F. (2016) The physiology and habitat of the last universal common ancestor. *Nat. Microbiol.* V. 1. http://doi.org/10.1038/NMICROBIOL.2016.116.

Whalen S.C., Reeburgh W.S. (2000) Methane oxidation, production, and emission at contrasting sites in a boreal bog. *Geomicrobiol. J.* V. 17. P. 237–251.

Whiticar M.J. (1999) Carbon and hydrogen isotope systematics of bacterial formation and oxidation of methane. *Chem. Geol.* V. 161. P. 291–314.

Yavitt J.B., Basiliko N., Turetsky M.R., Hay A.G. (2006) Methanogenesis and methanogen diversity in three peatland types of the discontinuous permafrost zone, boreal western continental Canada. *Geomicrobiol. J.* V. 23. P. 641–651.

Yavitt J.B., Yashiro E., Cadillo-Quiroz H., Zinder S.H. (2012) Methanogen diversity and community composition in peatlands of the central to northern Appalachian Mountain region. *North Am. Biogeochem.* V. 109. P. 117–131.

Zavarzin G.A. (1995a) Anti-market in nature. *Priroda.* V. 3. P. 46–60 (in Russia).

Zavarzin G.A. (1995b) Microbial methane cycle in cold environments. *Priroda.* V. 6. P. 3–14. (In Russian).

Zavarzin G.A. (1995c) Söngen psychrophilic cycle. *Ecol. Chem.* V. 4. P. 3–12.

Zavarzin G.A. (1997) Emission of methane from the territory of Russia. *Microbiology.* V. 66. P. 669–673.

Zavarzin G.A. (2001) Formation of the biosphere. *Vestnik Russ. Acad. Sci.* V. 71. No. 11. P. 988–1001.

Zavarzin G.A. (2002) Microbial geochemical calcium cycle. *Microbiology.* V. 71. No. 1. P. 1–17. Translated from Mikrobiologiya, V. 71. No. 1. P. 5–22.

Zavarzin G.A. (2003a) Evolution of the geospheric-biospheric system. *Priroda.* V. 1, P. 27–35 (in Russian).

Zavarzin G.A. (2003b) Formation of the system of biogeochemical cycles. *Paleontol. J.* V 2003. No. 6. P. 16–24 (in Russian).

Zavarzin G.A. (2004a) Microbes hold the sky. *Nauka iz pervykh ruk.* V. 2. P. 21–27 (in Russian).

Zavarzin G.A. (2004b) The future is selected by the past. *Vestnik RAN.* V. 74. P. 813–822 (in Russian).

Zavarzin G.A. (2006) Does evolution make the essence of biology? *Vestnik RAN.* V. 76. No. 6. P. 522–534 (in Russian).

Zavarzin G.A. (2008) *Microbial Cycles.* Russian Academy of Sciences, Moscow, Russia, Elsevier B.V. All Rights Reserved.

Zavarzin G.A. (2011) *Kakosfera. Filosofiya i Publitsistika.* Ruthenica, Moscow. 460 p. (In Russian).

Zavarzin G.A. (2015) *Selected Works.* MAX Press, Moscow. 512 p. (In Russian).

# 11 The Role of Astrobiology in Systems Thinking Education

*Julia Brodsky*

We stumble into the world we make possible as we lumber forward, with no or little insight or foreknowledge…We think that in physics we will find the foundations from which we can derive the world, the ultimate becoming. We cannot. The ultimate may rest on the foundations, but it is not derivable from them. This ultimate, an unknowable unfolding slips its foundational moorings and floats free.

**– Kauffman (2019)**

## 11.1 WHAT IS SYSTEMS THINKING?

The complexity of life is an undeniable reality. The term itself, derived from the Latin word "complexus," meaning "entwined," succinctly encapsulates the idea of multiple components interconnected in a way that renders them challenging to unravel. A quick internet search for images of "complexity" yields a plethora of visual representations, from spaghetti-like shapes to intricate diagrams, all purporting to depict the vast array of complex systems that permeate every aspect of our existence – from organisms and ecosystems to medicine and climate, engineering design and computer networks, organizations and societies, economics and education, the list goes on.

As humanity enters a new era marked by the advent of genetic engineering, emerging technologies, sophisticated AI, and dramatic ecological changes, we are faced with a world of complex, interconnected systems. As we become increasingly more technologically sophisticated and powerful, we must possess greater wisdom to avoid causing harm to ourselves and the Earth. Unfortunately, our intuitions about complex, nonlinear systems are often weak, leading to well-intentioned decisions that may have unintended consequences. Earth inhabitants, old and young alike, must develop a deeper understanding of these systems to make informed decisions.

The traditional approach to tackling complex tasks and questions involves breaking them down into smaller parts. However, this method can also make it difficult to understand the overall context of the problem. *Systems thinking*, on the other hand, is a way of recognizing the interconnectedness, interdependence, and cyclical nature of the world around us. Senge (1991) describes systems thinking as a discipline that allows us to see the whole picture, including patterns and relationships, rather than just isolated incidents. It provides tools and techniques to understand the underlying structures of complexity, making it possible to identify potential improvements with


DOI: 10.1201/9781003294276-11

minimal effort. Systems thinking also offers a new way of thinking about complex issues by expanding and reshaping our understanding.

Over the 20th century, the general principles of systems thinking have been developed and applied across various fields, ranging from physical and social sciences to engineering and management. In today's world, systems thinking is crucial for addressing a wide range of issues, such as combating global warming, preventing war, treating addiction, regulating advanced technologies, strengthening democracy, and making education more relevant for the future.

## 11.2 THE HISTORY OF SYSTEMS THINKING

The idea of understanding the world as an interconnected whole has been recognized by cultures around the world for centuries. Indigenous cultures have long observed the complex interdependencies in nature, as reflected in their traditional stories and beliefs. Ancient Chinese philosophy, as described in the classical Chinese text *Tao Te Ching*, credited to Tzu Lao (1998), poetically captures this holistic view of the world as the Tao, which is "undefined and complete," "formless," and "reaching everywhere and in no danger of being exhausted." Similarly, Indian Buddhist philosophy, which dates to the same period as Taoism, describes the world as a constantly changing and evolving system of interwoven causes and effects. It also recognizes human beings as complex systems that are an integral part of the world and advises its followers not to be deceived by the human tendency to desire a fixed reality. The concept of holism, which is synonymous with systems thinking, was developed by the Presocratic philosophers of ancient Greece, the Pythagoreans, and is derived from the Greek word "holon," meaning a universe made up of integrated wholes that cannot be understood by their parts alone (Seibert, 2018).

While Aristotle believed that naming and classification were essential, he also acknowledged, in line with the Presocratic tradition, that the whole is greater than the sum of its parts. However, this idea fell out of prominence during the 17th-century scientific revolution. One of the principal rules of Descartes' reductionistic approach involved "to divide each of the difficulties under examination into as many parts as possible, and as might be necessary for its adequate solution" (Descartes, 1637). Reductionism aims to simplify complexity to offer a rational explanation of phenomena, but it struggles to account for the emergence of new characteristics in the whole that are not present in its components.

In the late 19th century, John Stuart Mill pointed out that "no mere summing up of the separate actions of . . . elements will ever amount to the action of the living body itself" (Mill, 1859). Systems thinking emerged in the 1920s as a response to the limitations of scientific analysis. Instead of breaking down a system into its parts and using the properties of those parts to explain the behavior of the whole, systems thinking examines the properties of the system that arise from the interactions of the parts. Biologists in particular noticed how hierarchical levels of organization lead to increasing complexity in organisms, with new properties emerging at each level. Ludwig von Bertalanffy was one of the first to develop the concept of Systems Theory through his "Organismic Biology" theory (Bertalanffy, 1966) as early as the 1920s. The advancement of cybernetics, the study of control systems and

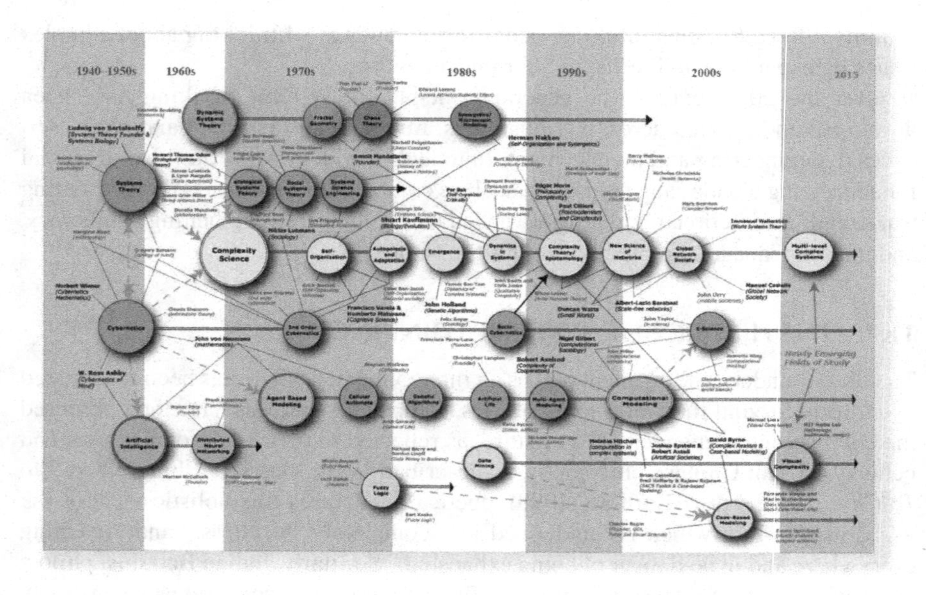


**FIGURE 11.1** "Leading figures and key areas of research in the complexity sciences."
*Photo courtesy:* Wikimedia Commons

information flow, further emphasized the importance of feedback loops. Figure 11.1 offers an overarching perspective on the historical development of systems thinking.

Jay Forrester of MIT Sloan School of Management, the founder of systems dynamics, introduced the term "systems thinking" and recognized that it should be taught to students from an early age (Forrester, 2016). Forrester's student Donella Meadows continued to investigate systems theory in her book, *Limits to Growth* (1972), to describe the relationship between human activity and the environment and to express her concerns about the long-term effects of exponential economic and population growth on natural resources and their implications for the future of humanity. Peter Senge, who was also a student of Forrester, has noted that from a young age, we are taught to "fragment the world" by breaking problems apart. While this may make complex tasks and subjects more manageable, we also lose our connection to the larger whole.

## 11.3 THE CHALLENGES AND BENEFITS OF TEACHING SYSTEMS THINKING

Most educators would agree that helping students discover interconnections between different concepts and subjects, adapt to the challenges of the rapidly growing complexity of their world, and become resilient in the face of uncertainty are critical objectives of modern education. A longitudinal study conducted by Waters Center for Systems Thinking (https://waterscenterst.org/) has confirmed the impact of systems thinking in schools. However, educational programs still do not place enough emphasis on helping students feel comfortable with complexity. While the Next Generation Science Standards (www.nextgenscience.org/) include cross-cutting concepts such as "stability and change" and "systems models," systems thinking has yet to become a fundamental aspect of modern school curriculums. This is unfortunate as enhanced systems thinking (Figure 11.2) will shape children's perspectives in all areas of their lives for years to come, from personal relationships and social interactions to business decisions and political involvement.

If our role as parents and educators is to prepare our children to thrive in a future world that is uncertain and fast-changing, we should focus on systems thinking as the tool of choice. It should be taught to students starting at an early age. It helps young students adapt to the challenges of the rapidly growing complexity of their world and become resilient in the face of uncertainty. It helps students develop a big-picture view of the world, teaches them to understand the behavior of nonlinear systems (with their tripping points and instabilities), and encourages them to consider any problem from multiple perspectives.

However, before designing an effective curriculum for teaching systems thinking, it may be beneficial to understand both the origins of our innate tendencies to think either analytically or holistically and the biases that can distort our understanding of systems behavior. By reinforcing children's natural propensities to see the world as a whole while making them aware of common misconceptions, we may achieve significantly greater progress than by simply presenting complexity as a "modern" skill to master, similar to computer programming.

In his book *The Master and His Emissary,* a neuroscientist McGilchrist (2019) examines the distinctions in the brain hemispheres concerning our capability to perceive the world as a unified whole where everything is considered in context. He suggests that the

**FIGURE 11.2** Habits of a systems thinker
*Photo courtesy:* Waters Center for Systems Thinking

left hemisphere breaks down and handles the world while the right hemisphere considers context and the overall human experience. McGilchrist explains that every animal needs to be able to eat without being eaten, which requires both focused and broad attention. He proposes that to solve this problem, evolution created two distinct but connected brain hemispheres that work together. Each hemisphere creates a unique perspective of the world. The left hemisphere creates a "re-presented" version of our experience, containing static, bounded, fragmented entities, grouped into classes, on which predictions can be based. McGilchrist argues that while both hemispheres are essential for survival and progress, the left hemisphere's language-based view has become dominant with negative consequences, including a decline in holistic thinking.

The ability to identify similarities among things and group them accordingly allows us to divide the interconnected world into categories. As a result, we tend to assign properties to objects based on their category rather than their interactions within the larger whole. Psychologists refer to this bias as "dispositional attribution," in contrast to "situational attribution," which explains an entity's behavior through its specific interactions. Recognizing an object as a category member allows us to make rapid predictions about it, increasing our evolutionary fitness. Pinker (2002) notes that, without this ability, the category "duck" would be useless in a world where "walking quacking objects" were no more likely to be a source of food than any other object. Even newborn babies may possess a few basic categories to aid in their learning process.

Every word we use splits our experience into categories that guide our expectations. When the mind categorizes objects that share features and behaviors, it envisions an unobservable "essence" that gives each category its identity. However, our language conditions the mind to see the world as full of "things" with fixed essences that determine their qualities, imposing invisible discontinuities between a seed and a flower or a caterpillar and a butterfly. Essentialist bias emerges at a very early age across different cultures. The essences of complex systems may be fundamental to our thinking, particularly when it comes to living systems and, occasionally, inanimate objects such as computers.

Another cognitive reason why targeted educational interventions may be necessary to help students understand complexity is "cognitive economy," a strong preference for simpler explanations. Research in psychology suggests that our brains, which are geared toward taking action, prefer simplified (but energy-efficient) *mental models* to more complex ones. Even scientists use the principle of parsimony, known as "Occam's razor," to gravitate toward simpler explanations, which may not always be justified by the phenomena under investigation, but rather reflect a deep-seated bias. Hence, even though we are surrounded by complex systems, our mental models of them are often inaccurate. For example, most of us can easily imagine the Pegasus flying gracefully in the sky, which is not physically possible due to the wings being either too short to lift the Pegasus off the ground or too heavy for the Pegasus to lift them off the ground.

It is important to identify the boundary of a system in order to understand its behavior within its environment. The boundary separates what is inside the system from what is outside. Changing our mental model of the system by moving its boundary can change our perception of the system and lead to more appropriate actions. For example, considering gut microbiota as part of the body can change our approach to health, and understanding the interdependence between parts of a forest ecology can motivate conservation efforts.

In the context of complex systems, we tend to underestimate the impact of seemingly insignificant events on the dramatic outcomes they may have set in motion (the "butterfly effect"). Similarly, it is easier for us to interpret large-scale social dynamics as the result of a powerful conspiracy rather than the confluence of economic, technological, regulatory, and demographic changes that are happening in plain sight. Due to the reciprocal flow of influence among the components of a system, the causes and effects of specific interactions are not closely related in time and space, and the results of any intervention in a system may be unexpected.

The inability to incorporate delayed feedback in our predictions is a common misconception that hinders our understanding of complex systems. For example, while touching a hot stove provides immediate visceral feedback, which facilitates swift learning, longer delays, such as those between catching an infection and getting sick, may take weeks, months, or even years. Since infections are caused by invisible bacteria and viruses, learning about them would require an accurate mental model of the complex phenomena in question. As a result of the different representations of the dynamics in our mental model, we may share the same static model of the system but disagree about the behaviors implied by it.

It is particularly challenging to understand our impact on a complex system when we are a part of it. For example, a seemingly functional "perpetual motion machine" was presented at the Paris Expo in the 1860s. It was a simple device consisting of a large wheel on a rolling bearing. Visitors took turns trying to stop it from rotating, but the wheel kept turning as soon as they let go. In reality, the machine had a spring-loaded mechanism concealed inside, which the visitors kept inadvertently rewinding while trying to stop the wheel. Such unforeseen side effects of our interventions in complex systems are often referred to as "unintended consequences." As Peter Senge observed, "Today's problems come from yesterday's solutions." Our interventions in the system will have many unexpected effects on the parts we were unaware of influencing. Examples of this abound in both history and popular culture, such as the story of the sorcerer's apprentice and the tale of King Midas.

Unintended consequences often result from an overarching focus on immediate improvements, which, combined with our inadequate mental models, often make the original problem even worse. For example, during the Great Plague of London, authorities encouraged the extermination of dogs and cats, believing they were spreading the disease. In fact, the plague was carried by fleas that were transported by rats, whose population was kept in check by dogs and cats. Another example is the "Cobra Effect," named after a sequence of decisions by the British government in colonial India (Newell and Doll, 2015). To reduce the risks associated with cobras, the government offered a bounty for every dead snake. The incentive appeared successful as the number of cobras started to decrease, but the number of dead cobras brought to the officials kept on rising. People started breeding cobras for the bounty. Having realized their mistake, the government shut down the reward program, and the breeders set the snakes free, expanding the cobra population far beyond the original one.

Combined with our failure to account for time delays, our limited ability to account for the consequences of our actions in complex systems may correct over-correcting the apparent discrepancies. Examples of this include getting burned in a shower after continuing to turn up the knob to make the water warm up and military pilots producing uncontrollable oscillations in their aircraft through their efforts to stabilize it by countering the effect of their previous commands (Skybrary).

It is therefore insufficient to teach about complexity as a topic in the school curriculum – we need to teach children how to think about it. Young students enjoy discovering interconnections in the world around them, such as examining the effects of their actions on family and friends, and then observing how those actions come back to them, often after substantial delays. Older children are eager to learn how remote astronomical events affect life on Earth and how climate change doomed ancient societies. Still older children may recognize the effects of inadequate healthcare policies on the national economy and education. A variety of tools, such as computer simulations and educational video games, which have become widely available in recent years, can provide new opportunities to gently introduce systems thinking concepts early on.

According to Uri Wilensky, the founder of Northwestern University's Center for Connected Learning and Computer-Based Modeling, it is vital to teach children about complexity from a young age to help them adapt to the increasingly complex world (Brodsky, 2021). He argues that understanding how complex systems change over time is crucial, as even small changes in a critical parameter can greatly impact the overall system. This can be seen in phenomena such as phase transitions, epidemics, and uncontrollable wildfires. Teaching systems thinking does not have to focus on abstract concepts and can instead be incorporated into various courses as a new perspective. Representing concepts in a way that is easy to understand is essential in making them accessible to everyone. For example, the adoption of Hindu-Arabic numerals greatly improved the ability to do multiplication and had a transformative effect on the world.

Wilensky uses agent-based models as his preferred method for helping students understand complex, nonlinear processes. These models are based on the idea that nearly every situation involves interactions between individual actors, known as agents. Each agent has a set of rules that govern their behavior and interactions with other agents, which leads to the emergence of the overall system's behavior. He has developed an educational tool called NetLogo, which uses these models to make complexity accessible to students of all ages, by helping them visualize complex systems without the use of mathematical formalism. His research (Jacobson and Wilensky, 2006, 2014) suggests that this method leads to a deeper understanding of complex systems compared to traditional science instruction. Additionally, it can help compensate for our innate biases and provide actionable insights without fundamentally altering our cognition. An example of this is a NetLogo forest wildfire model (https://ccl.northwestern.edu/netlogo/models/Fire) which reflects the perspective of a tree, with simple rules of whether it will catch fire based on the "burning status" of its neighboring trees. Even this seemingly simple model can be misleading to our intuition. This serves as a reminder that our intuition about complex systems is often incorrect.

## 11.4 ASTROBIOLOGY AS A UNIQUE TOOL FOR TEACHING SYSTEMS THINKING

Every time I turn on the faucet, I ask myself, how does the water flow on other planets?

> Should we define a human as a creature that is willing and able to change itself?
>
> I really loved the complexity and emergence topic as well as the Golden Record! The complexity and emergence topic gave me a name for the things I like to think about. The Golden Record gave me an opportunity to think about the complexity of our current human knowledge.
>
> (Students' quotes, Introduction to Astrobiology class, students age: 11–14 years old)

There is a growing interest in teaching the holistic approach to science among K-12 science education and parental communities. The exciting field of astrobiology provides a great example of a multidisciplinary perspective that emphasizes systems thinking, complexity, and emergence. By presenting familiar scientific phenomena in unusual settings and emphasizing the process over results, it helps students maintain their natural curiosity and develop critical thinking.

Astrobiology, which studies the emergence of life, has an immense potential to integrate concepts of complexity and emergence into STEM education for upper elementary and middle school students. Through examining the stellar life cycle, from cold hydrogen clouds to the collapse of stars, children can learn about how gradual accumulation can lead to sudden changes. Studying the nucleosynthesis within stars can help children understand the connections between star processes, the formation of new stars and planets, and ultimately, the emergence of life. Space science, dealing with enormous distances and durations, is literally about *the* "big picture" of the Universe as a whole.

Astrobiology allows students to take a systems view of the world by providing a framework to understand the interconnectedness of systems on a cosmic scale. It encourages students to adopt systems thinking approach and to recognize the complexity of the universe and how the whole is greater than the sum of its parts. Additionally, astrobiology provides a unique opportunity for students to understand the evolution of systems over time and the ways in which different disciplines come together to form a comprehensive understanding of the world. The interdisciplinary nature of astrobiology makes it an ideal field for introducing students to systems thinking.

Astrobiology brings together scientific disciplines that are frequently taught in isolation (Crawford, 2019; Staley, 2003). Through the lens of systems thinking, children see everything as deeply interconnected on every level – and always changing. Astrobiology, including cognitive astrobiology, helps students look at themselves from a different, extraterrestrial perspective and consider the potential diversity of forms that other forms of life and mind could take. It instills a sense of wonder about the world's many unanswered questions and encourages students to think about global problems from a systems perspective, such as how to maintain life on Earth and how to tackle complex issues facing humanity.

Astrobiology has a history of challenging our preconceived notions. For instance, the concept of a habitable zone has been reevaluated. For a long time, the possibility of life on the moons of Jupiter or Saturn was not taken seriously, because they were considered too cold. This is an example of how as surface dwellers, we tend to place a high value on life that exists on the surface. However, recent discoveries have shown that life can also exist underground or under ice, which is applicable to planets that are too hot, too cold, or lack protection from cosmic rays.

In our online educational program Art of Inquiry (www.artofinquiry.net) and the ASK program at the Blue Marble Space Institute of Science (www.bmsis.org), we teach systems thinking to preteens and teens by introducing the main concepts and helping them develop the habit of thinking holistically about the world. Based on our experience (Brodsky, 2019), there is no need to wait for high school or college to introduce ideas such as evolution, time delays, complexity, feedback loops, self-organization, or emergent behaviors. No advanced math background is needed to understand concepts such as natural feedback loops, fractal art, and computer simulations of emerging phenomena. Our observations show that young children are highly motivated by deep conceptual problems and are eager to take intellectual risks. Such problems stimulate imagination, enhance creativity, demonstrate the value of brainstorming, and encourage children to revisit their solutions and seek new insights. Along the way, students share their understanding with their peers and families, serving as ambassadors of systems thinking in their communities.

For example, to illustrate the concept of feedback, we use the Daisy World model (Lovelock and Watson, 1983), a well-known thought experiment in astrobiology, which demonstrates how a simple system can produce complex behavior. The Daisy World model is a simulation of a hypothetical planet that is home to two types of flowers – black and white daisies. As the temperature of the star increases over time, the populations of black and white daisies change in response, creating a feedback loop that affects the planet's surface temperature and the distribution of the flowers. This model helps to demonstrate the concept of self-regulation in complex systems and the potential for such systems to exhibit emergent behaviors. Additionally, we use simple tools like Loopy (https://ncase.me/loopy/), developed by Nicky Case, which allow us to connect entities and their properties with positive and negative feedback loops within an animated simulation.

We often begin our systems thinking instruction by identifying the boundaries and analyzing the behavior of systems within their environments. We use games such as "a system vs. a pile" to help students understand the difference between a collection of disjointed parts and a functional system. For example, students will inquire whether a Lego structure is a "system" and how the answer changes if the blocks can assemble and disassemble the structure on their own. This discussion serves as an introduction to self-assembling molecules, followed by computer modeling experimentation, which leads to the topic of self-organization. Many students find that the concept of self-organization in nature, rather than in human society, is a new and fascinating idea.

Our courses also help students to understand the unintended consequences of their actions. For instance, how does space travel impact our planet? In the 1950s, there was no space debris orbiting the Earth. However, now NASA tracks over half-a-million objects that are large enough to destroy a spacecraft. At this rate, we may eventually find ourselves trapped in a planet without internet satellites, GPS, or interstellar probes. By exposing students to these consequences, we teach them the importance of thorough analysis, reinforce the need for a broad perspective, and help them avoid costly mistakes.

We do not shy away from encouraging students to question the nature of fundamental physical concepts such as matter, space, and time, even if they may initially seem intimidating, as they have the potential to deeply engage students. We may start with seemingly simple questions, such as the following: What is your age on Mars? How does this change in perspective impact children's understanding of concepts such as year, month, and day? By further reflecting on the concept of time, it becomes easier for the students to focus on the role of time in systems behavior and to question assumptions they may have held.

After discussing space and time, we introduce the concept of motion in the universe (Brodsky, 2020). When students are asked if the Earth moves around the Sun, most of them will answer in the affirmative, but when asked if the Sun stays still, they will have to think harder. Many students have an ingrained idea that the Sun is at the center of the solar system and the planets move around it but may not have considered whether the Sun itself is in motion. The concept that everything in the universe is in constant motion can be a revelation to many students. They also learn that any object can be arbitrarily chosen as a "still" reference point. As students become more proficient in systems thinking, we move on to explore the interactions between planets and their host stars. It is noteworthy that many students expect celestial objects to remain unchanged throughout their lifetimes and are excited to learn about the life cycle of stars and planets.

We explore how complex structures and behaviors can arise from basic components at all scales, both small and large. We examine a variety of examples that are familiar to students, such as crystallization, neural networks, bee swarms, tornadoes, star formation, and the emergence of civilizations. We also emphasize the nonlinear nature of self-organization and explore the conditions that trigger and disrupt it. NetLogo provides an opportunity for computer-savvy students to program various agent-based complexity processes. The discussion of emergence prompts students to consider how simple chemical systems can give rise to more complex systems, such as living and thinking organisms, and how those interact and adapt over time. For example, self-organization in physics is demonstrated through Rayleigh-Bénard convection cells in viscous liquids and trade winds (Hadley cells). This knowledge can be applied later when studying the processes in planetary mantles and atmospheres. It is surprising and counter-intuitive to many students that self-organization can give rise to coherent actions that occur without a controller, either internal or external.

## 11.5 CONCLUSION

Our observations make it clear that students have a strong preference for understanding complex systems and the patterns and relationships within them, as opposed to simply memorizing facts. This type of learning often leads to a sense of excitement and accomplishment as students begin to recognize scientific concepts in diverse contexts and observe how systems change over time. Even young children enjoy making connections and building diagrams that help them understand cyclical phenomena in the natural world.

Unfortunately, the educational system may not be keeping pace with the rapidly changing world and may not be providing children with the skills they need to navigate it. Traditional schooling methods often fall short when it comes to fostering this type of systems thinking. The fragmented nature of the typical curriculum makes it difficult for students to grasp the big picture and develop the analytical tools needed to understand complex systems. This can hinder their ability to fully engage with and appreciate the interconnectedness of the natural world. The introduction of systems thinking via an exciting field of astrobiology helps to support academic engagement, mitigate socioeconomic disparities, and develop the mindset required to come up with new solutions for future challenges.

As an interdisciplinary field that combines knowledge from various sciences, astrobiology offers a unique approach to early science education. By emphasizing open discussion and the preservation of natural curiosity, this approach encourages children to ask questions and not be intimidated by them. It also helps children to be comfortable with the unknown and to appreciate the vastness of what is yet to be explored. In addition to fostering a growth mindset and building mental resilience, this approach encourages students to become active participants in the scientific process, showing them that their ideas are respected and can make a difference. Astrobiology provides an opportunity for children to make sense of their place in the world and to understand the interconnections between different systems, which is crucial to address future challenges.

The open-ended approach to education offered by astrobiology allows even those students with little prior knowledge to participate on an equal footing with their more informed peers. In astrobiology classes, students are often asked to work on group projects that rely heavily on systems thinking, such as redesigning the solar system to make it more hospitable to life, exploring AI scenarios, or developing an IQ scale for extraterrestrial life. Since there is no one "correct" answer, the teacher's role is to model different methods of inquiry and encourage students to consider a variety of solutions in novel situations.

As students work together to advance each other's understanding, they learn to value collaboration over competition. The lack of formal assessments also helps to reduce anxiety and replace fear with curiosity. The class asks profound scientific and philosophical questions that require little factual knowledge, actively inviting students to brainstorm and share their ideas. Our preliminary observations, based on collaboration with an educational nonprofit working with underrepresented children in New York and Los Angeles, as well as Ukrainian refugees, suggest that an astrobiology-based systems-thinking approach has the potential to address educational disparities.

It is our conviction that astrobiology education holds immense potential as a means of engaging students in STEM fields and developing essential systems thinking skills. By providing a comprehensive understanding of the workings of life on Earth and its place in the larger cosmic context, this approach to education can foster a sense of planetary responsibility and awareness in the next generation. With the ability to think critically about the interconnectedness of all living systems, these students will be better equipped to navigate the challenges of the 21st century (Brodsky, 2020).

I would like to end this chapter with another student's quote: "Astrobiology is unique. It is the study of us, and life among the stars. It brings together social studies, space studies, and biology. It brings us together!"

## REFERENCES

Art of Inquiry, *Online Educational Program for Young Explorers* (www.artofinquiry.net)

*Astrobiology Studies for Kids (ASK) Program at Blue Marble Space Institute of Science* (www.bmsis.org)

Bertalanffy, Ludwig Von, *Organismic Psychology and Systems Theory*, Clark University Press, 1966

Brodsky, Julia, *"Astrobiology, Complexity and Emergence" for Middle School*, AbSciCon, 2019

Brodsky, Julia, *Playing with the Unknown: Astrobiology for Young Students*, AbSciCon, 2020a (www.researchgate.net/publication/344301306)

Brodsky, Julia, *Space-inspired Systems Thinking for 10–13 yr Olds*, The Innovation-Medium, 2020b

Brodsky, Julia, *How Net Logo's Complex Systems Models Introduce the Ideas of Complexity to Millions Around the World*, Forbes, 2021

Crawford, I. A., Widening Perspectives: The Intellectual and Social Benefits of Astrobiology, Big History, and the Exploration of Space. *Journal of Big History*, III(3), 205–224, 2019

Descartes, Rene, *Discourse on the Method*, 1637.

Forrester, Jay W., Learning through System Dynamics as Preparation for the 21st Century, *System Dynamics Review*, 32, 187–203, 2016

Jacobson, M.J., and Wilensky, U., Complex Systems in Education: Scientific and Educational Importance and Implications for the Learning Sciences, *Journal of the Learning Sciences*, 15(1), 11–34, 2006

Jacobson, M.J., and Wilensky, U., *Complex Systems and the Learning Sciences*, The Cambridge Handbook of the Learning Sciences, Second Edition, Cambridge University Press, 2014, pp. 319–338

Kauffman, S. A. *A World Beyond Physics: The Emergence And Evolution Of Life*. Oxford University Press, 2019

Lao, Tzu. *Tao Te Ching: A Book About the Way and the Power of the Way*. Shambhala Press, 1998

Loopy (https://ncase.me/loopy/)

Lovelock, James and Watson, Andrew, *Biological Homeostasis of the Global Environment: The Parable of Daisyworld*, Tellus, 1983

McGilchrist, Iain, *The Master and His Emissary*, Yale University Press, 2019

Meadows, Donella et al., *Limits to Growth*, Potomac Associates, 1972

Mill, John Stuart, *System of Logic, Rationative and Inductive; Being a Connected View of the Principles of Evidence and the Methods of Scientific Investigation*, Harper and Brothers, 1859

NetLogo (https://ccl.northwestern.edu/netlogo/index.shtml)

NetLogo Forest Wildfire Model (https://ccl.northwestern.edu/netlogo/models/Fire)

Newell, Barry, and Doll, Christopher, *Systems Thinking and the Cobra Effect*, Our World, United Nations University, 2015

Next Generation Science Standards (www.nextgenscience.org/)

Pinker, Steven, *The Blank Slate: The Modern Denial of Human Nature*, Penguin Books, 2002

Seibert, M.K., Systems Thinking and How It Can Help Build a Sustainable World: A Beginning Conversation, *The Solutions Journal*, 9(3), pp 1-9, July 2018.

Senge, P.M., *The Fifth Discipline: The Art and Practice of the Learning Organization*, Doubleday, 1991

Skybrary (https://skybrary.aero/articles/pilot-induced-oscillation)

Staley, James T., Astrobiology, the Transcendent Science: The Promise of Astrobiology as an Integrative Approach for Science and Engineering Education and Research, *Current Opinion in Biotechnology*, 14, 347–354, 2003

Waters Center (https://waterscenterst.org/)

Wikimedia Commons (https://commons.wikimedia.org/wiki/File:Complexity_Map.svg)

# 12 Knowledge Gaps in Astrobiology

## *Can the Systems Approach Help Elucidate Them?*

*Vera M. Kolb*

## 12.1 INTRODUCTION AND BACKGROUND

### 12.1.1 A Brief Review of the Systems Approach versus Reductionist Approach

The systems approach to solving complex problems is superior to the reductionist approach, as we have shown in the primer on the systems approach, in Chapter 2 of this Guidebook. The primer has a substantial list of references on the subject. We do not repeat these references here.

We present first a brief rendition of the systems approach versus the reductionist approach. In the latter, one tries to ascertain how a system operates by focusing on the *individual parts* of the system. This approach builds the knowledge about the parts which constitute the system but are not necessarily fruitful in elucidating how the system *functions*. This creates a *knowledge gap* between the understanding of the nature of the individual parts of a system, which may be substantial, and their exact role in the functioning of the system, which may be lacking. The systems approach, which focuses on interactions between the parts, their organization, hierarchy, and the complex features such as autocatalysis and feedback control, is generally more successful than the reductionist approach in understanding the operation of the system.

The advantages of the systems approach as compared to the reductionist approach should not prevent us from realizing the importance of the reductionist approach. For example, consider a case in which we have identified the parts of the system but have not studied their individual properties in detail. If so, we may not understand how these parts can interact with each other to eventually bring about the functioning of the system. Because of this, we must keep studying the nature of the parts.

A generalized scheme of closing the knowledge gaps by the systems approach is shown in Figure 12.1.

DOI: 10.1201/9781003294276-12

**FIGURE 12.1** A generalized scheme of closing the knowledge gap by the systems approach. The cogwheels denote the system, and people walking over the gap show researchers bridging the gap.

*Photo courtesy:* Shutterstock

### 12.1.2 What Are the Knowledge Gaps?

It seems obvious what the knowledge gaps are. Nevertheless, we need to clarify what is meant by a particular knowledge gap. If confusion exists at such a basic level, it will extend to the systems approach as well. We give some general guidance for the clarification process. It is adopted and modified from the paper by Schuerger et al. (2018), titled "Science questions and knowledge gaps to study microbial transport and survival in Asian and African dust plumes reaching North America". While the subject of their paper is not in line with our astrobiology objectives, their methodology is very clear, and we can adopt it for our purposes. The authors first define key science questions for their study. Then, per each individual question, they list the knowledge gaps, which are about 10 per each or their science questions in their case. Following their methodology, we first need to state clearly what our science question is. Then, we describe the knowledge gaps which are preventing us from being able to answer fully our science question.

We now give an example of this methodology as applied to one astrobiology knowledge gap, namely that of the role of autocatalysis in the origin of life. We first need to understand what is meant by *autocatalysis*. Thus, our science question needs clarification of this term. It may appear that the term autocatalysis is clear, but researchers use this term in different ways. For example, several definitions of autocatalysis are presented in the paper by Hordijk, Hein, and Steel (2010), titled

"Autocatalytic sets and the origin of life". We present some of these definitions, which vary for different groups of scientists. For example, chemists use the term *autocatalytic reactions* for cases in which a molecule directly catalyzes its own productions. The term "*autocatalytic set*", however, does *not* mean a set of autocatalytic reactions. Instead, it means that a set of molecules and reactions are "*collectively autocatalytic*" in which all molecules help in producing each other through mutual catalysis in a closed, self-sustained system. More definitions of autocatalysis were presented; this is just one example. The lesson to be learned is that we must carefully examine and understand what different researchers mean by autocatalysis in their studies. But our search for exact definitions does not stop here. We also need to clearly define our scientific question which contains the term autocatalysis, such as the role of autocatalysis in the origin of life. We need to define the aspects of the origin of life to which autocatalysis will be applied. A review and a very clear step-by-step definition of these is presented by Lancet, Zidovetzki, and Markovitch (2018) in their paper "Systems protobiology: origin of life in lipid catalytic networks". Their paper also represents an exemplar of the systems application to the origin of life problem. More on this subject of autocatalysis in origins of life is covered in Chapters 7 and 13 of this Guidebook.

One of the major knowledge gaps in astrobiology is that of defining life, which we discuss in *Section 12.2*.

## 12.2 KNOWLEDGE GAPS WHICH IMPEDE OUR UNDERSTANDING OF LIFE AS A PHENOMENON

One of the main objectives of astrobiology is to explain the origins of life on the Earth and to search for the putative extraterrestrial (ET) life (Des Marais et al., 2008; Des Marais, 2019, pp. 15–25; Kolb, 2019, pp. 3–13, and the references cited therein). To pursue these objectives, we first need to agree on what life is. Unfortunately, although there are over 100 proposed definitions of life, none are universally accepted. This was reviewed and discussed in depth elsewhere (e.g., Pályi, Zucci, and Caglioti, 2002; Popa, 2004, 2010, 2015; Kolb, 2019, pp. 57–63, and the references cited therein). It appears that there is a *knowledge gap between our understanding of life and our ability to define it*. To remedy this problem, we applied systems approach to defining life (Kolb and Clark, 2023) and proposed a new definition which is systems-based. This definition is shown in the following subsection.

### 12.2.1 Our Definition of Life Which Is Systems-Based

Our definition of life (Kolb and Clark, 2023) comprises three parts, which are shown next. It is slightly modified from the original one.

Part 1: *We define life as it is on the Earth:*

***Life is a complex system comprising three subsystems – those of metabolism, information, and membrane – and is a part of two larger systems, those of the environment, which supplies nutrients and energy, and biosphere, which, in conjunction with the environment, enables evolution.***

By "complex" in this definition, we mean complexity by chemical diversity, organization, and by established networks.

Part 2*:* For any definition of life to be astrobiologically significant, it must include an explanation of the origin of life. We thus supplement the previous definition by adding the following sentence:

> *Life originated in our Solar System (and perhaps elsewhere) by prebiotic chemical evolution, which led from heaps of prebiotic chemicals to prebiotic proto-life systems, which then underwent transition to life by an unknown process, for which, however, several hypotheses exist.*

To make the previous two parts applicable to extraterrestrial life, we modify Part 1 and show the modification in bold:

> *Life is a complex system comprising of three subsystems – those of metabolism, information, and membrane,* ***or their functional equivalents****, and is a part of two larger systems, those of the environment, which supplies nutrients and energy, and biosphere, which, in conjunction with the environment, enables evolution".*

This definition appears satisfactory for both earthly and the ET life. However, it does not cover the borderline forms of life, which have some but not all properties of life. We present in the following subsection a condensed version of the in-depth discussion on this subject (Kolb and Clark, 2023).

### 12.2.2 Are Viruses Alive and Could a Systems Approach Help Answer This Question?

A *virus* is an infective agent whose genetic material (DNA or RNA) is located within a protein coat, which is known as a *capsid*. Virus also has in some cases an outside envelope consisting of lipids (e.g., Wikipedia, virus; Solé and Elena, 2018). Viruses cannot self-reproduce and are not considered alive based on the definitions of life which require self-reproduction as one of the criteria for life.

However, viruses are able to multiply within the living cells of their host, where they hijack the host's metabolic capacity to reproduce. Prior to infecting cell, viruses exist as particles termed *virions* (e.g., Wikipedia, virus; Solé and Elena, 2018). The life cycle of viruses is shown in Figure 12.1. This figure shows both the virion and the multiplication of viruses within the host.

The controversy about the status of viruses as being alive or not is illustrated in the following statement:

> viruses today are thought of as being in a gray area between living and nonliving: they cannot replicate on their own but can do so in truly living cells and can also affect the behavior of their hosts profoundly . . . viruses, though not fully alive, may be thought of as being more than inert matter: they verge on life.
>
> (Villarreal, 2004)

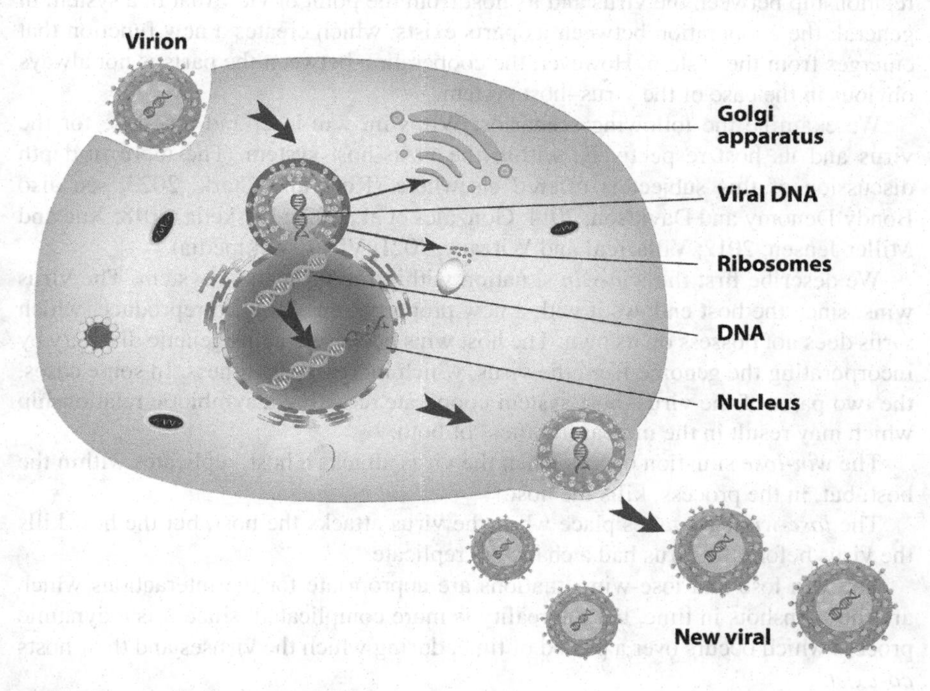


**FIGURE 12.2** Virus replication.
*Photo courtesy:* Shutterstock.

Often, the nonspecialists do not appreciate the difference between the *virion*, which is the virus particle outside the host, and the *virus* inside the host, where it can replicate. This is described well in the following citation:

> And here is the final reflection about what viruses are and whether they are alive: considering the virion particle as the true virus is equivalent to considering a grain of pollen as a redwood or an ovulum as a human. The virion would be equivalent to the germ line of the virus while the virus factory would equate to the somatic line.
>
> (Solé and Elena, 2018)

Since the virion has a potential to become alive under the right circumstances, this creates the gray area and controversy if virion is alive or not.

The controversy if viruses are alive may be resolved by considering the virus–host system. Within this system, the virus gains the ability to reproduce by seizing the metabolic capacity of the host and thus can be thought of as being alive.

Further examination of the virus–host system provides us also with some general insights into the functioning of systems in general. We start by examining the relationship between the virus and its host from the point of view that in a system in general, the cooperation between its parts exists, which creates a new function that emerges from the system. However, the cooperation between the parts is not always obvious in the case of the virus–host system.

We examine the following scenarios: Win-win, win-lose, and lose-lose for the virus and its host respectively, within the virus-host system. The more in-depth discussion of this subject is offered elsewhere (Kolb and Clark, 2023; see also Bondy-Denomy and Davidson, 2014; Gonzáles et al., 2021; Koskella, 2018; Xue and Miller-Jensen, 2012; Villarreal and Witzany, 2021; Virus, Wikipedia).

We describe first the *win-win* situation within the virus–host system. The virus wins, since the host endows it with a new property, the ability to reproduce, which virus does not possess on its own. The host wins because it gains genetic diversity by incorporating the genome from the virus, which increases its fitness. In some cases, the two parts of the virus–host system cooperate, creating a symbiotic relationship which may result in the increased fitness of both.

The *win-lose* situation occurs when the virus attacks a host, replicates within the host, but, in the process, kills the host.

The *lose-win* case takes place when the virus attacks the host, but the host kills the virus before the virus had a chance to replicate.

The win-lose and lose-win situations are appropriate for the interactions which are the snapshots in time, but the reality is more complicated, since it is a dynamic process which occurs over a period of time, during which the viruses and their hosts *co-exist.*

Our focus here is to bring up the win-lose and lose-win scenarios of the two parts of the virus–host system and examine if these should be considered in other systems. The win-lose and lose-win cases are not considered in the systems in general, since the parts of the system are antagonistic toward each other. Typically, the parts of the systems are considered to be only cooperative. In general, for a system to form, exist, and persist, the antagonism between its parts cannot prevail over their cooperation, since the latter is responsible for the functioning of the system. Because of this, the antagonism between the parts needs to be controlled in some fashion. Possibilities include silencing, neutralizing, or destroying the antagonistic parts or preventing them in some other fashion from interfering with the cooperation between the parts of the system. It could be that the parts of the prospective system are not always compatible and that some sort of selection occurs, otherwise the system would not form. Another scenario could be that the system which is formed has parts which do not contribute to the system's function and are thus neutral. These may be a burden on the system, and, as the system evolves, it may develop mechanism which restricts the inclusion of noncooperative parts.

In other cases, the malfunction of some parts within the system may impede or terminate the functioning of the entire system. Investigations of robotic systems may provide some clues along these lines, albeit the robotic systems are much simpler as compared to anything which is alive. This is addressed in Chapter 14 of this Guidebook.

We now further consider the scenario in which the system has neutral parts, which do not have an obvious function in the system but are not antagonistic toward system's formation and functioning. For example, a neutral part in a chemical system would be comprising chemically unreactive species, which seemingly just "sit there" and "do nothing". However, this may be an oversimplification, since the unreactive parts, such as most solvents, dilute the chemical system and slow down chemical reactions and absorb the heat when it is generated by the chemical system. Thus, they affect the chemical system although they are unreactive.

In conclusion of this section, the antagonistic and neutral properties of the parts of the system are complex. We need to be on a lookout for these and find out if they exist in a system. Otherwise, addressing only the cooperation between the parts of the system may impede our knowledge of the functioning of the system in general and prevent us in narrowing the knowledge gap between the interactions of the system's parts and the emergence of the system's function.

## 12.3 THE KNOWLEDGE GAP ABOUT THE TRANSITION FROM ABIOTIC TO BIOTIC (A-2-B) SYSTEMS

We start with a brief background about the transition from abiotic to biotic *(a-2-b)* systems. This transition represents one of the *widest* knowledge gaps in astrobiology. The currently accepted hypothesis is that life evolved from the abiotic matter, but the exact nature of this transition is poorly understood.

The transition to life is hypothesized to be the result of an evolutionary process of matter, starting from simple prebiotic compounds to more complex. This hypothesis is too broad, and its details are sparce. Still, there are fruitful attempts to address this knowledge gap, as shown in the selection of references with various approaches to the problem (e.g., Kolb, 2005, 2012, 2013, 2016; Perry and Kolb, 2004; Kolb and Liesch, 2008; Peretó, 2019; Jeancolas, Malaterre, and Nghe, 2020; Luisi, 2016; Banfalvi, 2021, Kitadai and Maruyama, 2018; Mason, 1991; Deamer, 2019, 2020; Fry, 2019, pp. 437–462; Krishnamurthy, 2017; Longstaff, 2015; Walker, 2015).

As the complexity of the prebiotic chemical systems grew, the so-called *transition zone* (*TZ*) between the advanced pre-biotic systems and the primitive biotic systems was reached (Perry and Kolb, 2004). Kolb and Liesch expanded on this in their paper titled "Abiotic, biotic, and in-between" (2008). Based on the *TZ* concept, there are three systems: The abiotic (*a*), the transition zone (*TZ*), and the biotic (*b*). The *TZ* system does not encompass the entire *a* and the entire *b*. Instead, it includes only parts of these systems: The advanced abiotic and the primitive biotic. Thus, the broader systems *a* and *b* need to be narrowed. This process can be visualized as follows.

The evolution of an abiotic system proceeds from a more primitive one, all the way to the advanced prebiotic one. The latter has a capability to evolve to life, since it is quite complex. On the other hand, the biotic system spans from the most primitive alive systems to the most advanced forms of life. We need to focus only on the primitive alive systems. *Within such a context, the TZ is hypothesized to exhibit selected properties of the advanced prebiotic and the primitive biotic systems.*

Let us first examine the chemical evolution from the primitive to the advanced chemical systems. This evolutionary process starts from the elementary prebiotic compounds from which the building blocks of life are synthesized (e.g., Miller and Orgel, 1974; Mason, 1991; Longstaff, 2015). These relatively simple building blocks are amino acids, sugars, nucleobases, and fatty acids. More complex chemical systems, such as primitive peptides (proteinoids), RNA elements, and rudimentary lipids, are made either from the simple building blocks or by different synthetic routes. Prebiotic chemical systems are coupled with the environment, since they require energy, catalysts, and chemical ingredients. Ultimately, chemical evolution led to the primitive metabolic, informational, and membrane-based compartmental subsystems, which led to the primitive pre-life (e.g., Banfalvi, 2021; Becker et al., 2019; Benner, Kim, and Carrigan, 2012; Cleaves, 2014, 2013; Pascal, Pross, and Sutherland, 2013; Patel et al., 2015). The systems approach to these events is covered in detail elsewhere (e.g., Kolb and Clark, 2023, and the references therein). Here, we focus on the *TZ*.

It is hypothesized that *coevolution* between the proto-metabolism, the primitive informational, and membrane-based compartmental systems led to the *TZ*, which then further evolved to life. Understanding this coevolution would be helpful in narrowing the knowledge gap for the *a-2-b* transition. However, by necessity, this needs to be done piecemeal by looking at the individual subsystems, their subsystems, and the constituent parts. For example, some simple chemical systems may be studied separately by modeling and computational means. One such study, by Wołos et al. (2020), demonstrated the emergence of an advanced feature of the system, namely autocatalysis at the prebiotic level. This study is very important since it shows that complexity can emerge from the simplest starting prebiotic materials. The buildup of complexity does not have to be gradual.

Next, we address some difficulties in studying the parts of the chemical systems, such as the elementary prebiotic compounds and primitive subsystems such as those of proto-metabolism, information, and membrane-based compartments. The evolution from simple to complex is dynamic in time and space. Some chemical systems evolve to more complex ones, while some others are decomposed. Not all chemical systems which are complex have the capability to evolve to life. Some are ephemeral. Thus, we need to look for trends in chemical evolution and focus on the evolved parts and then back-engineer their precursors which we believe were reasonably appropriate for the historic process of the emergence of life. Such back-engineering is not universally accepted. Some researchers agree with this idea and believe that life is a guide for prebiotic syntheses, such as those of nucleotides, which are key building blocks of life (e.g., Harrison and Lane, 2018), while others believe that such syntheses resulted from radically different pathways which are more feasible for the prebiotic chemical conditions (e.g., Patel et al., 2015). There is lots of guessing, intuitions, speculations, and hypothesizing involved. To build more substantial hypotheses, we need to go back and forth constantly and reevaluate the hypotheses considering new data.

It is believed that the three subsystems of the *TZ*, namely the primitive informational, metabolic, and membrane-based compartmental, co-evolved in some fashion. As one example, the coevolution between peptides and RNA was studied (Strazewski, 2019, pp. 409–420).

The coevolution could start with the primitive symbiosis-like prebiotic process in which there was a *merger* of different chemical systems (Monnard and Walde, 2015). While this concept was brought up in the context of prebiological compartmentalization, we can think about it broadly. While RNA-first, metabolism-first, and membrane-first hypotheses have various problems, which have been covered in depth elsewhere (e.g., Orgel, 2000, 2004; Peretó, 2012), one can hypothesize that two of these "firsts" systems evolved separately but then were joined by a primitive symbiosis. If so, the elucidation of the *TZ* may be easier. Instead of looking for the complicated and poorly understood process of coevolution of the *three* pre-life subsystems, one should explore the possibility that a symbiotic subsystem merger of only *two* such subsystems may have occurred at some point, and the third one followed. Further, the evolution may have started as a free exchange of valuable materials between the primitive pre-cellular compartments, such as coacervates, micelles, and vesicles (e.g., Monnard and Walde, 2015; Kolb, 2019, pp. 483–490; Kahana and Lancet, 2021; Deamer, 2000, 2017), reminiscent of the horizontal gene transfer (e.g., Gogarten and Papke, 2019, pp. 527–534). Such an exchange could continue over a period of time, even after a primitive genetic and metabolic systems were formed. This would be an evolutionary period of sharing valuable materials, rather than preventing sharing in a competition process. Thus, it may not be productive to stick to one of the "firsts" as separate entities. Our personal preference is the compartments first, and out of the options within compartments, we believe that initially coacervates were most feasible, as Oparin suggested in 1924 (Oparin, 1994, 1965, 1968; Kolb, 2019, pp. 483–490).

In conclusion of this subsection, it appears that the knowledge gap regarding the nature of the *a-2-b* transition and *TZ* can be elucidated and somewhat narrowed by the systems approach.

## 12.4 THE KNOWLEDGE GAP ABOUT THE UNDERSTANDING OF OUR LIFE ON THE EARTH AND CONCEIVING THE NATURE OF THE PUTATIVE EXTRATERRESTRIAL (ET) LIFE

One of the main objectives of astrobiology is the search for extraterrestrial (ET) life (Des Marais et al., 2008; Des Marais, 2019, pp. 15–25; Kolb, 2019, pp. 3–13, and the references cited therein). To search for such a life, we should have some idea about its nature. A generalized and universalized definition of life was proposed by Clark (2019a, pp. 65–74), which encompasses the diverse life on the Earth and is also applicable to the ET life. Clark's definition permits non-self-reproductive entities, such as the mule, children, eunuchs, and viruses to be considered as alive. This definition is applicable also to extraterrestrial environment:

> [W]e combine the two individual words of "life" and "form" into a single, all-representative word encompassing the minimum reproductive set of organisms, the Lifeform (Lf). . . . A Lifeform is a single organism, or a collection of specialist organisms, whose ability to reproduce is enabled by a set of indispensable yet modifiable instructions embedded in the Lifeform. . . . An Organism is any physical entity produced by a Lifeform that can, in a suitable environment, affect the flow and/or

conversion of energy to perform active functions guided by a subset of the Lifeform's instruction set. . . . By definition, an organism is "alive".

(Clark, 2019, pp 65–74)

Proponents of the ET life hypothesize that life is not limited to Earth but is a universal phenomenon and as such exists elsewhere in the universe. This view allows for the systems approach since one can consider the Earthly life and the ET life as subsystems of the life in the universe. These subsystems are not identical but are functionally equivalent. The reason is that our life is the result of chemical evolution on the early Earth whose environment is not likely to be the same across the habitable regions in our Solar System of the universe in general. Chemical evolution in different environments is expected to follow different paths.

A reasonable hypothesis about the nature of the ET life is that it should be similar to our own by its *functions*, rather than its detailed design. This means that the ET life should have the metabolism, information, and membrane-compartmental subsystems, which are *functionally equivalent* to those in our life. For example, the informational genetic biopolymers of life in general need to carry a repeating charge, and thus they should be polyelectrolytes. In addition, the building blocks of these biopolymers need to be regular in size and shape (Špaček and Benner, 2011).

At this time, it is believed that the best candidate for the ET life is Mars, albeit it could have been the past life. Much research is being done by looking for the precursors or remnants of life on Mars in the past and present space missions (e.g., Clark, 2019, pp. 801–817). Further, since at one time Mars and Earth were both habitable, life not only could have existed on Mars, but it may also have originated on this planet (Clark et al., 2021).

Recently, the equipment to search for such life has been proposed, named "Agnostic Life Finder" (ALF) (Špaček and Benner, 2011).

## 12.5 THE KNOWLEDGE GAPS IN THE SEARCH OF THE BIOSIGNATURES ON MARS

Astrobiology strategy for the search for life in the universe was laid out in the National Academy of Science, Engineering, and Medicine report of 2019. Life, present or past, can be detected via its biosignatures. The knowledge gaps in this area were discussed recently by Cabrol et al. (2021), in their paper titled "Addressing strategic knowledge gaps in the search for biosignatures on Mars", and the publication by the National Academies of Sciences, Engineering, and Medicine (2022), titled "Independent review of the community report from the biosignature standards of evidence workshop". It appears that the consensus has not been reached. However, the systems approach has been applied, which was helpful in elucidating this knowledge gap. For example, Cabrol and co-workers (2021) promote the consideration of coevolution of *life–environment system*, albeit life's existence (past/present) is still putative on Mars. Looking not just for life but also for the imprint that life had on environment is a positive example of the systems approach. The NAS report (2022) reviews potential biosignatures detection and evaluation, but, in this process, it takes into the account the input of not only scientists, but also of journalists and a broader community. It also re-evaluates the recommendations of one group of scientists by another, which, however, is equally qualified and knowledgeable about the subject. This approach illustrates the *systems approach* to *evaluation*, which uses additional evaluators rather than a narrow group.

The knowledge gaps in searching for biosignatures on Mars and elsewhere have been considerably narrowed by the recent work by Neveu et al. (2018), who introduced the ladder of life detection method, and by Schwieterman et al. (2018), who considered biosignatures on exoplanets. Both works are practical and rich in actual details.

## 12.6 KNOWLEDGE GAPS IN ASTROBIOLOGY EDUCATION

In general, the system's approach assumes that the parts of the system are identified and that we are knowledgeable about them. A lack of understanding of how a system functions is typically blamed on our lack of familiarity about the way the parts of the system interact with each other. However, the problem may be that some parts of the system are not included in the study. We illustrate this point on the example of the educational subsystem comprising the students. The parts of the system, namely students, are assumed to be automatically included in the system. This may not be the case, since there are underrepresented students who may not be included in the system. *We cannot assume that the missing parts of the system would not influence the rest of the system if included.* An assumption that the missing parts will behave in the same way as the recognized parts, and that the recognized parts will not be influenced by the parts that are missing, may not be correct. Therefore, if *parts of the systems are missing*, the systems approach may not give correct explanation for the system's function. This situation is depicted in Figure 12.3.

Much effort has been made to narrow such knowledge gaps in astrobiology education by including the underrepresented groups in the education pool (Nadkarni et al., 2020; Scalice, 2019). Nadkami *et al.*'s paper titled "Effects of astrobiology lectures on knowledge and attitudes about science in incarcerated population" and Scalice's chapter titled "Astrobiology-as-origins-story; Education and inspiration across cultures" describe recent efforts to include Native Americans and incarcerated population in the education pool.

## 12.7 SOME RECENT EVALUATIONS OF KNOWLEDGE GAPS IN ASTROBIOLOGY BY THE GROUPS OF EXPERTS

A recent paper by Müller et al. (2022), titled "Frontiers in prebiotic chemistry and Early Earth", brings up various knowledge gaps in current understanding of astrobiology, including some controversial ones. We select some illustrative examples.

The authors state that "The conditions, chemical components, and processes that favored the rise to life are intensely debated and remain unresolved". This statement summarizes serious and continuous knowledge gaps, which started at the inception of the scientific search for the origins of life and have remained to this date, despite much progress in the field.

**FIGURE 12.3** The knowledge gap in understanding the functioning of the system as caused by the missing information. The puzzle can be solved only if the missing data is filled.
*Photo courtesy:* Shutterstock

The energy requirement for life has been studied, and various forms of protobiological energies have been proposed. They include thermal, chemical, mechanical, and solar energies, as some examples. These are not universally applicable to all individual prebiotic chemical processes, which create both confusion and knowledge gaps. The authors recommended that a "systems chemistry" approach is needed to integrate these processes. We can then look at the chemical process utilizing the energy source which it needs, within the chemical system in which energy is a specific part.

A long-recognized knowledge gap about the emergence of homochirality remains a knowledge gap. It appears to be closing (Soai, Matsumoto, and Kawasaki, 2019; Blackmond, 2019, see also Chapter 7 of this Guidebook).

Müller et al. (2022) addressed the transition from chemistry to chemical evolution to Darwinian evolution and stated: "An important and controversial topic is the nature of chemical progression and primitive evolution prior to establishment of the Central Dogma of Molecular Biology, which involves DNA, RNA and polypeptide". The authors thus acknowledge the controversy about this topic. In our opinion, it is better to do so and look at the problem as a knowledge gap, rather than taking strong and polarizing stands which prevent the integration of knowledge.

Another serious knowledge gap is about the nature of the First Universal Common Ancestor (FUCA) and the Last Universal Common Ancestor (LUCA). We believe that both could be considered as a system which changes in time. The details of FUCA to LUCA transition remain to be fully understood and thus represent a knowledge gap.

## REFERENCES

Banfalvi, G. Prebiotic pathway from ribose to RNA formation. *Int J Mol Sci* **2021**, *22*(8), 3857. https://doi.org/10.3390/ijms22083857

Becker, SA; Feldmann, J; Wiedemann, S; Okamura, H; Schneider, C; Iwan, K; Crisp, A; Rossa, M; Amatov, T; Carell, T. Unified prebiotically plausible synthesis of pyrimidine and purine RNA ribonucleotides. *Science* **2019**, *366*(6461), 76–82. https://doi.org/10.1126/science.aax2747

Benner, SA; Kim, HJ; Carrigan, MA. Asphalt, water, and the prebiotic synthesis of ribose, ribonucleosides, and RNA. *Acct Chem Res* **2012**, *45*(12), 2025–2034. https://doi.org/10.1021/ar200332w

Blackmond, DG. The origin of biological homochirality. *Cold Spring Harb Perspect Biol* **2019**, *11*, a032540. https://doi.org/10.1101/cshperspect.a032540

Bondy-Denomy, J; Davidson, AR. When a virus is not a parasite: The beneficial effects of prophages on bacterial fitness. *J Microbiol* **2014**, *52*(3), 235–242. https://doi.org/10.1007/s12275-014-4083-3

Cabrol, N; Bishop, J; Cady, SL; Demergasso, C; Hinman, N; Hoffman, M; Kanik, I; Moersch, J; Noffke, N; Parro, V; Phillips, C. Addressing strategic knowledge gaps in the search for biosignatures on Mars. *Bull Am Astron Soc.* **2021**, *53*(4), 223. https://doi.org/10.3847/25c2cfeb.0eed7a57

Clark, BC. A generalized and universalized definition of life applicable to extraterrestrial environment. In *Handbook of Astrobiology*, Kolb, VM, Ed., CRC Press: Boca Raton, FL, **2019a**, pp 65–74.

Clark, BC. Searching for extraterrestrial life in our Solar System. In *Handbook of Astrobiology*, Kolb, VM, Ed., CRC Press: Boca Raton, FL, **2019b**, pp. 801–817.

Clark, BC; Kolb, VM; Steele, A; House, CH; Lanza, NL; Gasda, PJ; VanBommel, SJ; Newsom, HE; Martínez-Frías, J. Origin of life on Mars: Suitability and opportunities. *Life* **2021**, *11*(6), 539. https://doi.org/10.3390/life11060539

Cleaves, HJ. Prebiotic chemistry: Geochemical context and reaction screening. *Life* **2013**, *3*(2), 331–345. https://doi.org/10.3390/life3020331

Cleaves, HJ. Prebiotic synthesis of biochemical compounds: An overview. In *Astrobiology: An Evolutionary Approach*, Kolb, VM, Ed., CRC Press: Boca Raton, FL, **2014**, pp. 83–117.

Deamer, DW. Membrane compartments in prebiotic evolution. In *The Molecular Origins of Life: Assembling Pieces of the Puzzle*, Brack, A., Ed., Cambridge University Press: Cambridge, **2000**, pp. 189–205.

Deamer, DW. The role of lipid membranes in life's origin. *Life* **2017**, 7(1), 5. https://doi.org/10.3390/life7010005

Deamer, DW. *Assembling Life. How Can Life Begin on Earth and Other Habitable Planets?* Oxford University Press: Oxford, **2019**.

Deamer, DW. *Origin of Life: What Everyone Needs to Know*. Oxford University Press: Oxford, **2020**.

Des Marais, DJ. Astrobiology goals: NASA strategy and European roadmaps. In *Handbook of Astrobiology*, Kolb, VM, Ed., CRC Press: Boca Raton, FL, **2019**, pp. 15–26.

Des Marais, DJ; Nuth III, JA; Allamandola, LJ; Boss, AP; Farmer, JD; Hoehler, TM; Jakosky, BM; Meadows, VS; Pohorille, A; Runnegar, B; Spormann, AM. The NASA astrobiology roadmap. *Astrobiology* **2008**, *8*(4), 715–730. https://doi.org/10.1089/ast.2008.0819

Fry, I. The origin of life as an evolutionary process: Representative case studies. In *Handbook of Astrobiology*, Kolb, VM, Ed., CRC Press: Boca Raton, FL, **2019**, pp. 437–462.

Gogarten, JP; Papke, RT. Horizontal gene transfer in microbial evolution. In *Handbook of Astrobiology*, Kolb, VM, Ed., CRC Press: Boca Raton, FL, **2019**, pp. 527–534.

González, R; Butković, A; Escaray, FJ; Martínez-Latorre, J; Melero, Í; Pérez-Parets, E; Gómez-Cadenas, A; Carrasco, P; Elena, SF. Plant virus evolution under strong drought conditions results in a transition from parasitism to mutualism. *Proc Natl Acad Sci* **2021**, *118*(6). https://doi.org/10.1073/pnas.2020990118

Harrison, SA; Lane, N. Life as a guide to prebiotic nucleotide synthesis. *Nat Commun* **2018** *9*(1), 1–4. https://doi.org/10.1038/s41467-018-07220-y

Hordijk, W; Hein, J; Steel, M. Autocatalytic sets and the origin of life. *Entropy* **2010**, *12*(7), 1733–1742. https://doi.org/10.3390/e12071733

Jeancolas, C; Malaterre, C; Nghe, P. Thresholds in origin of life scenarios. *Iscience*, **2020**, *23*(11), 101756. https://doi.org/10.1016/j.isci.2020.101756

Kahana, A; Lancet D. Self-reproducing catalytic micelles as nanoscopic protocell precursors. *Nat Rev Chem* **2021**, *5*(12), 870–878. https://doi.org/10.1038/s41570-021-00329-7

Kitadai, N; Maruyama, S. Origins of building blocks of life: A review. *Geosci Front* **2018**, *9*(4), 1117–1153. https://doi.org/10.1016/j.gsf.2017.07.007

Kolb, VM. On the applicability of the principle of the quantity-to-quality transition to chemical evolution that led to life. *Int J Astrobiology* **2005**, *4*(3–4), 227–232. https://doi.org/10.1017/S1473550405002818

Kolb, VM. On the laws for the emergence of life from abiotic matter. In *Instruments, Methods, and Missions for Astrobiology XV* (Vol. 8521), **2012**, p. 852109. https://doi.org/101117/12.924817

Kolb, VM. Development of the algorithm for life for the search for extraterrestrial life. In *Instruments, Methods, and Missions for Astrobiology XVI* (Vol. 8865). International Society for Optics and Photonics, **2013**, p. 88650B. https://doi.org/10.1117/12.2021040

Kolb, VM, Ed. *Astrobiology: An Evolutionary Approach*. CRC Press: Boca Raton, FL, **2015**.

Kolb, VM. Origins of life: Chemical and philosophical approaches. *Evol Biol* **2016**, *43*(4), 506–515. https://doi.org/10.1007/s11692-015-9361-4

Kolb, VM, Ed. *Handbook of Astrobiology*. CRC Press: Boca Raton, FL, **2019a**.

Kolb, VM. Astrobiology: Definition, scope and a brief overview. In *Handbook of Astrobiology*, Kolb, VM, Ed., CRC Press: Boca Raton, FL, **2019b**, pp. 3–13.

Kolb, VM. Defining life: Multiple perspectives. In *Handbook of Astrobiology*, Kolb, VM, Ed., CRC Press: Boca Raton, FL, **2019c**, pp. 57–64.

Kolb, VM. Oparin's coacervates. In *Handbook of Astrobiology*, Kolb, VM, Ed., CRC Press: Boca Raton, FL, **2019d**, pp. 483–490.

Kolb, VM; Clark, BC III. *Astrobiology for a General Reader: A Questions and Answers Approach*. Cambridge Scholars Publishing: New Castle Upon Tune, UK, **2020**.

Kolb, VM; Clark, BC III. *Astrobiology: A Systems-Level Science*. CRC Press: Boca Raton, FL, **2023**.

Kolb, VM; Liesch, PJ. Abiotic, biotic, and in-between. In *Instruments, Methods, and Missions for Astrobiology XI* (Vol. 7097). International Society for Optics and Photonics, **2008**, p. 70970A. https://doi.org/10.1117/12.792668

Koskella, B. Resistance gained, resistance lost: an explanation for host – parasite coexistence. *PLoS Biol.* **2018**, *16*(9), e3000013. https://doi.org/10.1371/journal.pbio.3000013

Krishnamurthy, R. Giving rise to life: Transition from prebiotic chemistry to protobiology. *Acct Chem Res* **2017**, *50*, 455–459. https://doi.org/10.1021/acs.accounts.6b00470

Lancet, D; Zidovetzki, R; Markovitch, O. Systems protobiology: origin of life in lipid catalytic networks. *J. Royal Soc. Interface* **2018**, *15*(144), 20180159. https://doi.org/10.1098/rsif.2018.0159

Longstaff, A. *Astrobiology: An Introduction*. CRC Press: Boca Raton, FL, **2015**.

Luisi, PL. *The Emergence of Life: From Chemical Origins to Synthetic Biology.* 2nd ed., Cambridge University Press: Cambridge, **2016**, pp 129–154. www.cambridge.org/core/books/emergence-of-life/selfreplication-and-selfreproduction/6A5751130DE0264C4F017AAC73DB1FF6

Mason, SF. *Chemical Evolution, Origins of the Elements, Molecules and Living Systems.* Oxford University Press: Oxford, **1991**.

Miller, SL; Orgel, LE. *The Origins of Life on the Earth.* Prentice Hall: Englewood Cliffs, NJ, **1974**.

Monnard, PA; Walde, P. Current ideas about prebiological compartmentalization. *Life* **2015**, *5*(2), 1239–1263. https://doi.org/10.3390/life5021239

Müller, UF; Elsila, J; Trail, D; DasGupta, S; Giese, CC; Walton, CR; Cohen, ZR; Stolar, T; Krishnamurthy, R; Lyons, TW; Rogers, KL. Frontiers in prebiotic chemistry and early Earth environments. *Orig Life Evolut Biosph.* **2022**, *7*, 1–7. https://doi.org/10.1007/s11084-022-09622-x

Nadkarni, NM; Scalice, D; Morris, JS; Trivett, JR; Bush, K; Anholt, A; Horns, JJ; Davey, BT; Davis, HB. Effects of astrobiology lectures on knowledge and attitudes about science in incarcerated populations. *Astrobiology* **2020**, *20*(10), 1262–1271. https://doi.org/10.1089/ast.2019.2209

National Academies of Sciences, Engineering, and Medicine. *An Astrobiology Strategy for the Search for Life in the Universe.* The National Academies Press: Washington, DC, **2019**.

National Academies of Sciences, Engineering, and Medicine. *Independent Review of the Community Report from the Biosignature Standards of Evidence Workshop.* Report Series – Committee on Astrobiology and Planetary Sciences. **2022**. nap.nationalacademies.org

Neveu, ML; Hays, LE; Voytek, MA; New, MH; Schulte, MD. The ladder of life detection. *Astrobiology* **2018**, *18*, 1375–1402. https://doi.org/10.1089/ast.2017.1773

Oparin, AI. *The Origin of Life* (published in Russian) in **1924**. English translation by A. Sygne published in *Origins of Life: The Central Concepts*, Deamer DW; Fleischaker GP, Eds., Jones and Bartlett: Boston, MA, **1994**, pp. 31–71.

Oparin, AI. *Origin of Life*, **1938**. Translated by S. Morgulis. Republication of the original publication by Macmillan Company by Dover Publications Inc., 2nd ed., **1965**. Dover: New York.

Oparin, AI. *Genesis and Evolutionary Development of Life*, published in Russian in **1966**, English translation by Maass, E., Academic Press: New York, **1968**.

Orgel LE. Self-organizing biochemical cycles. *Proc Natl Acad Sci* **2000**, *97*(23), 12503–12507. https://doi.org/10.1073/pnas.220406697

Orgel, LE. Prebiotic chemistry and the origin of the RNA world. *Crit Rev Biochem Mol Biol* **2004**, *9*, 99–123. https://doi.org/10.1080/10409230490460765

Pályi, G; Zucci, C; Caglioti, L, Eds. *Fundamentals of Life*, Elsevier: Amsterdam, **2002**, pp. 15–55.

Pascal, R; Pross, A; Sutherland, JD. Towards an evolutionary theory of the origin of life based on kinetics and thermodynamics. *Open Biol.* **2013**, *3*(11), 130156. https://doi.org/10.1098/rsob.130156

Patel, BH; Percivalle, C; Ritson, DJ; Duffy, CD; Sutherland, JD. Common origins of RNA, protein and lipid precursors in a cyanosulfidic protometabolism. *Nat Chem.* **2015**, *7*, 301–307.

Peretó, J. Out of fuzzy chemistry: From prebiotic chemistry to metabolic networks. *Chem Soc Rev* **2012**, *41*, 5394–5403. https://doi.org/10.1039/C2CS35054H

Peretó, J. Prebiotic chemistry that led to life. In *Handbook of Astrobiology*, Kolb, VM, Ed., CRC Press: Boca Raton, FL, **2019**, pp. 219–233.

Perry, RS; Kolb, VM. On the applicability of Darwinian principles to chemical evolution that led to life. *Int J Astrobiol* **2004**, *3*(1), 45–53. https://doi.org/10.1017/S1473550404001892

Popa, R. *Between Necessity and Probability: Searching for the Definition and Origin of Life*, Springer-Verlag: Heidelberg, **2004**, pp. 197–205 (definitions of life are given in chronological order starting from 1855).

Popa, R. Necessity, futility and the possibility of defining life are all embedded in its origin as a punctuated-gradualism. *Orig Life Evol Biosph* **2010**, *40*, 183–190. https://doi.org/10.1007/s11084-010-9198-x

Popa, R. Elusive definition of life: A survey of main ideas. In *Astrobiology: An Evolutionary Approach*, Kolb, VM, Ed., CRC Press: Boca Raton, FL, **2015**, pp. 325–348.

Scalice, D. Astrobiology-as-origins-story, education and inspiration across cultures In *Handbook of Astrobiology*. Kolb, VM, Ed., CRC Press: Boca Raton, FL, **2019**, pp. 49–53.

Schuerger, AC; Smith, DJ; Griffin, DW; Jaffe, DA; Wawrik, B; Burrows, SM; Christner, BC; Gonzalez-Martin, C; Lipp, EK; Schmale III, DG; Yu, H. Science questions and knowledge gaps to study microbial transport and survival in Asian and African dust plumes reaching North America. *Aerobiologia*. **2018**, *34*(4), 425–435. https://doi.org/10.1007/s10453-018-9541-7

Schwieterman, EW; Kiang, NY; Parenteau, MN; Harman, CE, et al. Exoplanet biosignatures: A review of remotely detectable signs of life. *Astrobiology* **2018**, *16*, 663–708. https://doi.org/10.1089/ast.2017.1729

Soai, K; Matsumoto, A; Kawasaki, T. The origin and amplification of chirality leading to biological homochirality. In *Handbook of Astrobiology*, Kolb, VM, Ed., CRC Press: Boca Raton, FL, **2019**, pp. 341–354.

Solé, R; Elena, SF. *Viruses as Complex Adaptive Systems*, Princeton University Press: Princeton, **2018**.

Špaček, J, Benner, SA. Agnostic Life Finder (ALF) for large-scale screening of Martian life during in situ refueling. *Astrobiology* **2011**. https://doi.org/10.1089/ast.2021.0070

Strazewski, P. Coevolution of RNA and peptides. In *Handbook of Astrobiology*, Kolb, VM., Ed., CRC Press: Boca Raton, FL, **2019**, pp. 409–420.

Villarreal, LP. Are viruses alive? *Sci Am* **2004**, *291*(6), 100–105. Accessed July 10, 2021. www.jstor.org/stable/26060805

Villarreal, LP, Witzany, G. Social networking of quasi-species consortia drive virolution via persistence. *AIMS Microbiol* **2021**, *7*(2), 138–162. https://doi.org/10.3934/microbiol.2021010

Virus. *Virus – Wikipedia*. https://en.wikipedia.org/wiki/Virus

Walker, SI. Transition from abiotic to biotic: Is there an algorithm for it? In *Astrobiology: An Evolutionary Approach*, Kolb, VM, Ed. CRC Press: Boca Raton, FL, **2015**, pp. 371–379.

Wołos, A; Roszak, R; Żądło-Dobrowolska, A; Beker, W; Mikulak-Klucznik, B; Spólnik, G; Dygas, M; Szymkuć, S; Grzybowski, BA. Synthetic connectivity, emergence, and self-regeneration in the network of prebiotic chemistry. *Science* **2020**, *369*(6511). https://doi.org/10.1126/science.aaw1955

Xue, Q; Miller-Jensen, K. Systems biology of virus-host signaling network interactions. *BMB Rep*. **2012**, *45*(4), 213–220. https://doi.org/10.5483/BMBRep.2012.45.4.213

# 13 Life's Emergence by Protocellular Mutually Catalytic Networks

*Roy Yaniv and Doron Lancet*

## 13.1 OPARIN'S FORESIGHT

This year marks the 100th anniversary of the publication of Aleksandr Ivanovich Oparin's pioneering book *The Origin of Life* (Oparin 1924). The book was translated from Russian and elaborated on over several decades. The English version published 30 years later *The Origin of Life on the Earth* (Oparin 1957) was written with much broader knowledge of biochemistry and expresses surprising insights about systems facets of life's origin, such as those described in this book. Such startling vision is detailed in this section in which texts are extracted from chapter VIII, "The Origin of the First Organisms" in Oparin's book. The subsections given in the chapter are named after Oparin's original subchapters. The quoted texts, labeled **O**, are followed by a brief contemporary summary, including allusion to the Graded Autocatalysis Replication Domain (GARD) model, a Systems model for the origin of life that involves assemblies harboring networks of small molecule (Section 13.8).

DOI: 10.1201/9781003294276-13 

### 13.1.1 The Principle of Selection

**O** Attempt to apply the principle of natural selection to the evolution of separate molecules cannot be held to be satisfactory. However, we shall adopt a different approach if we try to imagine the possibility of the evolution of those systems which we postulated as being the starting point on the road to the development of living systems, that is, to the evolution of the drops of complex coacervates which have the properties of open systems and the network of interdependent reactions characteristic of such systems.

*Summary*: Evolution can take place in networks, not in single molecules.

**O** In the very origin of such individual multimolecular formations there was already inherent the necessity for their further progressive development. During the time when organic material was completely merged with its environment, while it was dissolved in the waters of the primaeval seas and oceans, its evolution could be considered as a whole. However, as soon as it became concentrated at definite points, in colloidal multimolecular systems, as soon as these formations became separated from the surrounding medium by a more or less clearly defined boundary and attained a certain individuality, new and more complicated conditions were at once created. The later history of any individual coacervate drop might differ substantially from that of another coexistent system. The fate of such a drop depended not only on the general conditions of the external medium, but also on the specific internal organization in space and time of the system in question. The details of this organization were peculiar to the particular drop and may have been somewhat different in other drops, each system having its own characteristic peculiarities.

*Summary*: This stresses that the future is not only determined by environment, but mainly by the differences among coacervate drops, i.e. molecular assemblies. In other words, the progression toward life is less influenced by, e.g., dry–wet cycles and more by the emergence of molecular assembly heterogeneity that breeds selection.

**O** Complex coacervates obtained artificially, by simply mixing solutions of two differently charged colloids, are . . . formations with a static stability. . . . This, however, was not the sort of stability manifested by the systems which played the decisive part in the evolution of matter on the way to the origin of life. This evolution could only proceed on the basis of interaction between the systems and the external medium . . ., i.e. on the basis of the formation of open systems. We must remember that the coacervate drops, which arose somehow in the primaeval hydrosphere, were immersed, not simply in water, but in a solution of various organic compounds and inorganic salts which were certainly capable of entering into the coacervate drop and interacting chemically with the substances of which it was composed. . . . [I]t will be clear to us that under these conditions . . . the drop could . . . assume the character of an open system.

*Summary*: Progress toward life happens when Coacervates become open systems, exchanging molecules with the external organic soup, i.e. are in the state of nonequilibrium.

**O** However, let us suppose that the drop arose . . . from the formation of a surface membrane, separating the collection of these substances from the external medium. Even so the molecules of the external medium must have passed selectively through the surface membrane of the drop or been adsorbed selectively by the compounds contained in it and reacted with them . . . the products of the reaction either being retained within the drop or passing out of it back into the external medium, . . . even at this primitive stage of evolution of our original systems two circumstances were manifest which were of great importance for the further development of matter. On the one hand the individual peculiarities of the physico-chemical organization of each separate coacervate drop imposed a definite pattern on the chemical reactions, which took place within that drop. The presence in a given drop of this or that compound or radical . . . inorganic catalysts such as salts of iron, copper, calcium, etc., the degree of concentration of protein-like substances and other substances of high molecular weight . . . all these affected the rate and direction of the various chemical reactions which occurred within the given drop, (and) imparted a specific character to the chemical processes which took place within it. (This) could not be without effect on its future. . . . The only systems which maintained themselves in existence . . . were those which had an individual organization based on chemical reactions which were favorable for their existence.

Thus, even at this stage of the evolution of matter there appeared a certain 'selection' of organized colloidal systems on the basis of the suitability of their organization to the function of preserving the uninterrupted interaction of the system and the surrounding medium under given circumstances. This 'selection' was, of course, of a very primitive kind and not directly to be compared with fully developed natural selection, in the strictly biological sense of the term. Nevertheless, the further evolution of organic systems was controlled by 'selection' of this sort and thus acquired a definite direction.

*Summary*: Coacervates, including those ensheathed by a membrane, attain individuality by having different chemical compositions and reactions, affording primitive selection. This is equivalent to combinatorial heterogeneity of GARD micelles (see Section 13.10).

### 13.1.2 Processes of Self-Renewal of the Systems

**O** The coacervate state and the organization of the processes taking place within the drop may, to some extent, exist independently of one another. However, . . . during the course of directed evolution these two aspects of the organization must afterwards have become more and more unified within the single system, because the existence of the system depended on a network of reactions carried out within it. while, conversely, the network was determined by the organization of the system as a whole.

For example, if the stability of the drop depended on the formation of strong surface layers and if these disintegrated spontaneously at a definite rate but could be built up again in the course of chemical reactions within the drop, then the stability of the drop would depend on the relative rates of disintegration and reconstruction of the surface layers. If the chemical reactions took place fast enough in the drop, with a corresponding fast rate of formation of the firm surface layers, then the dynamic stability of the drop might also be very great. In this case an increase in the rate of the chemical reactions within the drop would have favored its stability. The increased rate of reaction within a drop would increase its stability, and prolong its survival under such conditions.

Finally, if the surface layers themselves were very strong and stable but not associated with any chemical reactions within the drop, then such static colloidal systems would be excluded from the course of the evolutionary process. Accordingly, as a result of the directed evolution of the original systems, their stability took on a more and more dynamic character. The coacervate drops were gradually transformed into open systems the very existence of which, under the given conditions of the external medium, depended on the organization of the processes taking place within them. In other words, there arose systems in which there was a background of continuous processes of self-renewal, and which could preserve themselves and exist for a long period on the basis of constant interaction with the external medium. The origin of this capacity for self-preservation may be regarded as the first result of the directed evolution of our original systems.

*Summary*: Self-preservation via coacervate–environment interactions happens by the fact that certain inner compositions are preserved by interactions between assembly and the environmental molecules. This is a key feature of the GARD model, whereby homeostasis within the assembly is governed by mutual catalysis with exogenous molecules. In vesicular protocells, similar interactions among membrane, lumen, and environment allow synchrony of inner content and shell buildup in the realm of metabolic GARD. Significantly, self-renewal and self-reproduction are synonymous in the scope of GARD (see Sections 13.9 and 13.10).

### 13.1.3 The Origin of the Capacity of the Systems for Self-Preservation and Growth

**O** The second step forward in the same direction was the emergence of systems which could not merely preserve themselves, but could also grow, increasing their mass by drawing substances from the external medium. As was pointed out in the previous chapter, the stationary state of open systems is maintained constant, not because the free energy of the system is at a minimum as in thermodynamic equilibrium, but because the system is continually receiving free energy from the surrounding medium in an amount which compensates for the expenditure of free energy within the system. In such chemical open systems as the coacervate drops of the primaeval ocean would seem to have been, the intake of free energy was mainly due to the entry into the drop of organic compounds which were relatively rich in energy, and which underwent some sort of chemical reaction within the drop. When chemical reactions are taking place . . . in open systems . . ., the rate of the reaction is considerably greater in one direction than in the other, and it follows that there may exist in them a coordination of processes leading to an increase in the mass of the system. Such systems enjoyed an undoubted advantage in the process of directed evolution and therefore, owing to the action of 'selection', they came to occupy a predominant position in the general extension of organized formations. In the absence of any appropriate experiments, even with models, one cannot say anything definite about the nature of such growth in our original systems. They might have become larger in the form of uniform layers of coacervate, but they might also have become divided into separate drops. As we know, dispersion of this sort may be achieved, even in such static systems, by means of external influences such as simple shaking, which may lead to emulsification. The dispersion of the primaeval growing coacervate may also have occurred in this way. However, as these were of the nature of dynamic stationary systems the existence of which was bound up with the occurrence of processes within them, their dispersion may have been evoked by internal factors, . . . for instance, have occurred when the osmotic pressure, which was increasing rapidly owing to the hydrolysis of compounds of high molecular weight, became too great for the strength of the surface layer of the drop.

*Summary*: An added feature is the capacity to grow by net absorption of molecules. Growth naturally goes along with fission, e.g. by simple physical shaking or turbulence, a clear advantage in reaching the process of directed evolution. Oparin highlights the role of modeling in parallel to experimentation, and indeed the GARD modeling highlights the fact that homeostatic growth, while preserving composition, leads to assembly reproduction, evidence for the significance of modeling. Growth also ends up with input of high-energy compounds from the environment. This is related to contemporary concepts, whereby energy-rich molecules, such as phosphorous and sulfur compounds, are a key source of energy, parallel to photochemistry. In contrast, thermal energy (e.g., heat–cool cycles) is less specific; as in heterogeneous molecular assemblies, it will simultaneously affect a large number of reactions.

### 13.1.4 The Origin of the Highly Dynamic State of the Systems

**O** The position is radically altered if we include in the field of our investigations not merely one, but several open systems, existing simultaneously within a particular medium. This may be shown even by working with relatively simple models. For example, when there are several hydrodynamic stationary systems with common initial and final reservoirs, the greater part of the water will proceed through the system which enables it to pass through most quickly. In the case of several parallel, chemical open systems with a common external medium it is obvious that the main flow of substances will pass through the system which, by virtue of its internal organization (e.g. the presence of more efficient catalysts, etc.), provides the greatest over-all rate of chemical transformation. In this sense the chemical stationary system in which chemical processes occur fastest will have an advantage over other parallel chemical stationary systems so long as the increased rate of occurrence of the processes does not disturb the relationship of rates necessary for the self-preservation of the system.

In this connection we must bear in mind what was pointed out in the previous chapter, namely that, where there is a complicated network of chemical reactions, the attainment of the maximal rate by a process involves not merely the acceleration of one of the stages of the transformation but the establishment of a more effective relationship between all the parameters of the process. From what has been said it is clear that a dynamically stable coacervate drop capable of self-preservation and growth, which had acquired the ability to transform substances more quickly during its interaction with the external medium, would have a significant advantage over other drops which were immersed in the same solution of inorganic and organic compounds but in which the characteristic chemical processes proceeded considerably more slowly.

In the general mass of coacervates the relative proportion of such more dynamic drops would become greater and greater. There arose a special kind of competition among the drops, based on the speed with which reactions were accomplished within them and the rate of their growth. For this reason, the principle of the greatest speed (which must, nevertheless, be compatible with the existence of the stationary system as such) was a very important factor in the directed evolution of organized formations.

*Summary*: There can be competition among members of a coacervate population. This can happen by better catalysis, better networks, as well as better growth with preservation. This predicts the behavior in kinetic simulations of a multi-assembly reactor (see Section 13.10) in which the evolutionary process is observed within populations of GARD assemblies, showing the takeover of certain assembly proto-species.

### 13.1.5 The Origin of Systems Capable of Reproducing Themselves

> **O** It must, however, be pointed out that the capacity for self-preservation, and even for rapid growth, of the whole dynamic system did not imply the complete immutability of the system. On the contrary, the stationary drop of a coacervate, or any other open system, may be preserved as a whole for a certain time while changing continually in regard to both its composition and the network of processes taking place within it, always assuming that these changes do not disturb its dynamic stability. Changes of this sort were, in fact, a necessary part of the process of the emergence of life for they guaranteed the evolution of the initial systems. Without these changes no new material would have been provided for selection and the further development of the systems would have been frozen and brought to a standstill at some point.

*Summary*: The coacervates need to undergo continuous changes in composition and in order to manifest evolution. Their acquiring reproduction brings evolution, and this can be considered the origin of life. Indeed, GARD preserves composition and network upon growth, which is shown to be equivalent to reproduction. In parallel, GARD compositional mutations which lead to reaction changes allow selection and evolution.

> **O** Naturally, it was of the utmost importance that these changes should not overstep the bounds of the dynamic stability of the systems. Otherwise, any markedly unstable compounds which arose would be in constant danger of passing out of equilibrium and disappearing. Therefore, when there was rapid and massive growth of the original systems, selection took place, the only ones which were preserved for further evolution being those in which the network of reactions was so co-ordinated that there arose stationary chains of reactions which were constantly repeated or, even better, closed cycles of reactions in which the reactions always followed the same circle and branching only occurred at definite points on the circle leading to the constantly repeated formation of this or that metabolic product. Of course, these cycles must not be confused with the elementary cycles of chain reactions. This constant repetition of connected reactions, co-ordinated in a single network, also led to the emergence of a property characteristic of living things, that of self-reproduction. This may be taken as the origin of life. At this stage in the evolution of matter natural selection assumed its full biological meaning and formed the basis for the faster elaboration of higher and higher degrees of adaptation of living organisms to the conditions under which they existed, of the exact correspondence of all the details of their internal structure to their vital functions. In other words, there appeared that striking "purposefulness" of the structure of living bodies upon which we have already remarked.

*Summary*: Stationary chain of reactions in a single network leads to the emergence of self-reproduction, a property characteristic of living entities. Points on a circle lead to the repeated formation of a certain metabolic product. In the GARD formalism, the mere network, even without cycles, is sufficient for self-reproduction (Kahana, Segev et al. 2022).

> O The opinion is fairly widely held in contemporary scientific literature that the capacity for self-reproduction is to be found even in the chemical form of the motion of matter, that it can be a property of isolated molecules. Until comparatively recently many biologists regarded the constant formation of particular substances within the organism as being the result of the presence in the organism of readymade molds for those particular substances. These molds were supposed to ' multiply ' in some way and thus be responsible for the constancy of the composition and structure of the organism and for its reproduction. . . .
>
> The factual evidence of contemporary biochemistry was, however, radically opposed to this opinion and revealed a completely different mechanism of biosynthesis based on the constancy of certain sequences of biochemical reactions. . . . The constancy of the formation of the substances is simply a manifestation of the constancy of the sequences of the reactions. Here there is no self-reproduction of molecules in the literal sense of the term, no multiplication of them; here new molecules of exactly the same kind are repeatedly produced. The sequence of reactions on which this phenomenon is based does not depend on any single individual factor but is a manifestation of the whole organization of the protoplasm in its relationship to its environment. As we saw in Chapter VI, the biosynthesis of proteins constitutes no exception in this respect. Attempts to treat it as an autocatalytic process, in which one molecule of a given substance arises as a result of the catalytic activity of another of exactly the same sort, which was already present, have recently proved a complete fiasco. They showed that a single nucleic acid of tobacco mosaic virus completely freed from protein, when introduced into the plant, will evoke the formation in it of a specific protein which was not previously present in the plant. In this case there could be no question of any autocatalysis in the strictly chemical sense of the term. There was only definite co-ordinated interaction of all the processes of cells of the tobacco leaf, which were somewhat altered in character by the introduction of a new factor, the viral nucleic acid. The nucleic acid as an individual substance, a compound considered in isolation, could certainly not synthesize a protein by itself. It is only effective against the general background of the whole metabolism of the tobacco plant, as is confirmed by all the evidence at present available. The harmonious participation of a long series of catalytic systems is required for the biosynthesis of proteins, some providing the *energy* needed for the synthesis, some determining the strictly regular and constant relationship between the rates of the different reactions and, finally, some systems which control the spatial organization of the protein molecule in the process of its synthesis. Among these systems which determine the specific structure of the protein, nucleic acid plays a very important part, but it does not seem to be the sole determinant, it simply constitutes a part of the general organization of the living system. As has been pointed out above, nucleic acid itself also arises in the living organism in accordance with the same rules as the other components of the protoplasm, that is to say, on the basis of strictly co-ordinated, constantly repeated, catalytically induced exchange reactions. It is clear that no substance which forms a major component of protoplasm can be reproduced by a chance or easily attained relationship between the rates of reactions. It requires the absolutely constant, continually repeated chains and cycles of reactions which together comprise the network of the self-reproducing, living, open system. As we have seen above, the origin of such a system may be regarded, theoretically, as a result of the directed evolution of our original, dynamically stable, colloidal formations.

*Summary*: There are no single molecule reproducers. Reproduction relevant to the origin of life must be that of a multi-molecular system that self-reproduces as a whole. It requires the absolutely constant, continually repeated chains and cycles of reactions which together comprise the network of the self-reproducing, living, open systems. This is fully in accord with the GARD formalism.

**O** The living systems which were first formed already had all the features needed for their selection to be of the nature of purposeful ' natural selection ' in the biological sense of the expression. Further improvements in their internal organization, rationalization of their metabolism, therefore, went forward at a faster pace. As a direct result of this, all intermediate forms of organization were destroyed, swept from the face of the Earth by natural selection. This is why we have now no possibility of studying these forms directly and filling in, with factual material, the abyss which exists between the organization of the original systems and the organization of even the simplest of present-day organisms.

*Summary*: It appears that all intermediate forms of organization are extinct, swept away by natural selection. Therefore, it is impossible to bridge the chasm between the organization of the original systems and the organization of even the simplest present-day organisms.

### 13.1.6 The Evolution of Metabolism: The Origin of Enzymes

**O** Experiments with models which reproduce the phenomena in dynamically stable colloidal formations may, perhaps, play an important part in this connection. Studies of this sort are, however, still only beginning to be made and the results obtained from them are still very modest. Therefore, if we wish to formulate any sort of idea concerning the actual forms which developed during the course of evolution from the original systems to the first organisms, we must make as much use as possible of the data of comparative biochemistry (this is done more fully in the next chapter) and the results obtained from a study of the metabolism, or separate aspects of the metabolism, of isolated protoplasmic structures and collections of enzymic systems. In this way we may be able to reveal various features common to all living organisms and may try to form a mental picture of how these features could have arisen during the process of directed evolution of our original systems or in the earliest stages of the development of life.

*Summary*: In the future, it will be possible to form a picture of the early steps in life's origin by comparing enzymes from many species. This indeed has been partially fulfilled by the availability of DNA sequences that help define the Last Universal Common Ancestor (LUCA, Figure 13.5).

**O** During the interaction of our original colloidal systems with the medium surrounding them, and in the process of their later development, there must have been formed within them, not a single individual enzyme, but many specific catalysts. Their simultaneous activity determined the occurrence of some particular chain of chemical reactions or a whole network of reactions. On the nature of the organization of this chain or network depended the greater or less dynamic stability conferred by the network on the open system. The selection of systems was based on this stability, destroying those which had an "unsuccessful" combination of reactions and preserving for further

> evolution only systems with chains and networks which enabled them to survive for a long while under conditions of constant interaction with the external medium. It is obvious that it required a very prolonged and rigorous selection of a colossal variety of such systems for there to arise, at last, a chain consisting of more than 20 rationally concordant reactions such as take place in alcoholic fermentation. In principle, however, the origin of such a harmony between different catalytic reactions could quite well have occurred during the process of directed evolution and it seems that it must have come about at a comparatively early stage in the origin and development of life since the same basic collection of chains is common to, literally, all representatives of the living world which have been studied in this respect.

*Summary*: The intricacy of the early networks that led to today's life suggests that Darwinian evolution must have been in action very early on. This can only be true if self-reproduction appeared in metabolic network entities prior to the genetic apparatus.

## 13.2 FROM STRUCTURE TO ORIGIN

A living cell is an utterly complex molecular machine. Origin of life scientists hope that by understanding its intricacies, they will stumble upon an explanation for how it emerged. This aspiration is based on experiences with simpler systems. For example, by studying moon rocks, it is possible to come up with a mechanism of its formation by a cataclysmic planetary collision. However, this may not be true for entities whose complexity is many orders of magnitude higher. In such cases, it is necessary to employ divergent thinking, guided by an assessment of which features are fundamental and which are consequential.

It is widely agreed that the most crucial property of living entities is a capacity to generate similes (Chen and Walde 2010). This attribute underlies the process of Darwinian evolution and natural selection and must have been present very early in life's history. An often-stated conundrum is that life cannot reach even a most rudimentary level of complexity without self-replication, but that only sufficiently complex chemical systems can multiply (Lancet, Zidovetzki et al. 2018). The discovery of DNA, simple in design and seemingly capable of creating its own copies, has led to a wide belief that nucleic acids provide a solution for such an incongruity (Pace and Marsh 1985).

However, an examination of the structure of DNA and RNA reveals chemical intricacy that defies the notion that they were the first replicators. Even in the present-day setting, polynucleotides only undergo replication in the context of their cellular milieu. In a primordial chemically heterogeneous scenario, the spontaneous emergence of sufficient quantities of the required carbohydrate moiety and nitrogen base components is unlikely. Also, phosphodiester polymerization is thermodynamically unfavorable, and the repeated building of a long second nucleic acid strand on its first strand template would require intricate catalysis (Ertem and Ferris 1996, Bartel and Unrau 1999). For these and other reasons, it has been argued by quite a few authors that an RNA-like polymer may not have been the early seed of life (Shapiro 1984; Schwartz 1995).

The breakthrough discovery of Watson and Crick (Watson and Crick 1953a; Watson and Crick 1953b) led to the practical extinction of alternative hypotheses. The beauty of DNA replication was so compelling that any other "messier" pathway suddenly looked untenable. Yet, in view of the problems posed earlier, it is interesting to ponder on the untainted state-of-the-art of origin-of-life research as it existed prior to 1953. In the process, insight may be gained with respect to the dichotomy between the widely accepted RNA-first model in which sequence information reigns and an emerging, more seemingly debatable small molecules/lipid first model for life's emergence in which compositional information is the key (Figures 13.1 and 13.2).

## 13.3 OPARIN REVISITED

In his books (Oparin 1924; Oparin 1957), Oparin dealt with the problem of life's origin, for the first time, from a materialistic perspective uninhibited by the religious constraints prevailing in Western nations. The standard Oparin–Haldane theory

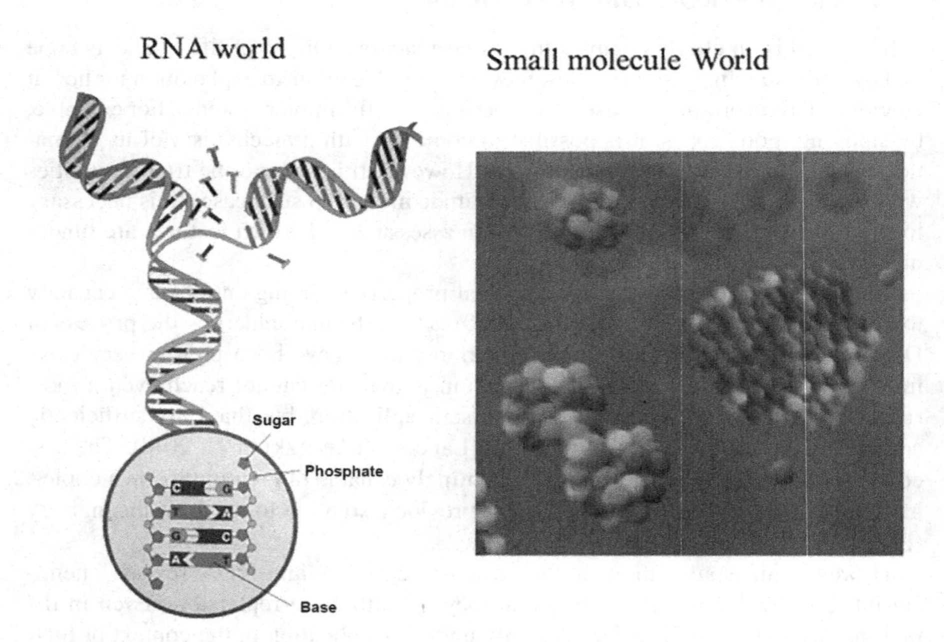


**FIGURE 13.1** The two dichotomous stereotyped "world" scenarios for life's origin. Both are likely to be over-simplifications, since prebiotic mixtures must have included diverse compounds, and not just one class of substances. Therefore, the two scenarios should be taken to represent opposing general concepts. On the one hand, the "RNA/DNA world" view (left) asserts that life began with individual covalently-strung large molecules that have a relatively complex structure. These generate progeny by non-trivial templating and strand separation. On the other hand, the "small molecule world" (right) is a scenario whereby there exists "non-covalent polymers", supramolecular assemblies of relatively simple compounds, accreted by weak non-covalent chemical forces, often hydrophobic interactions (Segre and Lancet 2000).

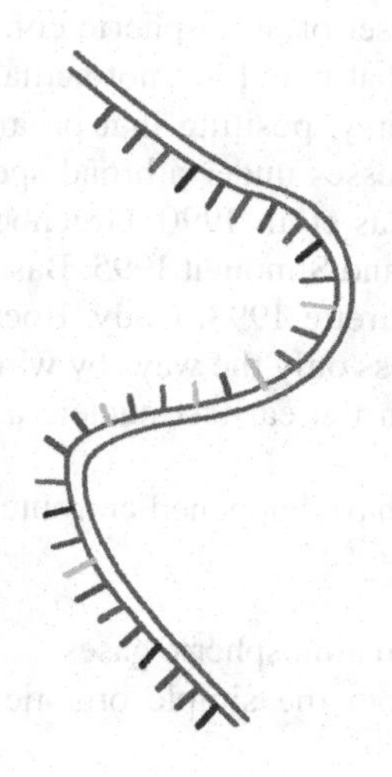
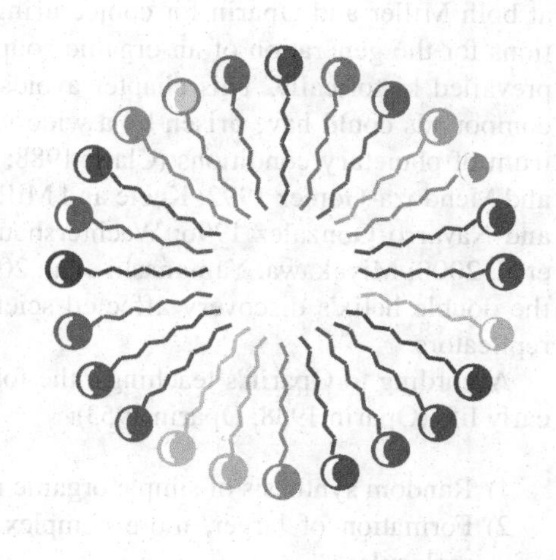


**FIGURE 13.2** In covalent biopolymers (left), information resides in the sequence of chemical units (monomers), while in non-covalent molecular assemblies (right), it rests in a "compositional vector" whose elements are the counts of the different molecular types in a given assembly (Lancet, Zidovetzki et al. 2018). In polymeric entities such as a single-stranded RNA, information is preserved by intricate second-strand construction (Figure 13.3) via templating and transmitted to progeny by energetically challenging strand separation. In the realm of small molecule assemblies, information is contained in chemical composition (a vector indicating the counts of every compound type). Compositional information is retained by homeostatic growth enabled by a mutually catalytic network acting upon energetically downhill monomer joining. Transmission to progeny takes place by simple fission energized by physical instability.

(Miller, Schopf et al. 1997) puts forward the production of organic molecules on the early Earth followed by chemical reactions that produced increased organic complexity. This process is perceived to have eventually led to organic life capable of reproduction, mutation, and selection. In this form, the theory has no specification of which molecules were first, or how self-replication came about. In fact, it predates, by two decades, the Watson–Crick breakthrough and is therefore unbiased by its implications.

The origin of organic molecules and the origin of replication are two separate problems. Actually, the first may be viewed as a prerequisite for the second. The groundbreaking experiments of Miller and Urey (Miller 1953), first published in the same year as DNA's structure, were specifically aimed at the first question.

They asked how, under primitive Earth conditions, a sufficient supply of organics was generated to warrant Oparin's "primordial soup", the milieu in which reactions leading to life are assumed to have occurred. Criticism has often been aimed at both Miller and Oparin for conjecturing a particular set of atmospheric conditions for the generation of an organic soup, conditions that may have not actually prevailed historically. This chapter avoids this controversy, positing that organic compounds could have arisen by a wide variety of processes under a broad spectrum of planetary conditions (Clark 1988; Chyba, Thomas et al. 1990; Greenberg and Mendoza-Gomez 1992; Keefe and Miller 1995; Leif and Simoneit 1995; Basiuk and Navarro-Gonzalez 1996; Wachtershauser 1997; Maurette 1998; Cody, Boctor et al. 2000; Miyakawa, Yamanashi et al. 2002). We address only the ways by which the double helix's discovery affected scientific views on the earliest nucleic acid replicators.

According to Oparin's teachings, the following steps have happened en route to early life (Oparin 1938; Oparin 1953):

1) Random synthesis of simple organic molecules from atmospheric gases
2) Formation of larger, more complex molecules from the simple organic molecules
3) Formation of coacervates – unique droplets containing the different organic molecules
4) Development of a capacity to take up molecules and discharge specific molecules and maintain a characteristic chemical pattern or composition
5) Development of "organizers" that allowed controlled reproduction to ensure that resultant split-related daughter cells have the same chemical capabilities
6) Beginnings of evolutionary developments so that a group of cells could adapt to changes in the environment over time

This clear-cut hierarchical scenario was considered for decades an acceptable view of how life began. In later years (Oparin 1976), Oparin had guessed that the hypothesized "organizers" in step 5 might include nucleic acids, but, clearly, he was not thinking in terms of a full-fledge transcription and ribosome-mediated translation apparatus, as these would have been too complex for the primitive stages invoked. Oparin had the insight to consider the possibility of simile generation based on a web of chemical interactions, namely what would nowadays be called a prebiotic metabolism-like network (Kauffman 1993; Morowitz 1999; Morowitz, Kostelnik et al. 2000). Self-replication was thought of as being related to the splitting of coacervate droplets, including its entire molecular repertoire, a process conceptually similar to modern cell division. Oparin was not influenced by the later realization that within each cell a component, namely DNA, was present and was capable of information storage and molecular self-replication. We note that terrestrial infall as well as local syntheses appear to have led to an enormous primordial chemical diversity, or 'messy chemistry' (Guttenberg, Virgo et al. 2017), inhabiting the prebiotic soup. The inclusion of amphiphiles had early evidence as demonstrated in extracts from the Murchison meteorite (Deamer 1985; Deamer and Pashley 1989). Indirect evidence for amphiphiles in a primordial soup is found in measurements of the vapor plumes of Enceladus, a moon of Saturn (Kahana, Schmitt-Kopplin et al. 2019). The most convincing evidence is based on detailed mass-spectrometric analyses of Murchison extracts that showed tens of thousands of organic compounds in this carbonaceous meteorites (Schmitt-Kopplin, Gabelica et al. 2010; Kahana, Schmitt-Kopplin et al. 2019). In parallel, supporting data comes from computationally derived networks of prebiotic reactions (Wołos, Roszak et al. 2020). Therefore, it is expected that diverse lipid molecules played a central role in delimiting coacervates en route to life, referred to by Oparin as protein–lipid coacervates (Oparin 1957).

## 13.4 THE NUCLEIC ACID TAKEOVER

In the post-1953 era, origin of life research underwent a revolution, not less significant than that which affected all of molecular biology. This revolution hinged on the notion that base-paired nucleic acid polymers might be capable of generating their own copies. Thus, the vaguer, but potentially more realistic scenario of a "mixed bag" ["garbage bag" in Dyson's terminology (Dyson 1999)] that undergoes self-reproduction has been largely abandoned. This view is in many ways reflected in a more molecularly detailed model, the Collectively Autocatalytic Set (CAS) paradigm (Kauffman 1986; Kauffman 1993), as described next.

The new scenario was simple and seemingly elegant. If only a single self-replicating base-paired nucleic acid could form in the soup, life would emerge. This is based on the notion that once self-replication is in place, selection and evolution would be jump-started and then gradual improvements could lead from a naked nucleic acid to a protocell. This scenario envisions that the additional paraphernalia needed for a full-fledged cell-like entity, such as metabolic pathways for the synthesis of hundreds of molecular components, membrane enclosures, passive and active transport molecules, energy-harvesting mechanisms and ATP-like free energy-rich compounds, primitive tRNAs and their associated synthetases, ribosomes, and protein enzymes would all somehow appear and join the original nucleic acid. Typically, it is not specified how the fact that a self-replicating nucleic acid has already turned up aids the appearance of the other molecular constituents. Also, little is said about how, if a single molecular replicator is present, it will recruit all the other components into the replication, selection, and evolution game.

The polynucleotide-first scenario brings to mind Marcia Brown's popular children story, based on an old tale, about three hungry soldiers who manage to cook up a fine soup by cajoling the necessary ingredients from a village of wary peasants. The pot initially contained only a stone immersed in water, and of course the soup emerged irrespective of it, solely because meat, vegetables, salt, spices, and all the rest had been stealthily added. The early evolution moral is that the "recruitment by nucleic acids" picture may be somewhat simplistic and that alternative, Oparin-style scenarios need to be explored. In the latter, the hypothesized early entities would include a diverse repertoire of components right from the start, and full-fledged double-helix-based mechanisms would arise much later. Replication could emerge as a capacity of an entire molecular ensemble, so that that gradual improvement, specifically by Darwinian evolution, would apply to all components in unison and not solely to a hypothesized core information carrier.

## 13.5 TEMPLATES AND CATALYSTS

One of the most popular means of portraying the chicken and egg nature of early evolution is asking what came first: Nucleic acid-based information storage molecules needed in order to synthesize proteins or globular protein catalysts without which the DNA/RNA machinery would come to a grinding halt. One way to answer the question is to ask which of the two sets of chemical phenomena appears more basic. The RNA world protagonists' answer is that since ribonucleic acids are capable of both template-based replication and catalysis, RNA must have been the primordial mover. However, a view derived from basic chemistry may lead to a different answer. This alternative view probes the very basic definition of a templating reaction. As seen in Figure 13.3, all a templating strand is doing when directing the incorporation of a properly base-paired nucleotide into a growing second strand is a special case of rate enhancement. While this reaction does not conform to the strict definition of a multiple turn-over enzyme reaction, it is nevertheless a selective catalysis phenomenon. One may envisage a very primitive system in which the perfect Watson–Crick structure has not yet evolved and in which the control over a replication-related reaction is in the hands of a more unwieldy set of organic catalysts. Addressing the chemical nature of templating in such a general perspective leads to the notion that sets of mutually catalytic substances may be, in principle, sufficient for primitive forms of self-replication.

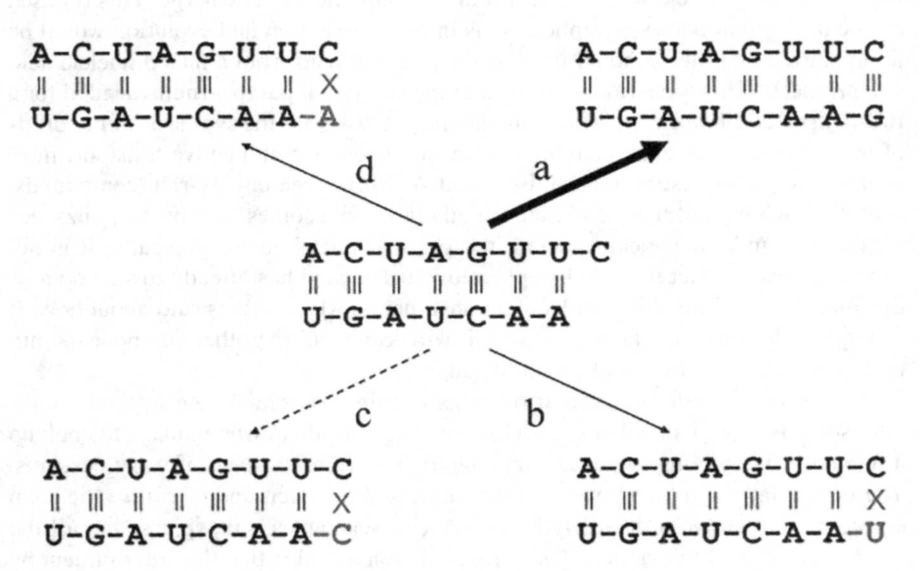


**FIGURE 13.3** We propose that templating is a derived concept, and the more basic notion is catalysis. The elementary step of adding a base to an extending second strand in a double stranded RNA biopolymer may be viewed as four competing reactions, one which involves legitimate Watson–Crick pairing (a) and the other three (b, c, d) forming illegitimate pairing hence with decreasing kinetic efficacy. The top strand may be considered to be an "enzyme" that catalyzes chemical reaction a in favor of the side reaction reactions b, c, and d which are much slower.

It is possible to envisage how through a long, gradual process of selection and evolution the double helix would emerge from its "bag of catalysts" ancestor. Another way to view this is to say that a fairly complex network resembling present-day metabolism is more basic, therefore likely to have been earlier than a double helix-based genetic apparatus. Indeed, Oparin's original paper (Oparin 1924; Oparin 1967) makes the argument that

> "slowly but surely, from generation to generation, over many thousands of years, there took place an improvement . . . directed towards increasing the efficiency of the apparatus for absorption and assimilation of nutrient compounds . . . (and) the ability to metabolize".

## 13.6 PRIMORDIAL INFORMATION STORAGE

Oparin's first article was written before Avery's discovery of DNA's role as a genetic information store (Avery, MacLeod et al. 1944; Lederberg 1994), and much before Watson and Crick uncovered the mechanism by which this is made possible. To nucleic acid supporters, Oparin's description of how biological information was stockpiled and transmitted looks rather untenable. He describes how bits of protoplasm-like "gel" would be "broken down and give rise to new pieces . . . constructed or organized just like their parent . . . and therefore their structure was inherited by them from the gel from which they were formed". The essence of this argument is that the most important aspect of replication is bequeathing the overall structure and function of a molecular ensemble through a capacity to propagate an internal chemical consistency or chemical composition, following a simple process of physical split.

While, in fact, a purist's point of view would assert that this is true also for a present-day living cell. Accordingly, the molecular templating of DNA is just one of a large set of mechanisms through which it is assured that after a physical split, the two cellular halves would be capable of replenishing their molecular repertoires en route to the next split. Parallel mechanisms would be protein synthesis, the assembly of ribosomes, the protein-mediated copying of the centrioles, and the production of new lipid bilayers for organelles such as mitochondria and the Golgi apparatus. From a more encompassing angle what happens upon cell division and subsequent growth is that the proteome, transcriptome, and metabolome remain largely unchanged when comparing the predecessor and the progeny.

## 13.7 THE COLLECTIVELY AUTOCATALYTIC SET (CAS) MODEL

The origin of life is accepted by many to be the point at which a self-copying chemical entity emerges, thus affording Darwinian evolution. A popular view is that this entity was RNA, a long polynucleotide that self-copies its sequence by Watson–Crick base-pair templating. However, with the aforementioned profound criticism regarding RNA, first an alternative model is sorely needed. The Stuart Kauffman innovation posits that certain sets of simple organic molecule of diverse types would show a sufficient degree of mutual catalysis to fulfill "catalytic closure". This is a condition at which each of its constituents is synthesized by a reaction catalyzed by one or more of the set members. The reactants are other set members or external "food". It was contended, and later formally shown, that such a molecular ensemble, driven by a mutually catalytic network, would constitute a Collectively Autocatalytic Set (CAS), capable of self-reproduce in its entirety (Kauffman 1993). What is being copied is the relative amounts of constituents rather than sequence of individual biopolymers. Every contemporary living cell is such a supramolecular catalytic set. Self-reproducing CAS has thus emerged as a legitimate systems chemistry alternative for self-copying RNA as the early seed of life.

## 13.8 THE GARD MODEL

While the words of Oparin vividly describe how structure and function might have been passed down the generations, a more specific and rigorous description would be needed to convince a modern molecular biologist that the fission of a molecular assembly is sufficient to carry out the information transfer over many generations. To this end, 25 years ago, we advanced the Graded Autocatalysis Replication Domain (GARD) model (Segre, Ben-Eli et al. 2000; Segre and Lancet 2000; Markovitch and Lancet 2014; Lancet, Zidovetzki et al. 2018; Kahana and Lancet 2021). GARD is an elaboration on Kauffman's CAS model, adding chemical kinetics equations, and allowing to trace the fate of molecular assemblies along the time axis, thus verifying the occurrence of self-reproduction (Lancet, Zidovetzki et al. 2018; Kahana, Segev et al. 2023). The essence of this model relates to the capacity of a daughter molecular assembly, derived by a physical split from its ancestor, to withstand the "trauma" of fission (e.g., the loss of scarce but essential molecular species). GARD molecular assemblies are shown to be capable of undergoing a complex replenishment process during growth, thus assuring homeostatic growth and a capacity for self-replication upon fission (Segre, Ben-Eli et al. 2000; Segre, Shenhav et al. 2001). Computer simulations, which embody a set of rigorously defined kinetic and thermodynamic constraints, demonstrate the capacity of such assemblies to grow homeostatically, i.e., while preserving their compositional information (Segre, Ben-Eli et al. 2000), a prerequisite for Oparin-style recovery from fission (Haldane 1929; Oparin 1938; Kolb 2016). The specific compositions that fulfill the conditions for homeostasis, hence compositional reproduction, are termed *composomes*. Thus, GARD assemblies appear to alleviate the chicken-and-egg problem, harboring within them both replicable information and a metabolism-like network.

In a modern-day cell, this process is known to depend crucially on protein synthesis, based on information inscribed in RNA and DNA. Such a central dogma pathway is, however, so complex that it is unlikely to have been present in very early protocells (Morowitz, Heinz et al. 1988; Morowitz, Kostelnik et al. 2000; Shapiro 2000). The RNA world concept asserts the existence of very few template-replication biopolymers, assisted by an ill-specified set of catalysts, potentially RNA itself, to make proteins. The GARD scenario claims that early on, none of the three components of the central dogma may have existed and that the only resemblance of early replicating and evolving entities to present-day cells was in matters of general design.

Specifically, GARD assumes a network of mutually catalytic, relatively simple organic molecules. Most likely candidates for soup molecules to spontaneously form dynamically long-lived networks are amphiphiles (lipids) which assemble together spontaneously to form micelles and vesicles, kinds of coacervates. A key advantage for these molecules is that their accreting together is governed by nonspecific hydrophobic interactions among amphiphile tail groups. This phenomenon also allows a great measure of promiscuity: As long as hydrophobic tails are present, head groups of diverse structures can form the assemblies.

In the basic GARD model, the most rudimentary form of rate enhancement needed for the establishment of a mutually catalytic network is an enhancement of amphiphile entry into the molecular assembly (Figure 13.4), resulting in a behavior that resembles heterotrophic cellular homeostatic growth (Segre, Ben-Eli et al. 2000, 2001; Lancet, Zidovetzki et al. 2018; Lancet, Segrè et al. 2019; Kahana, Lancet et al. 2022). More advanced versions of GARD also include catalyzed covalent bond formation events that lead to generating synthesis of novel molecules including oligomerization and to an initial emergence primitive metabolism (Shenhav, Bar-Even et al. 2005; Lancet, Zidovetzki et al. 2018). The capacity of lipids exerting stereospecific recognition and catalysis on other molecules has extensive experimental support, as reviewed (Kahana and Lancet 2021; Kahana, Maslov et al. 2021). In this view, simple catalysts (enzyme mimetics) underlie the emergence of primitive catalytic networks and the eventual appearance of polynucleotides that subserve the

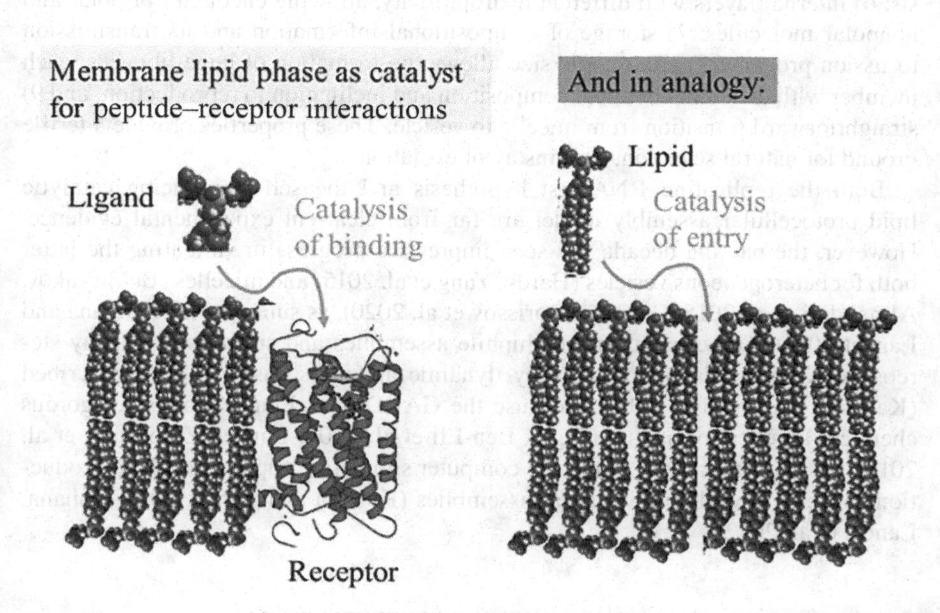


**FIGURE 13.4** Left: Membrane lipid phase as catalyst for peptide–receptor interactions (Sargent and Schwyzer 1986). Right: In analogy, neighboring lipids in a micelle or a bilayer may catalyze the entry/exit of another molecule into the amphiphilic assembly. Such a relatively simple notion serves as a core of the GARD model as described in the text and elsewhere (Segre, Ben-Eli et al. 2001; Lancet, Zidovetzki et al. 2018).

making of better catalysts, e.g., folded polypeptides necessary to generate highly effective enzymes.

## 13.9 PROTOCELLS

Although Oparin did not write on protocells, it is clear that his profound view of metabolism within coacervates and his keen interest in uptake and secretion of organic compounds across the environment–interior boundary suggest that the coacervates have a strong resemblance to protocells (see Section 13.1). A modern statement contends that amphiphilic molecules might be the first player in the evolution from molecular assembly to cellular life (Deamer and Dworkin 2005; Walde 2006). A step from vesicle toward protocell might be to develop self-reproduction capacity coupled with the metabolic system (Sakuma and Imai 2015). The latter challenge is addressed by the GARD model, as it proposes how metabolism and reproduction are intertwined.

Especially attractive in the realm of GARD is the idea that the first protocells were nanoscopic micelles. These miniscule amphiphile assemblies have a series of advantages that lend support to their suggested role in early life: 1) Fast and spontaneous formation of multi-molecular assemblies; 2) promiscuity for admixture of a large diversity of amphiphiles; 3) disposition for growth and fission; 4) extremely high concentrations of components; 5) fluidity that permits effective mutual catalysis; 6) internal layers with different hydrophilicity, allowing encounter of polar and nonpolar molecules; 7) storage of compositional information and its transmission to fission progeny; 8) nanoscopic size allows the formation of huge libraries, each member with different chemical composition and inclination to reproduction; and 9) straightforward transition from micelle to vesicle; These properties provide a fertile ground for natural selection, a mainstay of evolution.

Both the replicating RNA-first hypothesis and the self-reproducing catalytic lipid protocellular assembly model are far from clear-cut experimental evidence. However, the passing decade has seen impressive progress in validating the latter both for heterogeneous vesicles (Hardy, Yang et al. 2015) and micelles (Bukhryakov, Almahdali et al. 2015; Colomer, Borissov et al. 2020), as summarized (Kahana and Lancet 2021). The capacity of amphiphile assemblies and surfaces to portray stereospecific molecular recognition by dynamic aptamers has also been described (Kahana, Maslov et al. 2021). Because the GARD model encompasses a rigorous chemical kinetics formalism (Segre, Ben-Eli et al. 2000; Lancet, Zidovetzki et al. 2018), the near future will likely see computer simulation support for the reproduction capacity of early protocellular assemblies (Kahana and Lancet 2019; Kahana, Lancet et al. 2022).

## 13.10 PROTOCELLULAR DARWINIAN EVOLUTION

The original CAS model is only rarely analyzed with the assumption that the mutually catalytic network is spatially delimited (Hordijk, Naylor et al. 2018). But it is clear that for Darwinian evolution to be observed, compartmentalization is essential.

In other words, the lipid GARD implementation of CAS is by definition compartmentalized, since the amphiphilic assembly is assumed to have a delimited volume to grow and to fission. In other words, GARD is defined as a protocell. This is true whether the model is applied to a vesicular configuration or a micellar one (Kahana and Lancet 2021). Thus, the GARD embodiment affords a process in which each protocell begets fission-generated progeny and under the proper conditions shows compositional reproduction with mutations (Lancet, Zidovetzki et al. 2018). Therefore, it is expected that along the time axis, protospecies will emerge and undergo natural selection, e.g., based on the speed and precision that will take place (Figure 13.5). Such an evolutionary process has been observed in kinetically rigorously simulations of populations of GARD assemblies, showing the takeover of certain assembly protospecies (Markovitch and Lancet 2014), as well as simulations in which compositional mutations may increase the fitness of certain compositions (Inger, Solomon et al. 2009; Lancet, Zidovetzki et al. 2018). Simple arithmetic predicts that on the surface of our planet's oceans, $>10^{33}$ spontaneously generated micellar protocells will appear (Lancet, Zidovetzki et al. 2018). Based on the micellar size and the magnitude of the environmental organic molecular repertoire, combinatorics suggest that each member of the formed micellar "library" will have a different composition (Figure 13.6). While a huge majority of these micelles will be non-reproducing, a significant number will be reproduction-prone (Kahana, Segev et al. 2023). Such a phenomenon is akin to what has been defined as selection for replication (Nowak and Ohtsuki 2008). A prediction is that selection can take place on the basis of randomly formed diversity, even prior to natural selection. Such a phenomenon can take place in diverse environmental conditions.

Our GARD simulative analyses recently discovered a fundamental phenomenon: In a mutually catalytic network, stemming from the CAS school, reproduction is a dynamic attractor (Kahana, Segev et al. 2023). This demonstrates that even when the network is initially far from reproduction, the dynamical properties end up "pulling" the complex system toward better and better self-reproduction capacities. This observation strongly augments the probability that randomly seeded assemblies will develop to be reproducers.

## 13.11 CONCLUSIONS

As we proceed into the post-genomic era, emphasis is being shifted from DNA sequences to the more global molecular interplay in the realm of systems biology and systems chemistry (Brown and Okuno 2012). Similarly, it is hard to envision primordial cellular systems without considering some form of mutually interacting sets of compounds. It is possible, however, that delineating the progression from a simple protocell to the "Last Universal Common Ancestor" (LUCA) of terrestrial life, including reproducing and encoding biopolymers, would be as convoluted as outlining how human beings evolved from a lowly worm.

The uncovered central dogma of contemporary biology brought with it tremendous progress with respect to our understanding how present life is molecularly constructed. Paradoxically, it may have changed the course of prebiotic evolution

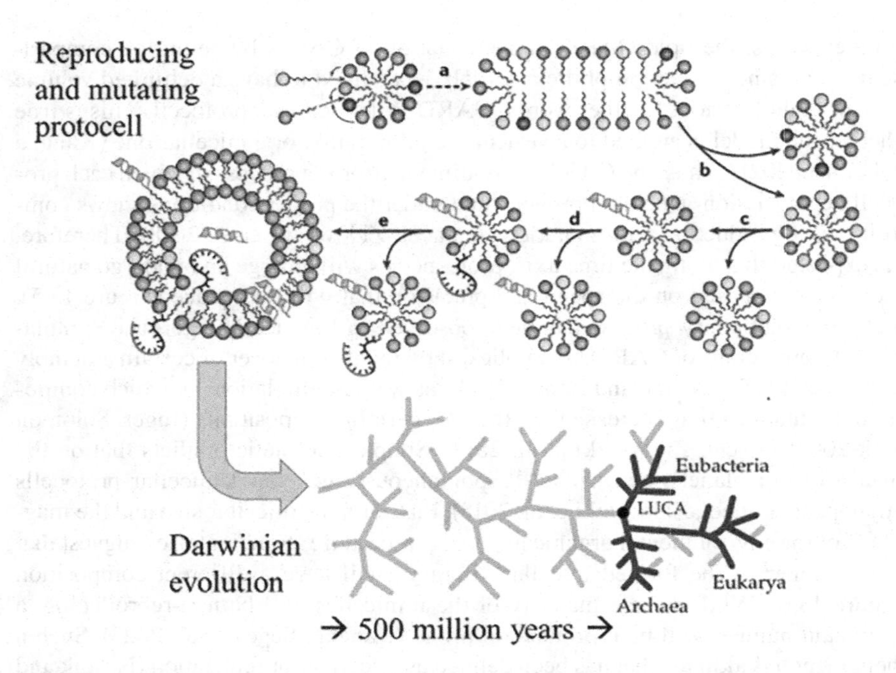


**FIGURE 13.5** Top: Micellar reproduction and initial evolution. (a) Mutually catalytic networks prevailing in a micelle based on rate enhancement interactions between micellar lipids and environmental free lipidic monomers, whereby rates of entry are enhanced by lipids within the assembly (Segre, Ben-Eli et al. 2000; Lancet, Zidovetzki et al. 2018). Along catalyzed growth, micelles assume distinctive compositions (composomes) at which homeostatic growth prevails. (b) A subsequent step of random fission marks the occurrence of micellar self-reproduction, upon which compositional information is passed to the next generation (Segre, Ben-Eli et al. 2000; Markovitch and Lancet 2014; Lancet, Zidovetzki et al. 2018). (c, d) Homeostatic growth and fission entail compositional reproduction (curved arrows) with mutations (straight arrows), involving both non-covalent accretion as in (a) and covalent lipid modifications, as experimentally occurring in certain catalytically reproducing heterogeneous micelles (Colomer, Borissov et al. 2020) and supported by GARD simulations with covalent modifications (Shenhav, Bar-Even et al. 2005). These reactions could, among others, involve catalyzed oligomerization of micelle-attached amino acids and nucleotides (α-helical peptide; short folded RNA). e) Some lipid modifications, such as a conversion of single-chain to double-chain lipids by another covalent reaction, would result in a micelle-to-vesicle transition. At this point, the vesicular lumen could be populated by both catalyzed uptake and catalyzed covalent lysis. The sequence of events depicted here portrays a gradual progression from micelles that undergo purely compositional inheritance to more elaborate vesicular protocells that embody sequence-based replication and core metabolism (Lancet, Zidovetzki et al. 2018). In the shown progression, each step constitutes a self-reproducing assembly capable of reproduction. Natural selection will thus eliminate all cases in which reproduction efficacy goes below a threshold. Those that reproduce well will be present in many similar copies, i.e., as a species. Bottom: A suggested progress from an early protocell as described (Kahana and Lancet 2021) to the last Universal Common Ancestor (LUCA), employing a published pre-LUCA progression (Cornish-Bowden and Cárdenas 2017). In this described scenario, the grey tree structure represents Darwinian evolution in which each node is an early species, all currently extinct. LUCA itself is assumed to have complexity approaching that of present-day prokaryotes (Mushegian 2008). The figure is adapted jointly with modifications from both Kahana and Lancet (2021) and Cornish-Bowden and Cárdenas (2017).

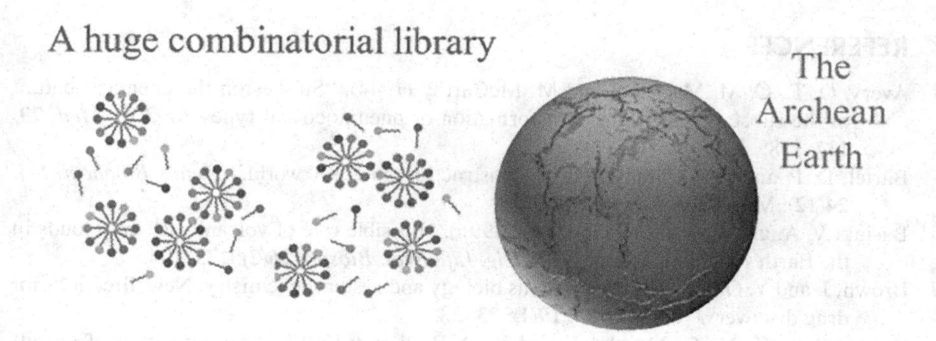


**FIGURE 13.6** It is worthy to consider the impact of the substantial volume available for early chemistry. Computations performed with underestimation (Lancet and Shenhav 2009; Lancet, Zidovetzki et al. 2018) show that our planet's top 1m of water bodies can contain $10^{33}$ (decillion) micelles. With a molecular repertoire of 100 amphiphiles, as used in most GARD simulations, it turns out that a typical micelle will appear in ~$10^{35}$ compositional combinations. This means that, by and large, each micelle will have a different composition. Such an astronomical number of spontaneously emerging micelles suggest a very high reproduction probability planet-wide, so that at some locales, micellar reproduction will appear. This depiction provides quantitative routes for addressing the question of life's probability in the universe (De Duve 2011).

research in a somewhat less favorable fashion. Time will tell whether proponents of the elegant RNA-first scenario are right or wrong. Yet, considering much simpler molecular mechanisms that could lead to early reproduction will surely be beneficial for the cracking of the most important open question in "low energy" science – how life emerged from inanimate chemical compounds.

The models described in this chapter open the door to a rigorous and convincing alternative for RNA as the first primordial entity with potential to reproduce. A reproduction scenario such as the CAS-GARD model, beginning with a spatially bordered supramolecular assembly harboring a large number of catalytic events, has a great advantage. This is because the three pillars of the origin of life, information, metabolism, and containment, appear simultaneously. The alternative scenario also has strong explanatory power: The spontaneously arising, relatively simple network that can reproduce, mutate, and be selected by fitness paves a road to a very early form of Darwinian evolution (Kahana and Lancet 2021). This will greatly alleviate the paradox of how to attain the unimagined complexity of the central dogma components without genes and ribosomes.

## ACKNOWLEDGMENTS

We thank Omer Markovitch for helpful discussions and grant support from the Minerva Foundation.

## REFERENCES

Avery, O. T., C. M. MacLeod and M. McCarthy (1944). "Studies on the chemical nature of the substance inducing transformation of pneumococcal types." *J. Exp. Med.* **79**: 137–158.

Bartel, D. P. and P. J. Unrau (1999). "Constructing an RNA world." *Trends Biochem. Sci.* **24**(12): M9–M13.

Basiuk, V. A. and R. Navarro-Gonzalez (1996). "Possible role of volcanic ash-gas clouds in the Earth's prebiotic chemistry." *Orig. Life Evol. Biosph.* **26**(2): 173–194.

Brown, J. and Y. Okuno (2012). "Systems biology and systems chemistry: New directions for drug discovery." *Chem. Biol.* **19**(1): 23–28.

Bukhryakov, K. V., S. Almahdali and V. O. Rodionov (2015). "Amplification of chirality through self-replication of micellar aggregates in water." *Langmuir* **31**(10): 2931–2935.

Chen, I. A. and P. Walde (2010). "From self-assembled vesicles to protocells." *Cold Spring Harb. Perspect. Biol.* **2**(7): a002170.

Chyba, C. F., P. J. Thomas, L. Brookshaw and C. Sagan (1990). "Cometary delivery of organic molecules to the early Earth." *Science* **249**: 366–373.

Clark, B. C. (1988). "Primeval procreative comet pond." *Orig. Life Evol. Biosph.* **18**(3): 209–238.

Cody, G. D., N. Z. Boctor, T. R. Filley, R. M. Hazen, J. H. Scott, A. Sharma and H. S. Yoder, Jr. (2000). "Primordial carbonylated iron-sulfur compounds and the synthesis of pyruvate." *Science* **289**(5483): 1337–1340.

Colomer, I., A. Borissov and S. P. Fletcher (2020). "Selection from a pool of self-assembling lipid replicators." *Nat. Commun.* **11**(1): 176.

Cornish-Bowden, A. and M. L. Cárdenas (2017). "Life before LUCA." *J. Theoret. Biol.* **434**: 68–74.

Deamer, D. W. (1985). "Boundary structures are formed by organic components of the Murchison carbonaceous chondrite." *Nature* **317**(6040): 792–794.

Deamer, D. W. and J. P. Dworkin (2005). "Chemistry and physics of primitive membranes." *Preb. Chem.*: 1–27.

Deamer, D. W. and R. Pashley (1989). "Amphiphilic components of the Murchison carbonaceous chondrite: Surface properties and membrane formation." *Orig. Life Evol. Biosph.* **19**: 21–38.

De Duve, C. (2011). "Life as a cosmic imperative?" *Philos. Trans. Royal Soc. A Math. Phys. Eng. Sci.* **369**(1936): 620–623.

Dyson, F. (1999). *Origins of Life.* Cambridge, Cambridge University Press.

Ertem, G. and J. P. Ferris (1996). "Synthesis of RNA oligomers on heterogeneous templates." *Nature* **379**(6562): 238–240.

Greenberg, J. M. and C. X. Mendoza-Gomez (1992). "The seeding of life by comets." *Adv. Space Res.* **12**(4): 169–180.

Guttenberg, N., N. Virgo, K. Chandru, C. Scharf and I. Mamajanov (2017). "Bulk measurements of messy chemistries are needed for a theory of the origins of life." *Philos. Trans. Royal Soc. A: Math. Phys. Eng. Sci.* **375**(2109): 20160347.

Haldane, J. B. S. (1929). "The origin of life." *Rat. Ann.* **148**: 3–10.

Hardy, M. D., J. Yang, J. Selimkhanov, C. M. Cole, L. S. Tsimring and N. K. Devaraj (2015). "Self-reproducing catalyst drives repeated phospholipid synthesis and membrane growth." *Proc. Nat. Acad. Sci.* **112**(27): 8187–8192.

Hordijk, W., J. Naylor, N. Krasnogor and H. Fellermann (2018). "Population dynamics of autocatalytic sets in a compartmentalized spatial world." *Life* **8**(3): 33.

Inger, A., A. Solomon, B. Shenhav, T. Olender and D. Lancet (2009). "Mutations and lethality in simulated prebiotic networks." *J. Mol. Evol.* **69**: 568–578.

Kahana, A. and D. Lancet (2019). "Protobiotic systems chemistry analyzed by molecular dynamics." *Life* **9**(2): 38.

Kahana, A. and D. Lancet (2021). "Self-reproducing catalytic micelles as nanoscopic protocell precursors." *Nat. Rev. Chem.* **5**(12): 870–878.

Kahana, A., D. Lancet and Z. Palmai (2022). "Micellar composition affects lipid accretion kinetics in molecular dynamics simulations: Support for lipid network reproduction." *Life* **12**(7): 955.

Kahana, A., S. Maslov and D. Lancet (2021). "Dynamic lipid aptamers: Non-polymeric chemical path to early life." *Chem. Soc. Rev.* **50**(21): 11741–11746.

Kahana, A., P. Schmitt-Kopplin and D. Lancet (2019). "Enceladus: First observed primordial soup could arbitrate origin-of-life debate." *Astrobiology* **19**(10): 1263–1278.

Kahana, A., L. Segev and D. Lancet (2022). "Protobiotic network reproducers are compositional attractors: Enhanced probability for life's origin." Available at SSRN 4317204.

Kahana, A., L. Segev and D. Lancet (2023). "Attractor dynamics drives self-reproduction in protobiological catalytic networks." *Cell Rep. Phys. Sci.* **4**(5).

Kauffman, S. A. (1986). "Autocatalytic sets of proteins." *J. Theoret. Biol.* **119**(1): 1–24.

Kauffman, S. A. (1993). *The Origins of Order: Self-Organization and Selection in Evolution*. Oxford, Oxford University Press.

Keefe, A. D. and S. L. Miller (1995). "Are polyphosphates or phosphate esters prebiotic reagents?" *J. Mol. Evol.* **41**(6): 693–702.

Kolb, V. M. (2016). "Origins of life: Chemical and philosophical approaches." *Evol. Biol.* **43**(4): 506–515.

Lancet, D., D. Segrè and A. Kahana (2019). "Twenty years of 'lipid world': A fertile partnership with David Deamer." *Life* **9**(4): 77.

Lancet, D. and B. Shenhav (2009). "11 compositional lipid protocells: Reproduction without polynucleotides." *Protocells*: 233.

Lancet, D., R. Zidovetzki and O. Markovitch (2018). "Systems protobiology: Origin of life in lipid catalytic networks." *J. Royal Soc. Interface* **15**(144): 20180159.

Lederberg, J. (1994). "The transformation of genetics by DNA: An anniversary celebration of Avery, MacLeod and McCarty (1944)." *Genetics* **136**(2): 423–426.

Leif, R. N. and B. R. Simoneit (1995). "Confined-pyrolysis as an experimental method for hydrothermal organic synthesis." *Orig. Life Evol. Biosph.* **25**(5): 417–429.

Markovitch, O. and D. Lancet (2014). "Multispecies population dynamics of prebiotic compositional assemblies." *J. Theoret. Biol.* **357**: 26–34.

Maurette, M. (1998). "Carbonaceous micrometeorites and the origin of life." *Orig. Life Evol. Biosph.* **28**(4–6): 385–412.

Miller, S. L. (1953). "A production of amino acids under possible primitive earth conditions." *Science* **117**: 528–529.

Miller, S. L., J. W. Schopf and A. Lazcano (1997). "Oparin's 'origin of life': Sixty years later." *J. Mol. Evol.* **44**(4): 351–353.

Miyakawa, S., H. Yamanashi, K. Kobayashi, H. J. Cleaves and S. L. Miller (2002). "Prebiotic synthesis from CO atmospheres: Implications for the origins of life." *Proc. Natl. Acad. Sci. USA* **99**(23): 14628–14631.

Morowitz, H. J. (1999). "A theory of biochemical organization, metabolic pathways, and evolution." *Complexity* **4**: 39–53.
Morowitz, H. J., B. Heinz and D. W. Deamer (1988). "The chemical logic of a minimum protocell." *Orig. Life Evol. Biosph.* **18**: 281–287.
Morowitz, H. J., J. D. Kostelnik, J. Yang and G. D. Cody (2000). "The origin of intermediary metabolism." *Proc. Nat. Acad. Sci. USA* **97**(14): 7704–7708.
Mushegian, A. (2008). "Gene content of LUCA, the last universal common ancestor." *Front. Biosc.-Landmark.* **13**(12): 4657–4666.
Nowak, M. A. and H. Ohtsuki (2008). "Prevolutionary dynamics and the origin of evolution." *Proc. Nat. Acad. Sci.* **105**(39): 14924–14927.
Oparin, A. I. (1924). *Proiskhozhdenie zhizny.* Moscow, Izd, Moskovski Rabochi.
Oparin, A. I. (1938). *The Origin of Life.* New York, MacMillan.
Oparin, A. I. (1953). *The Origin of Life.* New York, Dover Pub.
Oparin, A. I. (1957). "The origin of life on the earth." In *The Origin of Life on the Earth,* 3rd ed. New York, Academic Press.
Oparin, A. I. (1967). The origin of life. In *The Origin of Life*, ed. J. D. Bernal. London, Weidenfeld and Nicolson: 199–234.
Oparin, A. I. (1976). "Evolution of the concepts of the origin of life, 1924–1974." *Orig. Life* **7**(1): 3–8.
Pace, N. R. and T. L. Marsh (1985). "RNA catalysis and the origin of life." *Orig. Life Evol. Biosph.* **16**: 97–116.
Sakuma, Y. and M. Imai (2015). "From vesicles to protocells: The roles of amphiphilic molecules." *Life* **5**(1): 651–675.
Sargent, D. F. and R. Schwyzer (1986). "Membrane lipid phase as catalyst for peptide-receptor interactions." *Proc. Natl. Acad. Sci. USA* **83**(16): 5774–5778.
Schmitt-Kopplin, P., Z. Gabelica, R. D. Gougeon, A. Fekete, B. Kanawati, M. Harir, I. Gebefuegi, G. Eckel and N. Hertkorn (2010). "High molecular diversity of extraterrestrial organic matter in Murchison meteorite revealed 40 years after its fall." *Proc. Nat. Acad. Sci.* **107**(7): 2763–2768.
Schwartz, A. W. (1995). "The RNA world and its origins." *Planet. Space Sci.* **43**(1–2): 161–165.
Segre, D., D. Ben-Eli, D. W. Deamer and D. Lancet (2001). "The lipid world." *Orig. Life Evol. Biosph.* **31**(1–2): 119–145.
Segre, D., D. Ben-Eli and D. Lancet (2000). "Compositional genomes: Prebiotic information transfer in mutually catalytic noncovalent assemblies." *Proc. Natl. Acad. Sci. USA* **97**(8): 4112–4117.
Segre, D. and D. Lancet (2000). "Composing life." *EMBO Rep.* **1**(3): 217–222.
Segre, D., B. Shenhav, R. Kafri and D. Lancet (2001). "The molecular roots of compositional inheritance." *J. Theoret. Biol.* **213**(3): 481–491.
Shapiro, R. (1984). "The improbability of prebiotic nucleic acid synthesis." *Orig. Life* **14**(1–4): 565–570.
Shapiro, R. (2000). "A replicator was not involved in the origin of life." *Life (IUBMB)* **49**: 173–176.
Shenhav, B., A. Bar-Even, R. Kafri and D. Lancet (2005). "Polymer GARD: Computer simulation of covalent bond formation in reproducing molecular assemblies." *Orig. Life Evol. Biosph.* **35**: 111–133.
Wachtershauser, G. (1997). "The origin of life and its methodological challenge." *J. Theoret. Biol.* **187**(4): 483–494.
Walde, P. (2006). "Surfactant assemblies and their various possible roles for the origin (s) of life." *Orig. Life Evol. Biosph.* **36**: 109–150.

Watson, J. D. and F. H. C. Crick (1953a). "Genetical implications of the structure of deoxyribonucleic acid." *Nature* **171**(964–967).

Watson, J. D. and F. H. C. Crick (1953b). "A structure for deoxyribose nucleic acid." *Nature* **171**: 737–738.

Wołos, A., R. Roszak, A. Żądło-Dobrowolska, W. Beker, B. Mikulak-Klucznik, G. Spólnik, M. Dygas, S. Szymkuć and B. A. Grzybowski (2020). "Synthetic connectivity, emergence, and self-regeneration in the network of prebiotic chemistry." *Science* **369**(6511): eaaw1955.

# 14 Prototyping the Colonizer Cell
## *Combining Biology and Systems Engineering*

*Benton C. Clark*

**NOTE:** In this chapter, we include key references to relevant scientific papers, many of which are recent and advanced. In our examples of the general characteristics and functioning of biological systems which are well-known and for which reference material is readily available (e.g., Wikipedia), we do not provide published references to preserve readability. This also applies to other areas of common knowledge, such as for basic engineering systems.

### 14.1 INTRODUCTION

How did the global biosphere come about? To answer this question, we must begin at the origin of life. Among the hundreds or thousands of diverse environments on the early planet, there must have been a convergence of environmental properties to produce a setting, most likely dynamic, where the physical and chemical conditions were favorable to the formation of the first replicators. Such an environment is termed "urable" (Deamer et al., 2022) and could occur on Mars (Clark et al., 2021), on Earth, within ocean worlds, or on innumerable exoplanets. Such settings have more restrictive constraints on their physical and chemical properties than the widespread so-called "habitable" environments, which generally include all those where liquid $H_2O$ occurs and interacts with suitable chemical environs, whether geological or atmospheric, to produce a nutrient-solvated milieu.

The minimal nutrients required for life as we know it include a large variety of essential organic compounds, variously containing the elements carbon, hydrogen, oxygen, nitrogen, phosphorus, and sulfur (the CHONPS elements). In addition, a number of micronutrients seem essential to life forms. Most notable are ions of alkaline earth elements (especially $Mg^{2+}$ and $Ca^{2+}$), $K^+$, and transitions elements ($Fe^{2+}$, Ni, Zn, etc.). From these basic ingredients are constructed a bewildering variety of mostly organic molecules which interact in metabolic networks; informational macromolecules; as well as membranes and other structures in a harmonious way that results in a reproductive unit – the biological cell. At the simplest prokaryotic level, the chief components are DNA, RNA, proteins, plasma, and membrane.


DOI: 10.1201/9781003294276-14

As discussed in Chapter 6 herein (Gull and Pasek, 2024) and elsewhere (Todd, 2022), the availability and types of compounds that earliest life drew upon may have been significantly different than today's most widespread forms.

Furthermore, earliest life may have been primitive and different. For example, life may have begun with RNA rather than DNA (Joyce and Orgel, 1993; Robertson and Joyce, 2012); or, initially with thioesters (de Duve, 2000) rather than organic phosphates as the energy currency of the cell; or, with cosmic-derived amphiphiles as proto-membranes (Deamer and Pashley, 1989) rather than the ester- or ether-linked phospholipid membranes of contemporary species of bacteria and archaea.

As research in prebiotic chemistry, phylogenetic analyses, and primitive metabolisms continues to make exciting advances, it becomes ever more feasible to contemplate more primitive beginnings for the first globally competent cells. From these conjectures, with advances in modern lab techniques, it is also becoming possible to conceive relevant simple forms from which to physically construct prototypes of hypothetical earliest life forms.

Safety is also an important task in all laboratory work but will be especially so for the engineering of new self-replicating entities which potentially could escape control. It should therefore be the aim of this research not only to concentrate on elucidating the potential earliest evolution of cellular life, but also to steadfastly assure that these artificial organisms are never introduced into the wild, nor could they propagate if inadvertently released.

## 14.2 PROTOCELLS AND THE MACROBIONT

The urable (Deamer et al., 2022) cauldron from which life was born is envisioned as a special setting in which replicators arose and evolved, ultimately forming the first protocells. We term this setting the macrobiont (Clark and Kolb, 2020) because it is macroscopic in scale (meters to km), yet populated by microscopic entities which are the beginnings of biology. These protocells evolve to more sophisticated and competent macrobiont (MB) cells, which multiply and diversify. This, however, is not generally sufficient to form a biosphere because the MB environment is very likely to be subject to change and decay, especially over longer periods on geologic time scales. Furthermore, the urable environment to create a macrobiont may have necessarily been a rare juxtaposition of dissimilar environments, neither of which were necessarily representative of the more globally ubiquitous and potentially habitable environments.

### 14.2.1 Candidate Macrobionts

There are several scenarios for the origins of life. As listed by Smoukov et al. (2023, Preface), there are at least 25 different starting points that have been proposed (e.g., RNA World versus Metabolism-First versus Zinc World versus others), plus at least 14 different settings where it could have taken place. And at least 15 different theories of the trigger that created chirality have been proposed.

For simplicity of discussion, we consider two opposite settings: On land, from Darwin's speculation about "warm little ponds", versus the suboceanic hydrothermal vent hypothesis. An intermediate scenario is hot springs, which have not only access to the products of hydrothermal processing of minerals to release important nutrients, but also access to the surface for favorable solar UV-driven reactions and to facilitate the wet--dry cycling that can drive forward the difficult polymerizations (e.g., amino acids into proteins; nucleobases into DNA and RNA). These hypothetical macrobionts are all significantly different, but some tend to be special and relatively rare compared to the general occurrence of habitable bodies of water on the planet. Examples are shown in Figures 14.1–14.3.

In any case, there must have been transient cases where at least some of the progeny had become pre-adapted to external environments (e.g., along shorelines/mudcracks for ponds or ocean water/tidal flats for hydrothermal vents or effluent streams for hot springs). These special external environments provide the selection pressures for the rise of a subset of colonizer (CZ) cells from the predominant MB cells.

### 14.2.2 Colonizer Cells as the Bridge to LUCA

According to current knowledge, the most primitive version of life that can be inferred from available phylogenetic data has been termed the Last Universal Common Ancestor (LUCA). However, this organism was significantly more advanced than would have been necessary for the first colonizer cells.

As soon as one or more of these CZ cells escape and successfully reach one case of a widespread habitable environment, a global biosphere can begin to take root. Example: If life arose in a well-stocked Darwinian pond, its shoreline would be subjected to dilution by precipitation runoff and/or stream input, which mimics the more numerous but less-nutrient-rich ponds and lakes. Or, if life arises in a deep-sea hydrothermal vent, the MB cell mutants which experience the generic ocean interface will be subjected to the selection pressures of an enormous potentially habitable environment.

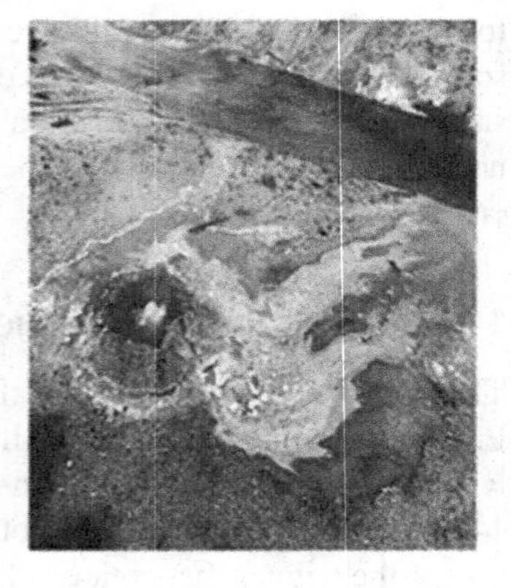

**FIGURE 14.1** Microbiological diversity from hot springs in Yellowstone National Park.
*Photo credit:* Author

**FIGURE 14.2** Episodically interconnected warm little ponds in Yellowstone National Park.
*Photo credit:* Author

**FIGURE 14.3** Lost City hydrothermal field.
*Photo courtesy:* Public domain, National Science Foundation; https://commons.wikimedia.org/wiki/File:Lost_City_(hydrothermal_field)02.jpg

Once the colonization of the planet begins, there can be an explosion of microbial populations. For example, Cole et al. (1993) found populations of $10^{10}$ microorganisms per liter in anoxic regions in 20 stratified lakes. In even more nutrient-rich environments, such as the intestines, population densities of bacteria can reach $10^{15}$ per liter (Ohland and Jobin, 2015).

Furthermore, the current number of ponds and small lakes on Earth is hundreds of millions (Downing et al., 2006), and the total area of land covered with freshwater reaches ~2% (Clark and Kolb, 2020). Using the scaling relationship between volume and area for lakes on Earth (Cael et al., 2017), this results in over $10^{16}$ liters of freshwater (not counting lakes larger than 2 km in diameter, which would increase the total by another factor of $>10^2$). These may be significant overestimates if the Hadean was devoid of continents and only volcanic islands protruding above a global ocean. However, this and other factors (such as latitudinal climate effects) would not reduce these cell populations in volumes of potentially nutrient-rich waters by more than 3–5 exponents out of a total of more than 25 (i.e., $10^{10} \times 10^{16} = 10^{26}$).

Additional exponents are added when accounting for the number of generations available for Darwinian evolution. Although the macrobiont might last only a millennium or less, the colonizer cell would have much more time to evolve to LUCA. For example, even if the CZ cell were sluggish compared to today's organisms, with a long reproduction time of, say 88 hours, then 5 more powers of 10 would be added to the total number of starting organisms in the number of new generations per 1,000 years (assuming stationary phase conditions, i.e., static overall population). This length of time can be short, of course, on the time scale for significant geological modifications.

Even with modest reproduction times of weeks or months, these enormous numbers of potential generations of modest CZ organisms provide the necessary springboard from which a variety of descendants can arise and periodically share DNA via horizontal gene transfer (HGT) (e.g., Soucy et al., 2015) to ultimately give birth to the progenitors of much more highly developed and competent cells.

Methods of HGT between microorganisms are varied, as summarized in Figure 14.4.

"Transformation" merely requires transiently leaky membranes which allow DNA segments to be released or taken up from the surrounding fluid.

The more complex "Conjugation" mechanism requires special bridges between organisms, which in modern cells is performed by hair-like pili extensions that enable transfers of DNA when two microbes "mate." These may not yet have been invented in the earliest forms of the colonizer cells, although the type IV pili are able to also provide motility to a microbe by twitching movements and therefore provide the additional competitive advantage of motility.

Bacteriophage viruses may also have arisen early (Prangishvili, 2015) and provided a vector for horizontal gene transfer between different strains of microbes that had arisen by the method termed "Transduction."

Because these more common environments are colonized but not necessarily each *stricto sensu* identical in terms of bioavailable nutrients and physical conditions, there will be additional selection pressures that result in a degree of global-scale diversity. From this vast multiplicity, there will be competition which can lead to ever-more-sophisticated means of survival and accelerated reproduction rates to

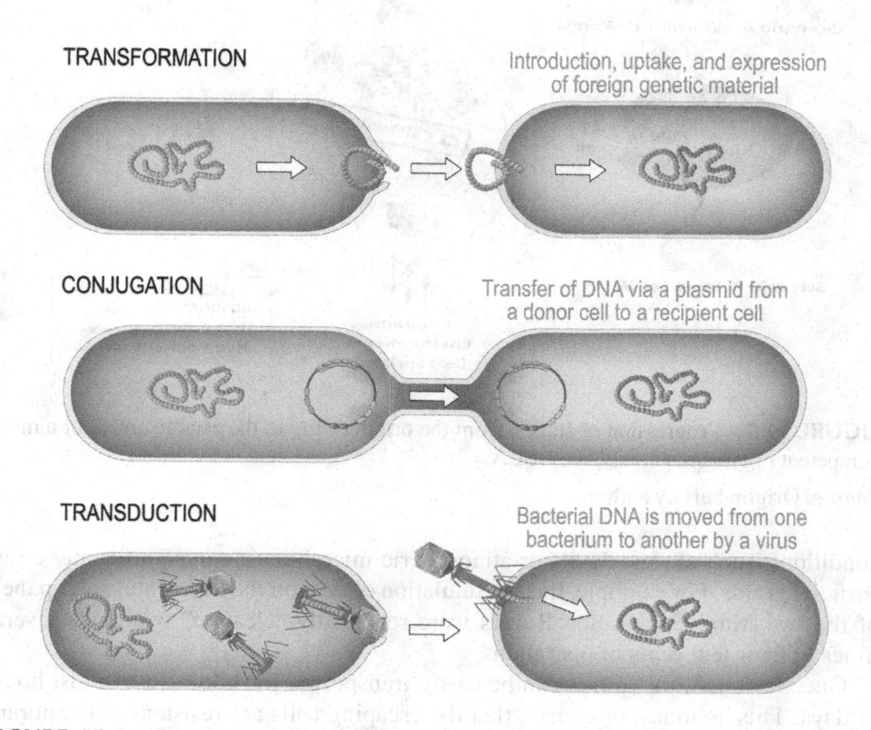


**FIGURE 14.4** The three modern methods by which procaryotic cells can interchange genetic information.

*Photo credit:* Shutterstock

out-compete the initial colonizing life forms. In our view, an essential intermediate precursor was the ecosystem created by CZ cell emigrants from the special cauldron in which life was born. The overall hypothetical scheme from protocell to LUCA is as shown in Figure 14.5.

## 14.3 THE COLONIZER CELL

Here, we hypothesize that a global biosphere takes hold only if the colonizer (CZ) cell arises and escapes the macrobiont. Modes of escape include transport by water currents if the earliest cells are hosted in an oceanic environment (vents or tidal flats). If in an environment exposed to the planetary surface, such as the warm ponds or hot springs, escape can be mediated by temporary overflow and transport by rivulets. The escape can also occur if they become isolated in dry conditions along a shoreline or in mudflats, followed by eolian lofting and dispersal. Highly dynamic

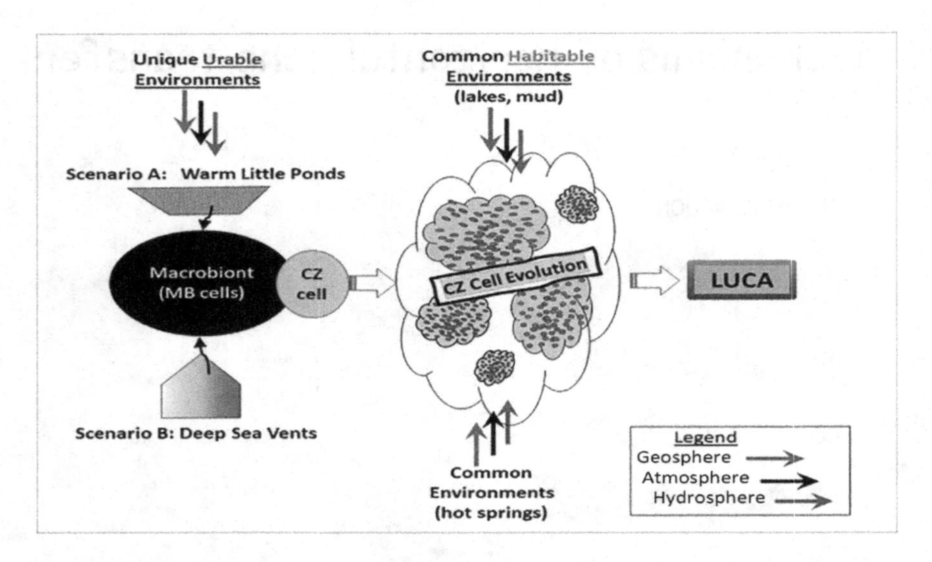


**FIGURE 14.5** Progression of stages, from the origin of life to the establishment of a most-competent organism, namely, the LUCA.

*Source:* Original art by author

conditions, such as dust devils or atmospheric microbursts, can significantly erode surficial grains. For example, the accumulation of dust on the horizontal solar panels of the two Mars Exploration Rovers were repeatedly "cleaned" overnight several times in their few years of operation.

Once levitated, the grains can be easily transported for kilometers in just hours or days. This assumes, of course, that the escaping cells are resistant to temporary desiccation and are able to avoid the lethal pre-ozone solar UV by shielding by regolith grains to which they are attached and/or transport during nighttime. As with all forms of microscopic life, their intrinsically high population numbers can overcome low probabilities of successful dispersal, as attested by the widespread occurrence of all manner of microorganisms in surface soils to which they are not necessarily pre-adapted.

We further hypothesize that because the LUCA as derived from phylogenetic studies is highly sophisticated (Kirschning, 2022), with over 1,000 gene families (Ouzounis et al., 2006), the CZ cells themselves could have been much more rudimentary, yet capable of colonizing the planet on the global scale (ponds to lakes, rivers, and potentially to an ocean). This provided an opportunity for the expansion of the population of CZ cells as well as exposures of subsets to a wide diversity of selection pressures from a multitude of nonidentical habitats to enable the rise of LUCA-class cells. The subsequent intermingling of cells with somewhat diverse capabilities and weaknesses under conditions in which horizontal gene transfer can occur could enable the rise of cell lines with some optimum combination of reproductive performance versus resilience to erratic changes in the environment.

### 14.3.1 Irreversible Effect of the Colonizer Cell

One of the key characteristics of any system is its tipping point, at which the system changes irreversibly and cannot assume its prior state. The history of the tipping point as a scholarly concept rather than a metaphor was reviewed by Milkoreit et al. (2018). These authors revealed that this concept came from the early works of the chemist Hoadley in 1884 and the mathematician Poincaré in 1885. According to Poincaré, a system undergoes a qualitative change during bifurcation, which is a well-accepted concept in fields such as mathematics, physics, ecology, environmental, evolutionary, and systems science.

While the systems approach to astrobiology and the emergence of life which it addresses are covered in a recent book (Kolb and Clark, 2023), the specific applications of the tipping point concept to the emergence of life are mostly general.

We bring up here a fruitful application of the tipping point concept to the example of the CZ cell. When life emerges, it must be able to become established and to perpetuate to survive. A single alive cell will not be able to accomplish this. Once life emerges, the tipping point for the abiotic chemical evolution is reached. This means, for example, that the newly established life may "eat" the founding organic chemicals, which then will not be able to further evolve. Further, once cellular life emerges, the tipping point toward life will make all proto-life cells obsolete, since they will not be able to compete successfully in survival and reproduction with the CZ cells.

Our approach follows that of Gladwell (2002), whose influential book *The Tipping Points* applied this concept to multiple fields. In short, he proposes three rules about reaching the tipping point: The Law of the Few, the Stickiness Factor, and the Power of Context. In his book, Gladwell amply supports these rules. Here, we consider example applications of the rules to the colonizing cells.

The Law of the Few in the context of the colonizing cell simply means that such cells will spread not by a sheer number of them, but by a few such cells that are uniquely able to achieve colonizing. Thus, they have a special fitness for it. Many other cells will have very low success in colonizing and will become obsolete. It is by the power of biological exponential growth that a few original constituents can overpower the others and "takeover" the system.

A Stickiness Factor example is that the colonizing cell which escapes the original pond and finds a new setting will stick to the new setting and thus will prevail. While many cells may escape and potentially reach the new setting, only some will literally "stick" to it. The ones which do not will not survive or thrive. The Power of Context in our example means that the colonizing cell will stick to and flourish in the new setting only if such setting provides the correct context, i.e., an inhabitable environment for it.

The tipping point in our example means that life can establish and spread successfully from colonizing cells. The Tipping Point will be the emergence of the new quality of competent colonizing cells, thus enabling the establishment of a widespread colonization and the earliest beginning of a global-scale biosphere. Only when there are both breadth and depth in the biosphere will it become virtually indestructible, except in only the most challenging of circumstances (e.g., stellar evolution which eliminates the possibility of liquid water, such as the sun's growth leading up to the Red Giant phase).

### 14.3.2 Approaches to Conceptualizing the Colonizer Cell

As seen in Figure 14.6, the prokaryotic cell is much simpler than its eukaryotic descendants.

It is fortunate that, for a variety of reasons, a diverse and widespread array of prokaryotes have survived to this very day in spite of the voracious appetites of the animals and other eukaryotes which opportunistically consume them. Thus, we have before us examples of simpler life whose gene streams are still available in spite of their forerunner species being totally extinguished (as far as we know at this point in time). These cells serve as our window into the past and provide the only sound jumping-off point from which to judge, hypothesize, and simulate their less proficient ancestors.

There are three fundamentally different approaches that are being taken to derive the properties of the earliest cells. Great contributions have already been made through phylogenetic tracing to gain insights to the capabilities of LUCA from the "root" of the tree of life. From this, researchers have derived characteristics of the LUCA, such as that it was an anaerobic chemolithoautotrophic microorganism, with mesophilic or thermophilic compatibility (Weiss et al., 2016).

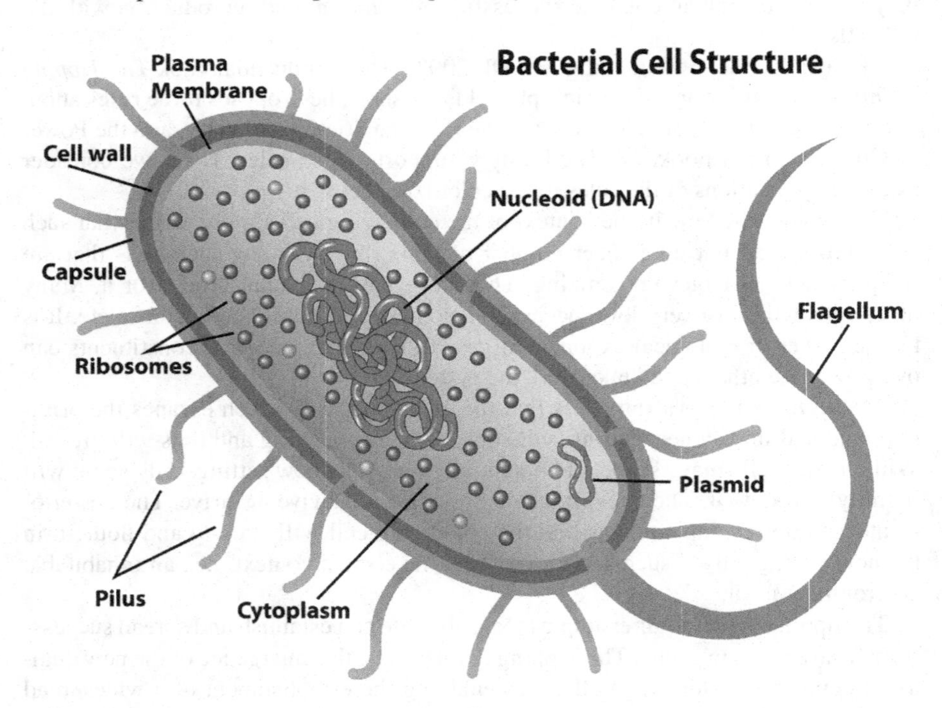


**FIGURE 14.6 (A AND B)** Prokaryotic cells are relatively simple compared to eukaryotic cells.

*Photo courtesy:* Shutterstock

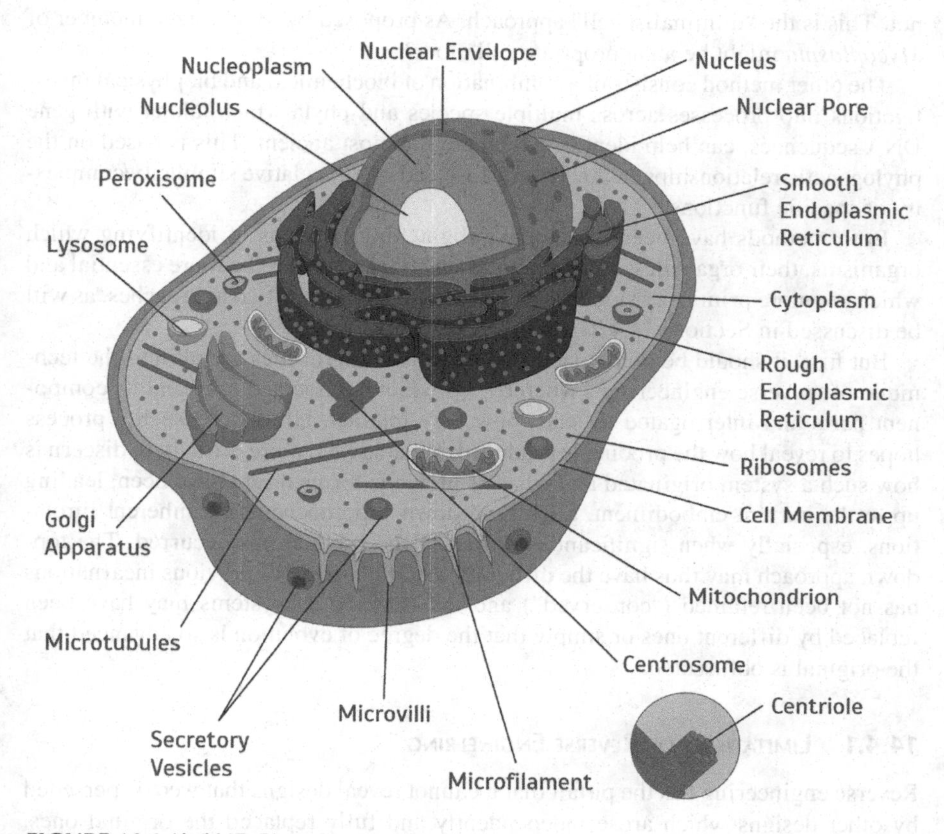


**FIGURE 14.6 (A AND B)** *(Continued)*

A second approach is to investigate the properties of a minimalist cell, as discussed in the next section.

And a third approach is a bottoms-up approach, which we will consider subsequently, by the exploration of prebiotic chemical evolution (PCE) to investigate natural chemical pathways that would generate molecular assemblages that provide the cornerstones of life as we know it: Amino acids which can become polymerized to form proteins; nucleoside bases that can be polymerized to form RNA and DNA; and lipids or other suitable molecules to form the membranes that would envelop the cellular contents.

## 14.4 TOP-DOWN APPROACHES TO THE MOST PRIMITIVE CELL

There are two fundamentally different top-down methodologies to investigate what could be a most primitive cell. In the first, a simple procaryote that exists in nature is selected and then perturbed by inactivating one gene at a time (gene "knock-out" technique) or by the use of a broader-based transposon mutagenesis approach to find out which genes are essential to cell survival and reproduction and which are not. This is the "minimalist cell" approach. As proposed by Morowitz, a member of *Mycoplasma* might be an appropriate cell model.

The other method consists of a combination of biochemical and biophysical investigations into processes across multiple species and phyla which, along with gene DNA sequences, can help identify which are the most ancient. This is based on the phylogenetic relationships in the Tree of Life and on the relative simplicity/complexity of specific functions.

Both methods have been extremely enlightening in terms of identifying which organisms, their organelles, and metabolic functions appear to be more essential and which are more primitive. Much has been learned from these two approaches, as will be discussed in Sections 14.4.2 and 14.4.3, respectively.

But first, it should be recognized that these two approaches are akin to the technique of "reverse engineering", whereby a product is dissected to reveal its component parts and interrogated to learn how they interact. Ultimately, such a process hopes to reveal how the product is produced. What is even more difficult to discern is how such a system originated and what its previous forms might have been, leading up to the current embodiment. Such "top-down" approaches have inherent limitations, especially when significant evolution of the product has occurred. The top-down approach may thus have the difficulty that the history of previous incarnations has not been retained ("conserved") and that specific subsystems may have been replaced by different ones or simply that the degree of evolution is so advanced that the original is obfuscated.

### 14.4.1 Limitations of Reverse Engineering

Reverse engineering has the pitfall that it cannot reveal designs that were superseded by other designs which arose independently and fully replaced the original ones. This is evident in all developments invented by human intelligence, as new technologies come to the forefront (in availability and affordability).

For example, the reverse engineering of a contemporary mobile smartphone from the original hard-wire relay-switched telephone networks simply does not work. All the technologies involved are different. The smartphone is powered by an onboard battery based on lithium-ion technology; the original phone system was powered by electrical generators located remotely. The mobile phone communication system is wireless, based on the emission and reception of electromagnetic waves at radio frequencies; the original electrical communications were transmitted as pulses (telegraph) or modulated voltages (telephone) along a wiring network interconnected at both ends by switches or magnetic relay boxes. All phone systems require the use of microphones and speakers to transduce between sound wave energy and electrical energy, but the detailed technologies for these conversions have been different and evolved over time.

In the case of the CZ cells, the most likely circumstance is that their primitive metabolism has been replaced by more capable ones. We need only look at how modern technological systems have driven out their forerunners. The abacus was replaced by tooth-geared mechanical calculators; the slide rule approximator was replaced by electronic analog versions, only to be replaced by digital electronic central processing units (CPUs). Although it may be of interest to investigate which bamboo or marble is better suited for the optimum abacus, it does not lead to the invention of the CPU. In fact, the modern CPU which not only performs calculations but also provides logic functions, sequences instructions, and reads and writes memory banks only became practical with the replacement of the vacuum tube diodes and amplifiers by microscopic impurity-atom-doped semiconductor-based equivalents. Painfully obvious it is that one cannot back-engineer a modern CPU to derive the "ancient" vacuum tube (or "thermionic valve"), which is shown in Figure 14.7. Nor can the abacus be envisioned by disassembly and study of even the simplest electronic calculator. One could not expect to back-engineer the design of a conventional self-winding mechanical watch by taking apart a modern electronic watch.

However, life did evolve sequentially, whereas the electronic watch was not evolved faithfully from the mechanical watch. Neither the controls, nor shape, and

**FIGURE 14.7 (A AND B)** Electron tubes and discrete parts (resistors, capacitors) replaced by highly integrated circuits utilizing solid-state semiconductor chips.

*Photo courtesy:* Shutterstock

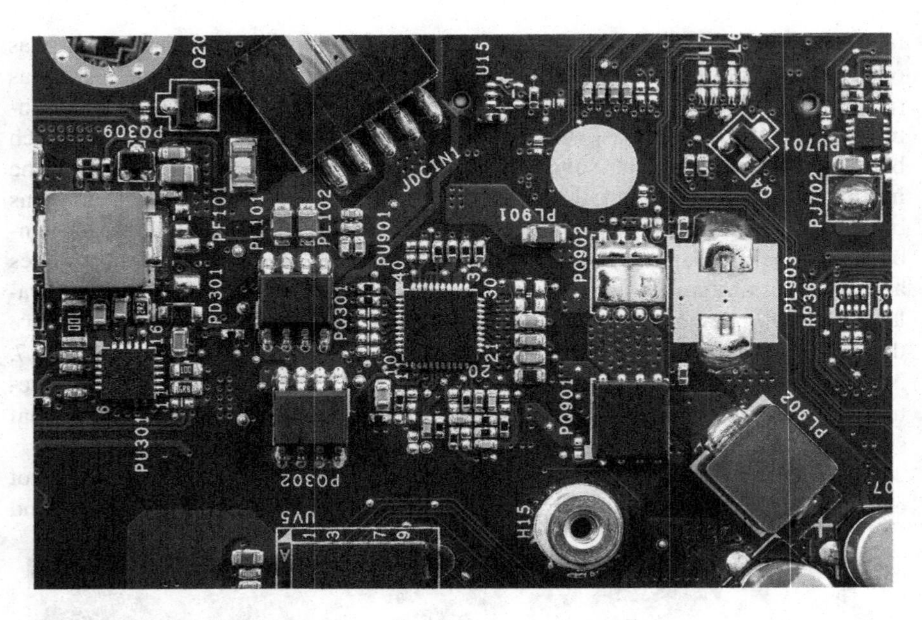


**FIGURE 14.7 (A AND B)** (Continued)

not even the wristbands are the same. Fortunately, there is some basis for the expectation that the mechanisms of some modern organisms may indeed recapitulate, in parts at least, the history of its origins.

Because certain portions of the DNA sequences are actually highly conserved (i.e., have not substantially changed), they apparently do go back to the beginning. The question is what "beginning" do they preserve. It is known, for example, that much is preserved that predates the branching between archaea and bacteria. This means that these preserved sequences were part of the LUCA genome.

But that does not guarantee what may have predated LUCA. We hypothesize that earlier genomic specifications and metabolic functions could have been less complex, albeit perhaps not as efficient, robust, or versatile. Yet, the earlier process(s) were sufficient to promulgate life in certain highly suitable environments. We presume that such environments could be somewhat rare, but nonetheless sufficiently abundant that they could be interconnected, biologically, by natural actions such as wind transport or hydrological linkages.

Likewise, although the earliest automobiles were steam- or electric-powered, these gave way to fundamental changes. For decades thereafter, the vast majority of all modern automobiles were powered by the combustion of petrochemicals with air and in piston/crankshaft-driven engines. No matter how detailed the study of even a minimalist car today, it could not be derived that early wagons were pulled by animals or powered by a steam engine.

Telecommunication by microwaves cannot be back-engineered to derive the telegraph to communicate over wires. An ocean cruise ship has virtually no back-connectivity to the birchbark canoe, other than the principle of floating by volumetric displacement of more than an equivalent mass of water and a streamlined shape and high aspect ratio to minimize drag and thus lessen the energy required to locomote.

The automobile has evolved since its early inceptions, as exemplified in Figure 14.8. First, the wheel had to be invented, then the multi-wheeled wagon. Steam-engine powered wagons were built but were too impractical. Although electric-powered wagons first appeared two centuries ago, they were also impractical because the rechargeable battery had not yet been invented. But, by the late 1880s, rechargeable batteries were sufficiently capable that enabled the demonstration of an electric vehicle with a top speed of 20 mph and 50 miles range.

About the same time, Carl Benz was able to obtain a patent for a gasoline-engine-powered vehicle, albeit the early versions were with less than 1 horsepower. Electric vehicles were popular because of no hand-cranking, no exhaust smell, and quiet operation. However, gasoline eventually became more widely available and affordable, and the combustion-based engines became more powerful and lightweight.

The earliest automobile versions were barebones designs, with no enclosure except perhaps a sunshade and not even windshields or doors. Fast-forward to today's automotive wonders, with its climate-control system complete with concert-quality musical reproduction, plus access to information sources and built-in charging stations for your smartphones and other accessories. Steering and braking are power-assisted. Wheels have rubber tires. Windows are button-powered up and down. Lighting systems can be set on automatic to sense night or fog. Windshield wipers can automatically respond to raindrops.

**FIGURE 14.8** (a) Evolution of the wheel and wagon, from carriage to race car. (b) Example of reverse-engineered concept.

*Photo credit:* Shutterstock

Modern automobiles have one or more computers, performing functions such as monitoring/controlling emissions, performing anti-lock braking, providing the security system, enabling cruise control, and providing warnings. Especially important now are the safety features. Highways enabling high speeds also can result in fatal crashes and collisions. But these accidents are no longer always fatal because of energy-absorbing protective structures and automatically deployed cushion bags. Accidents also occur less often, because of computerized radar and visual imaging systems that warn drivers of other vehicles in blind spots, or of dangerous closing speeds with obstacles. The modern automobile also includes advanced navigation and emergency communication functions. There are electrically heated seats and steering wheels. Aesthetics are enhanced by a deep-color, shiny, weather-resistant finish on a sleek, aerodynamic design. There are even cup holders.

Virtually no subsystem has been left untouched by modern innovations. Consequently, there are far more things that can go wrong in the modern automobile. However, improvements in engineering implementations and design have also vastly improved the reliability of each and every subsystem, down to the sub-sub-sub-subsystems and their component parts. In spite of this, among some tens or hundreds of thousands of a particular new model of automobile produced, there emerge flaws that must be corrected by "Recall" for remedial action. Likewise, the eukaryotic cell, with all its plurality of implementations and sophistications, can sometimes become the enemy of the multicellular organism it creates by becoming ineffective or defective or, worst of all, uncooperatively reproducing itself to dangerous levels (i.e., carcinogenic).

In short, the challenge is the possible absence of primitive remnants of metabolic pathways that became obsolescent by being outcompeted by more efficient ones. A central dogma of prebiotic evolution for the origin of life (OoL) has been, however, that many of the organic species or families that exist today in cells or in the natural environment are all that is or was available, ever. Thus, the fabled "silicon giraffe" has been discounted by the better understood limitations of silicon diversity compared to carbon (Lambert, 2019) and doped-semiconductors notwithstanding, with the possibility that an enterprising engineering team might yet construct a disconcertingly realistic robotic giraffe.

We must therefore acknowledge that a top-down approach can help elucidate some constructs that may ultimately shed a strong light on what the CZ cell might have been, in spite of the conjecture that LUCA was far more exquisitely developed than the putative CZ cell needed to have been. Also fruitful are the various discoveries of more simplistic metabolic pathways that accomplish the same ends as a more widespread pathway albeit with less efficiency or less robustness to adversity. Such simplistic pathways may be rare and occur in only novel organisms but should be sought out.

### 14.4.2 The Minimalist Organism

Important research has been underway in "synthetic biology" to systematically use genetic modification techniques to disable genes individually, in minimalist cells, to determine which genes are critical and hence deduce the minimum number of genes needed for the modified cells to survive and reproduce. Depending on the criteria, such cells seem to survive and reproduce through the action of less than 500 genes (Koonin, 2000; Gibson et al., 2010; Hutchison et al., 2016; Gibson et al., 2019; Glass et al., 2017).

The cells being studied are members of the genus *Mycoplasma*, a group of related bacteria which are generally strongly parasitic on multicellular eukaryotes. As shown in Figure 14.9, these members of bacteria are extraordinarily small and simple, although they can support either an aerobic lifestyle or an anaerobic one (facultatively).

This reverse engineering is enlightening but may not succeed in fully probing deeper than LUCA in equivalent complexity, i.e., down to the CZ cell, which in our hypothesis is likely to be significantly simpler and more primitive. There are at least two reasons for this. The test subject, the *Mycoplasma genitalium*, is in fact a pathogenic bacterium already highly dependent on a rich, favorable environment to

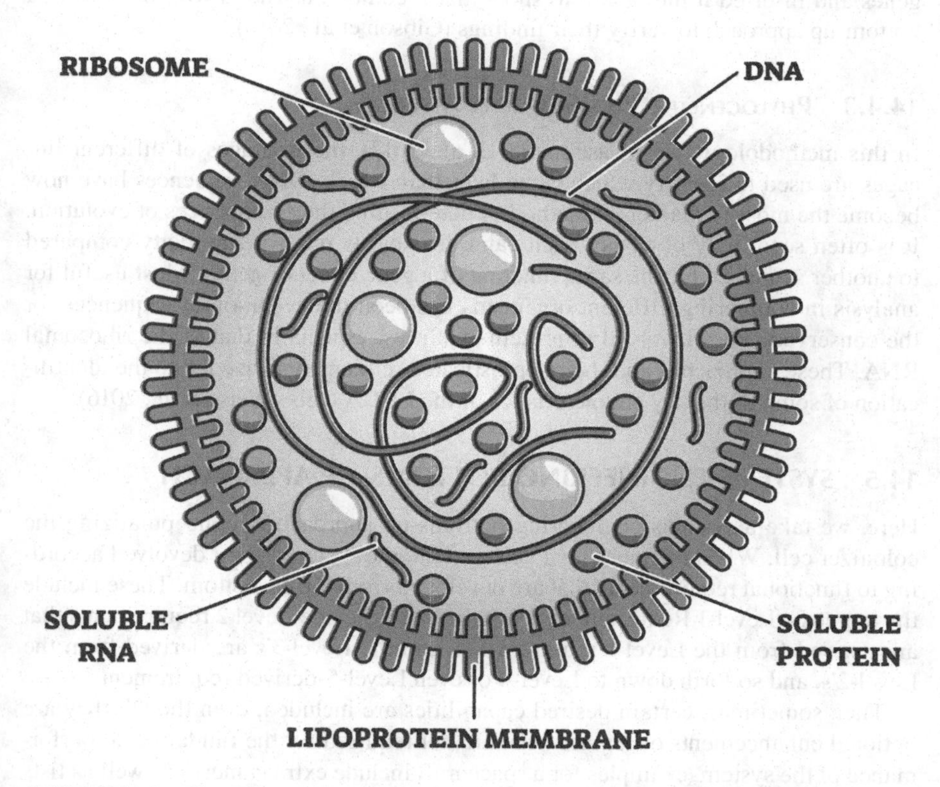


**FIGURE 14.9** The *Mycoplasma* genus comprises naturally minimalist organisms.
*Photo credit:* Shutterstock.

survive. Thus, the SP-4 growth medium needed to cultivate the mycoplasma cells (Hutchison et al., 2016 supplementary material) includes significant concentrations of all 20 biological amino acids, none of which this organism can synthesize itself. SP-4 medium also contains beef heart infusion, peptone, fetal bovine serum, and CMRL 1066 solution. The latter provides the amino acids, plus a dozen vitamins, a dozen additional organics (including glucose and NAD), and a solution of the needed inorganic salts (cations of Mg, Ca, Na, and K, as various chlorides, sulfates, or phosphates). Other than the salts, organic contents of this growth medium do not seem consistent with the composition of an abiotic pond, stream, or sea. Thus, in the sense of being akin to LUCA (an autotroph) or much less to CZ cells, this syn3.0 cell is dependent on factors beyond what these genes can accommodate.

Surprisingly, the functions of about one-third of the 373 genes that have been retained to enable a synthesized version of this so-called *syn3.0* cell line to be constructed are not known, a finding that has been termed "astonishing" (Glass et al., 2017). Additional studies on other species of mycoplasma give about the same result of ~400–500 genes (Glass et al., 2017). Although the deletion of genes from an existing organism is a top-down approach, the investigators also then artificially constructed a genome using DNA sequence manipulation to provide these identified genes and inserted it into a cell to show that it could construct a working cell in a bottom-up approach to verify their findings (Gibson et al., 2010).

### 14.4.3 Phylogenetics Plus Simplicity Approach

In this methodology (e.g., Nascimento et al., 2017), the heritages of different lineages are used to identify which came later than which. DNA sequences have now become the most fundamental method of determining these sequences of evolution. It is often seemingly obvious which came first by its relative simplicity compared to another sequence for the same function of a gene or set of genes. Most useful for analysis in comparing different organisms has been the evolution of sequences for the conserved genes involved in protein synthesis, especially that of the ribosomal RNA. These approaches and their sophisticated refinements have led to the identification of some of the key characteristics of the LUCA cells (Weiss et al., 2016).

## 14.5 SYSTEMS ENGINEERING BOTTOMS-UP APPROACH

Here, we take a systems engineering bottoms-up approach to conceptualizing the colonizer cell. When engineering a new system, the system is first devolved according to functional requirements that are developed from top-to-bottom. These include the topmost "Level-1 Requirements", but then also include Level 2 requirements that are derived from the Level-1 requirements, and the Level-3's are derived from the Level-2's, and so forth down to Level-4 or even Level-5-derived requirements.

Then sometimes, certain desired capabilities are included, even though they are optional enhancements or features that are not required for the fundamental performance of the system (examples for a spacecraft include extra cameras as well as lists of submitted names that are inscribed on a medallion, as a means to involve "public participation" in the spacecraft's mission of exploration).

We then take an overall design concept and perform detailed design and implementation mainly from the bottom to top. There are also "middle-out" activities which include, for example, mathematical simulations (models) which simulate the overall system and which can identify unrecognized synergies or conflicts between or within subsystems.

First and foremost is the development of the overview engineering block diagram, with subsystem high-level requirements included therein. Figure 14.10 shows a typical high-level systems engineering block diagram. In the context of astrobiology, these subsystems have been analyzed with respect to a planetary rover, the human physiological systems, and the macrobiont (Kolb and Clark, 2023, pp. 33–43, 52–62, and 90–98, respectively).

It is useful and potentially insightful to also compare these generalized subsystems with those which are analogous to the functions of a living cell of the type envisioned for the colonizers, as summarized in Figure 14.11 and discussed in sections which follow. Although not all is grounded in fundamental research, the use of unrestrained considerations can lead to potentially fruitful hypothesis formulations.

### 14.5.1 Energy Subsystem

A system performs functions that would not ordinarily occur in a natural environment. It is intrinsically out-of-equilibrium with the external environment, and to be

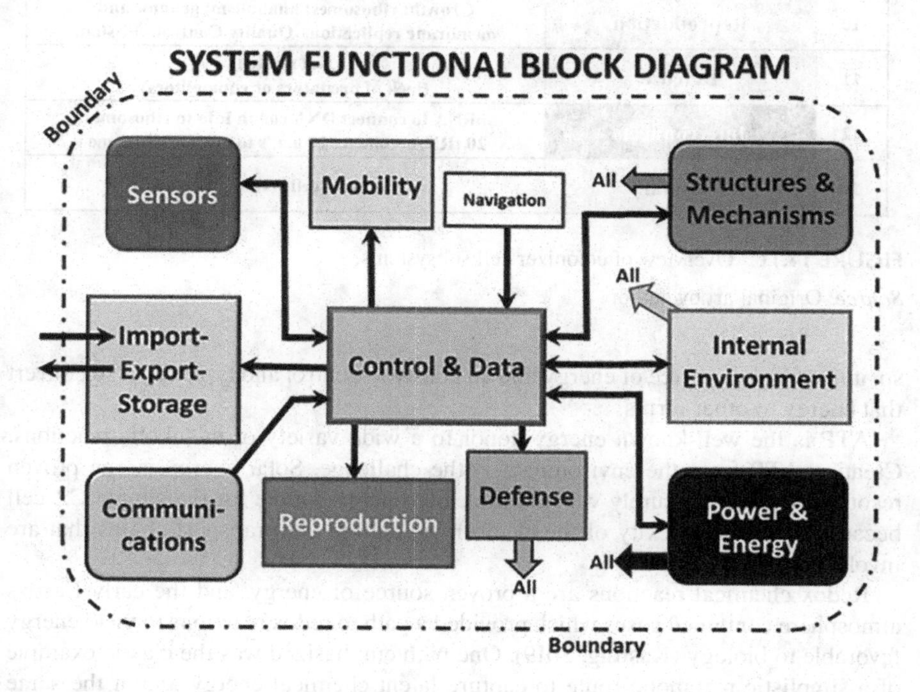


FIGURE 14.10 Generic block diagram of key system functions as subsystems.

*Source:* Original art by author

| | Subsystem | Colonizer Cell |
|---|---|---|
| 1 | Energy | Capture (chemical); Condition (form ATP); Maintain (ATP); Distribute (ATP); Redox (NADH) |
| 2 | Control and Data | Data: genome (DNA). All RNA regulators; Protein repressors, inducers, enzymes, and activators |
| 3 | Sensors | None needed? |
| 4 | Structual/Mechanical | Membrane confinement; DNA coiling; Cytoskeleton (not needed?); Pili (needed?) |
| 5a | Import | CHONPS plus Fe and transition elements. see text for other elements (Mg, K, Ca) and molecules |
| 5b | Export | Active transport outward of $Na^+$ and toxic ions and molecules; Catabolic deconstruction into subunits |
| 5c | Storage | Metastable equilibrium concentrations of ATP, NADH, acetyl CoA, other metabolites |
| 6 | Communications | Not required |
| 7 | Internal Environment | pH, redox (Eh), osmolarity, ion concentrations (see Import/Export) |
| 8 | Mobility | Not required |
| 9 | Navigation | Not required |
| 10 | Reproduction | Growth: ribosomes; anabolism; genome and membrane replication. Quality Control. Fission. |
| 11 | Defense | Not needed (lack of predators or competitors) |
| 12 | Interconnects | mRNA to connect DNA codon info to ribosomes; 20 tRNAs connect 20 a.a.'s to mRNA/ribosome |
| 13 | Boundary | Membrane (cell wall not needed) |

**FIGURE 14.11** Overview of colonizer cell subsystems.

*Source:* Original art by author

so, it must have a source of energy and an ability to control and typically also convert that energy to other forms.

ATP is the well-known energy donor to a wide variety of metabolic reactions. Creating ATP from the environment is the challenge. Solar energy is one proven resource, but almost surely was not the early energy source for the simple CZ cell because of the complexity of the electron generation and transport chains that are involved.

Redox chemical reactions are a proven source of energy, and the early Earth's atmosphere contained gases which provided a path to redox reactions to yield energy favorable to biology (Kasting, 2019). One path emphasized was the classic example of a simplistic metabolic route to capture latent chemical energy and at the same time to fix atmospheric carbon dioxide and hydrogen gas into essential organic molecules. This is provided by the metabolic Wood–Ljungdahl pathway (aka the reductive acetyl-coenzyme A pathway). At the same time, this pathway captures the energy from this overall reaction by converting attendant ADP to ATP (Varma et al., 2018). It is also this pathway which has been implicated as the primitive methanogenic metabolism of LUCA (Weiss et al., 2016).

Converting ATP back to ADP is an excellent way to add chemical energy to a receptive chemical reaction. However, the cell does not readily store ATP for future use because of its relatively poor stability. Instead, glycogen and glucose are synthesized to provide the primary energy storage functions (the cellular batteries) because their subsequent catabolic reactions can produce the ATP when needed.

### 14.5.2 Control and Data

Without some sort of controls, a system cannot function. It either follows a constant script or an adaptive script that is capable of varying its routine. In human-developed systems, this function is today generally provided by a computer which can implement built-in sequences through various outputs. However, if the system is to be responsive to the state of the external environment, it needs information, i.e., "data" about that environment. It also needs to monitor the internal environment sufficiently to maintain the health, safety, and sustainability of the system. Data are memorized for future interpretation, setting the need for temporary and permanent memory subsystems.

Control in the cell is especially maintained through the production of enzymes. Their catalytic activity performs the functions of a "leaky switch", in that they can strongly "turn on" a particular reaction pathway, or strongly inhibit another, but do not switch between zero and maximal production rates. Nonetheless, their actions are sufficient to meet the functional needs of the systems or a subsystem. Without enzymes, and a method for their control, a cell simply could not properly function.

Biological systems, as we know them, have an equisitite storehouse of control information, viz. their DNA, which employs a triplet code of four characters for specifying any given protein polymers from the 20 or so amino acids that make them up. In addition, the codes for starting and stopping the protein sequence and for overall functioning of the system are provided in the DNA sequence. In many respects, this is analogous to the "native code" in a computer, such that its interpretation requires an understanding of how the code is translated into actionable instructions.

Part of that interpreting function is accomplished by identifying important stretches of the overall DNA sequence (its "genes") and making copies into much smaller RNA polymer molecules. A large variety of small RNAs (sRNA) provide the necessary regulatory functions. These controls are somewhat like individual apps that can be called upon when necessary, such as when experiencing changing environments.

To make proteins, the single-messenger mRNA version must find a ribosome and become the template for which individual transfer molecules (tRNA) form a bridge between each triplet code and the corresponding amino acid molecule. Whether a given protein manufacturing sequence is activated depends on special protein "transcription factors" which bind to a special promoter region nearby the gene to be expressed. Because some proteins can behave as internal sensors and hence monitor the metabolic processes underway, they can affect these processes by their transcription activities.

Proteins can have other functions – some are structural components, some are transporters, some are chaperones (assisting in helping other proteins to assume their 3-D shape), etc. Gene regulation thus becomes the means by which a cell controls its own internal processes as well as any responses needed to external environments.

Because of the amazing range in capabilities of proteins, they are at the heart of implementing the overall control within the cell, through judicious use of the "data" stored in the DNA sequences, as implemented by special sRNA or proteins. In some ways, the cellular control subsystem combines the attributes of the digital and analog computer in the sense that enzymes are not perfect digital switches, and there is delay in the feedback loop when a need for adjustment is sensed before the response molecules can be generated and reach their intended location.

It is difficult to hypothesize any alternative or better way to control a complex, interrelated series of chemical reactions than by use of a stockpile of apps or subroutines, which a gene or group of genes comprise through the use of protein-based enzyme switches and regulators. These provide not only the instruction set but also the means for the feedback of processes and conditions which require adjustment in order for the cell to perform its overall functions, some of which are paramount for survival and some for the essence of life, i.e., reproduction.

### 14.5.3 Sensors

Advanced bacteria can sense various environmental factors and respond by movement. In some cases, they can even detect small gradients in such factors as temperature or nutrient concentrations. These sensors are useful especially when the microbe can move toward or away from a given stimulus, depending on whether it is beneficial to increase or decrease the level of exposure (e.g., toward food, or away from too hot or too cold temperatures). Even if gradients cannot be determined, the simple act of moving or not could be sufficient to provide a benefit to the organism.

However, thermotaxis or chemotaxis responses may not be essential, especially if the organism does not have motility capabilities (see Section 14.5.8). Our most primitive CZ cell need not necessarily be equipped with sophisticated sensory capabilities.

### 14.5.4 Structural and Mechanical

Every system needs its structural support to maintain its shape and integrity against external forces. The boundary (Section 14.5.13) membrane and possible cell wall are structural elements. In the simplest microbes, an internal structural component is difficult to image, but a cytoskeleton anchored by the outer membrane does exist, albeit much less substantial than the complex fibrous protein structures within eukaryotic cells. The lack of membrane-defining organelles or a central nuclear compartment minimizes structural needs for the simplest prokaryotes.

Mechanisms may also not be needed, given that many microbes are adequately compatible with their local environment, and those which can become flagellated have that as an option. Locomotory capability may not arise until later advancements. Fabrication of a flagellar subsystem requires the action of many genes, although a stepwise evolution of a few basis genes has been shown to provide a path to its development (Liu and Ochman, 2007).

The CZ cellular structure and mechanisms can be simple, but accomplishing the task of reproduction may invoke certain needs as discussed later (see Section 14.5.10).

### 14.5.5 Import/Export/Storage

The system boundary, in combination with active subsystems, must be able to sufficiently control the entry of materials and energies, as well as to dispose of excess materials and energies that could compromise the operation of the system. Storage of materials and energy needed in the future also enables a system to perform at peak efficiency and activity.

The boundary of the cell as a system is its semipermeable membrane. This provides a way not only to juxtapose key elements tightly (e.g., DNA and ribosomes), but also to provide for and maintain differentials in cytoplasm compositions versus the external aqueous medium. In modern cells, there are highly developed means of combining passive semi-permability (e.g., ion channels) and active transport (e.g., by locally staged transporter specialist proteins) to separately regulate the import and export of key ions and molecules. Among these, capture and import of dissolved $H_2$ can be one of the most important activities for early life. Organic molecules that could serve as feedstocks may be available in the environment from UV- or electric discharge-induced synthesis, and/or from the infall of exogenous material such as meteorites from carbonaceous chondrites, interplanetary dust particles, and comets.

Likewise, since early enzymes may have been even more dependent on metal ion cofactors than today, the acquisition of favored ions and the avoidance or export of interfering ions may be critically important. An important question, however, is to what extent these active, passive, and storage processes were necessary, compared to the natural abundances of various key cations in the primitive biosphere. Based on the needs of current microbes, as well as availabilities in the more reduced Hadean atmosphere (which affects the solubility of many ions), these micronutrients can include not only $Mg^{2+}$, $Ca^{2+}$, $K^{+}$, as well as many transition elements, including $Fe^{2+}$, Mn, Co, Ni, Cu, Zn, etc., but also the much more rare V and Mo (for certain nitrogenases, if natural abundances of nitrite, nitrate, and ammonium salts are too low for N acquisition). Other key elements are available naturally in suitable abundance, generally including P as phosphate (Walton et al., 2023; Gull and Pasek, 2024) and S as sulfates or sulfides.

But does a colonizer cell require these more advanced but complex capabilities or could it survive and continue to propagate and evolve without them? This query actually involves multiple questions, because each ion type may be different: Essential but rare, versus common but toxic.

Storage of needed ingredients has many forms. An obvious example is the phosphorus atom. Once it becomes incorporated into ADP, there is an intrinsic storehouse of two-thirds of the P atoms needed to create ATP. Another example is key metals.

Once a metal ion becomes bound or ligated as a cofactor into a metalloprotein molecule, that enzyme can be reused over and over. Many organic compounds can be stored as reserves. An example is the adenine core molecule that is incorporated into several key metabolites.

A final mechanism that should be considered is the conversion of molecules to something more useful, less detrimental, or which is more easily exported. Useless or antagonistical products can be a hindrance or even fatal to a cell's surival or reproductive prowess. These conversion processes and recycling processes are of special importance when they manifest as a metabolic cycle in which intermediate products are re-used, and the end product is identical to the starting substrate of the cycle but with net results (e.g., creation or consumption of ATP; likewise for NADH).

### 14.5.6 Communications

Transferring information outside a system is the basis of our category of communications (transfer of information *inside* a system is covered in our category of Interconnects, Section 14.5.12). Although microbes do not have well-developed communication techniques such as visual or electronic signaling, they do have ways of communicating with one another. As put by Ohland and Jobin (2015), "Owing to their ruthless requirements for survival, bacteria have also developed cooperative mechanisms such as horizontal gene transfer, biofilm formation, and quorum sensing to ensure the fitness of their own community as a whole". Witzany (2019) has given examples that communication is such a universal component that it should be considered one of the key characteristics of living systems.

Although a group of CZ cells may not initially have such techniques available or fully developed, they may eventually evolve such capabilities. Alternatively, HGT may actually have been an essential feature of the precursor protocells in the macrobiont.

### 14.5.7 Internal Environment

Unlike their macroscale homeothermic multicellular brethren, microscopic organisms have little hope of regulating their own thermal environment. This is because the thermal conductance of thin layers is high, even if the high conductivity of $H_2O$ is offset by the lower conductivities of fats and carbohydrates. Thus, their metabolic functions must be optimized for the predominant range of external temperatures, resulting in organisms which can be classified as psychrophiles (down to freezing temperatures), mesophiles (near room temperature), thermophiles, or hyperthermophiles (up to boiling). Although many microbes are reproductively active over a range of ~20°C, none seem to be able to grow well over the entire temperature range of liquid water (~ 0–100°C for 1 bar atmospheric pressure, i.e., Earth). The least challenge to the CZ cells would be if the macrobiont or the fringes in which the CZ cells arose would be akin to temperate climates, rather than hydrothermal extremes. However, it is useful to note that even hot springs typically have overflow streams which are much cooler and that suboceanic hydrothermal vents warm the surrounding water only for a short distance from their points of discharge. Thus, it seems likely that the CZ cells need not be other than mesophilic. This calls into question why LUCA phylogenetic inferences are for optimum metabolism at higher temperatures. One potential answer would be an evolutionary bottleneck when global environments temporarily became hot (Morales and Delaye, 2020; Gogarten and Deamer, 2016; Gogarten-Boekels et al., 1995).

Microorganisms often do have mechanisms for homeostasis with respect to other physical as well as chemical challenges from the external environment. To what degree would the CZ cells need such modern capabilities?

Thus, although some so-called *extremophilic* microorganisms are able to thrive at either low pH or others at high pH, the majority of organisms are best adapted to a more neutral pH, between 6 and 8. This encompasses the most common pH values that result when water interacts with igneous minerals to produce alteration products such as clays and salts. Hence, it is the most common range in bodies of $H_2O$ on Earth (and any basaltic- or felsic-rich planet).

Although there are examples where certain synthesis reactions of PCE experiments proceed better at higher temperatures, or extremes of pH, these can be taken into account when analyzing potential scenarios for the macrobiont, but the CZ cells themselves need not necessarily to be pre-adapted to such environments. Thus, there are reasons to consider that CZ cells are adapted to the more common terrestrial conditions and need not have internal controls for these particular properties.

Another major consideration would be to what extent a CZ cell would need to control its own internal composition. Modern cells do so to control ionic strength to avoid osmotic stress with the external ionic strength. Generally, most important is cellular regulation of $Na^+$ and $K^+$ ion concentrations. Although $Na^+$ is typically more common than $K^+$ in natural waters, due to mineral weathering reactions and subsequent solubilities, modern cells use a combination of selective permeability of their membrane as well as an active Na/K ion pump, powered by ATP, to export $Na^+$ and import $K^+$. As discussed in Section 14.5.5, there is actually a whole host of possible regulation of the internal environment's array of various types of ions and their concentrations.

### 14.5.8 Mobility

At the microorganism level, the ability to move around, using metabolic energy, is termed "motility." The earliest organisms, such as CZ cells, could have nonetheless survived and prospered without the need for active means of movement, as provided by coordinated pili or by a flagellum. The latter is especially complex and unlikely because it involves a sophisticated motor mechanism.

Microorganisms are readily transported to new environments by natural geological, hydrological, and atmospheric forces, such as gravity, stream flow, and wind currents. These forces of transport could have been more than adequate to enable the movement of a subset of organisms across local, regional, and eventually continental and even global scales. Depending on the occurrences of strong weather events, the spreading of organisms and their enhanced progeny could occur on rapid time scales compared to those factors reshaping a planet.

### 14.5.9 Navigation

Microbes with sensing receptors can choose a direction of motion by modulating the activity of their flagellar motors. Using the "tumble" mode to change orientation and the "run" mode to translate more or less linearly, they can navigate optimally even in the case of multiple desired conditions. For example, with chemotaxis, pH taxis, thermotaxis inputs (their equivalent of sensors, short-term memory, and processing), they can seek optimum conditions (Hu and Tu, 2014). Here again, we cast doubt on whether a bare-bones microbe such as a CZ cell will have evolved, or have needed to evolve, the combination of capabilities required for this activity in its earliest existence.

### 14.5.10 Reproduction

The reproduction of a cell involves two separate activities. First, the cell must increase in size through internal growth. Second, to birth the new cell, it may either expel a portion of itself (budding) or split into two roughly equal parts (binary fission). Even the latter can be controlled more by the physics of the instability of a larger object, allowing it to split into two or by a deterministic sequence of events and internal constructs driven by a biological plan, such as the use of a contractile ring, as shown in Figure 14.12. Eto et al. (2022) have developed a synthesis system for phospholipid growth suitable for use in constructing synthetic cells. Kohyama et al. (2022) have demonstrated the incorporation of a contractile division ring function inside lipid vesicles using purified versions of the relevant proteins in a synthetic biology experiment. Overall, however, even for Archaea, the reproductive cycle involves significant complexities (Lindås and Bernander, 2013).

Part of the process of growth is through the normal production of ribosomes, which involves of course making copies of the rRNA from the DNA and manufacturing the ribosomal proteins by previous ribosomes. Not only the rRNA but also a myriad of non-coding RNAs (ncRNA) must be made to perform various functions, especially for controls. The membrane needs to be increased in extent, incrementally, which is again facilitated by ribosomes and also the metabolic production of its special ingredients.

And not only must copies of RNAs be made in excess, the DNA genome itself must also be totally replicated and done so with high accuracy. This is "arguably the most fundamental biological process", but its mechanisms of initiation vary among several groups within the Archaea (Ausiannikava and Allers, 2017). And although there is one main DNA strand, there can be one or many more plasmids which also need to be replicated. Furthermore, could the CZ cell have mainly been composed of multiple, fairly large DNA molecules?

Although there may be simple ways to do this, somewhat analogous to the splitting of one large bubble into two or more smaller bubbles, modern cells have a relatively elaborate and definitive sequence of actions to accomplish this. Making sure each daughter cell has its own separate but complete set of genomic elements (nucleoid and plasmids) requires specific mechanisms for accomplishing the sorting.

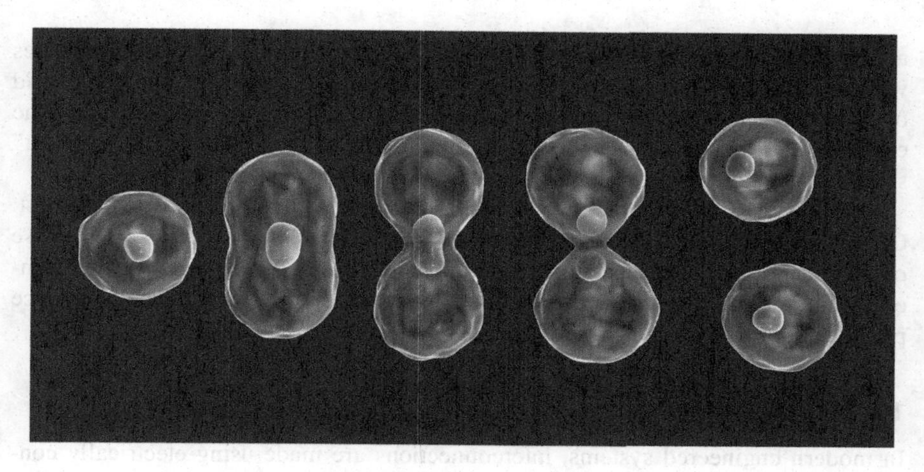

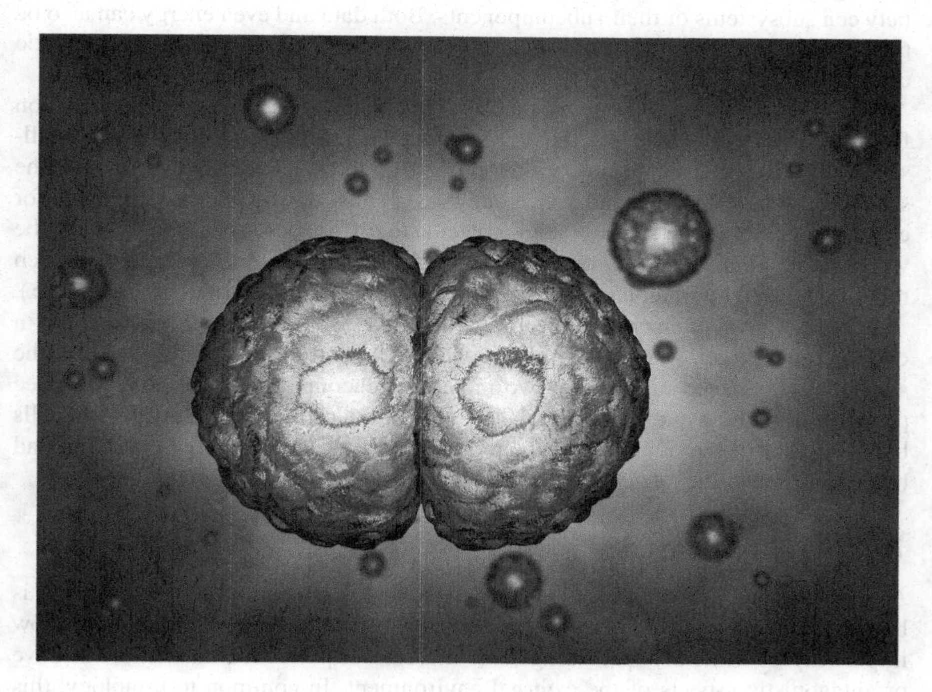

**FIGURE 14.12** Cell division (a) by physics versus (b) by biology.
*Photo courtesy: Shutterstock*

### 14.5.11 Defense

This function includes any measures undertaken to protect the system against outside antagonistic influences or agents. Such capabilities can obviously be a significant additional burden on any cell, given that there are so many intrinsic vulnerabilities within it or any complex system. However, for a planet with a virgin landscape devoid of other life forms of any type, the threats resolve only to the vicissitudes of the environments themselves. This is taken into account in the "Internal Environment" subsystem, which must provide means for its own maintenance.

In an environment devoid of preexisting life forms, this threat would not exist. Only when the CZ cells themselves spin off descendants which attack or otherwise outcompete the original inoculation would this be a threat. However, this would generally be due to cells which are even more competent in the environment and hence provide additional robustness toward the spread of the nascent biosphere.

### 14.5.12 Interconnects

In modern engineered systems, interconnections are made using electrically conducting wires, which can be variously used to distribute energy or to transfer data between subsystems or their subcomponents. Both data and even energy can also be transported between units using electromagnetic fields – as radio waves or magnetic coupling.

However, in biological systems, much of the connectedness is by simple diffusion from the source to the target. This is yet another advantage of cells being small-sized. Whereas diffusion is a very slow process on the macroscale, it varies with the second power of distance and hence can be very fast on the microscale. Thus, for example, the time for a small protein molecule to diffuse across a small pond can be weeks or months. In reality, on the macroscale, of course, mixing times are much more influenced by water currents due to other forces (wind, thermal gradients, etc.).

A typical protein enzyme molecule will diffuse from one end to the other of a one micron-size cell in only 10 milliseconds (Milo and Phillips, 2015, p. 213). The statistically averaged rate of any two molecules encountering one another is once per second! (Schulz and Jørgensen, 2001). One of the primary consequences of cells being smaller in size is that their interconnection times are shorter in duration and hence help speed the rate of reproduction.

### 14.5.13 Boundary

A "system" has a boundary, albeit one which is not uniformly fully contiguous because it not only keeps critical components contained together but also must allow import/export while also providing some measure of protection against the dilutive or antagonistic aspects of the external environment. In common terminology, this is a wall, barrier, box, etc. In biology, the molecular barrier that is constructed is termed a "membrane."

Our CZ cell needs a selectively permeable membrane. Possibly derived originally from amphiphilic compounds sourced from meteorites (Deamer et al., 2002), its great importance in the earliest stage of chemical evolution may lead us to assume that a phospholipid membrane arose prior to the development of the colonizer cell. In archaea, a single phospholipid layer can constitute the basis of a membrane, whereas in the Bacteria domain, it is commonly a double-layer lipid. The CZ cell may or may not also have a cell wall, which can provide additional structural support, protection, and some filtering action.

It has also been often pointed out that a boundary is needed to bring together the components of the metabolic system in order to achieve the higher concentrations that are needed for effective reaction times. The importance of compartmentation of the nucleic acids has also been stressed (Strazewski, 2019).

### 14.5.14 Summary of Subsystems

From the survey given, it is seen that not all subsystems are needed, at least not to the extent they are in the most highly competitive organisms and complex ecosystems of the modern world. On the other hand, the energy subsystem is complex, including the capture and conversion of non-chemical energy into the highly useful chemical energy of the phosphate bond, as an energy-donating participant in innumerable essential metabolic reactions.

In the next section, we will consider various useful techniques and approaches that are often used by experienced systems engineers and project managers when undertaking a major new project. These "Lessons Learned" are passed around after hard-won experiences, usually from problems, shortcomings, or failures in the history of any given organization which must grapple with repeated new and often very challenging projects – many of which are the next evolutionary step beyond the last successful one.

## 14.6 Lessons Learned from Engineering, as Applied to Colonizer Cell Research

Concurrent with defining the goals of a particular project is the assembling of experienced and motivated teammates who can work together as well as independently to address the varied aspects of the project, but with the overarching goal of integrating their results into a unified whole. This includes researchers who have extensive overall experience merged with members new to the field who are self-motivated by the challenges ahead. This includes those who are "technically oriented" or "inordinately motivated" or "experimentalists" or "analysts" or "resourceful", as some examples. Diverse backgrounds, areas of expertise, and methods of approach are components of a successful team, as portrayed in Figure 14.13. Particularly important are those who devote the time and effort to pursue options down to dead-ends or find innovative solutions or systematically characterize a path of investigation no matter where it leads. Dedication to the tasks pays off in the end.

The technical backgrounds and skills of the team members must span a wide range of disciplines, including those in Figure 14.14. Some members will have multiple skills, knowledge, or backgrounds, which may be of special value because of their broad points of view. The combination must be both multidisciplinary and interdisciplinary (Kolb, 2019; Repko, 2012).

Unlike the project to develop the atom bomb, where the principals and laboratories and equipment were sequestered at Los Alamos until the final results were achieved, the "team" for this endeavor will not be one assembled in a single place. They will not be under a single leadership but hopefully will freely share information of not only their successes but also their failures, so that progress can continue forward with significant momentum. Even the Manhattan Project needed the plutonium production from the nuclear reactors at Hanford and the uranium-235 from the gaseous diffusion plants at Oak Ridge, each of which drew upon their own scientific and engineering talents and expertise.

**FIGURE 14.13** Teams of scientists using a variety of techniques are needed to conduct leading-edge research and development for elucidating the origin of the biosphere.

*Photo courtesy:* Shutterstock 691541095

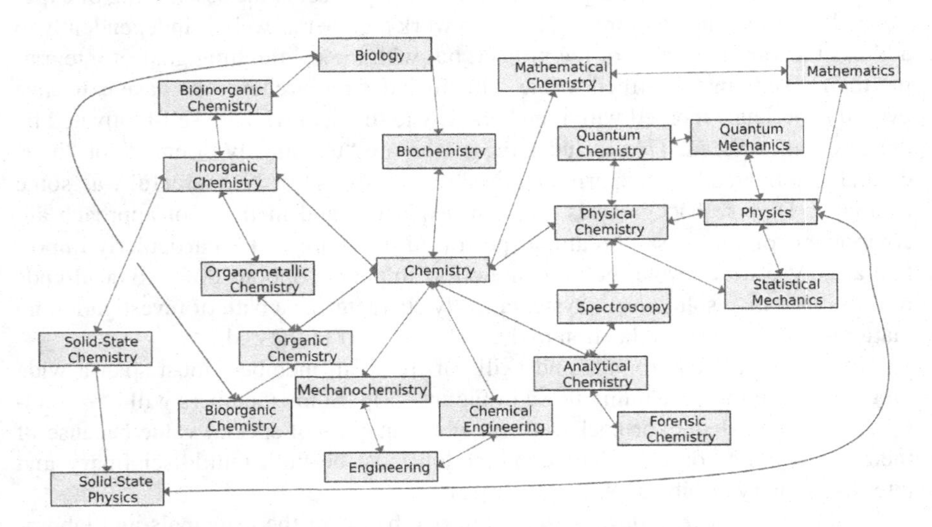


**FIGURE 14.14** Many disciplines must be applied to this challenge.

*Artwork courtesy:* Shutterstock

In contrast, elucidating and experimenting with entities, subsystems, and components of the putative primitive colonizing cells will require the contributions of countless researchers, situated in research laboratories around the world. Although there may be master strategies and goals proposed for special research emphasis, it is not unlikely that certain important breakthroughs may come about from serendipitous discoveries. Unforeseen contributions from independent lines of inquiry are sure to also be found to be of value or even essential. A few possible approaches to the ultimate goal of construction of a plausible colonizer cell-like system are provided in Figure 14.15.

In pursuing these goals, it is useful to consider how some of the lessons learned from decades of application of systems engineering can apply to this project. These lessons are some of the generalizations from all types of engineering projects, from developing and manufacturing a million new-model automobiles to a one-of-a-kind spacecraft based on only modest heritage from a previous spacecraft. Indeed, many may also trace back to construction projects of the ancients from the pyramids in Egypt to the aqueducts of Rome.

### 14.6.1 Lesson 1: Subsystems Evolve Independently

Once a given system is successfully designed, built, tested, and placed into service, it typically becomes obvious that various of its subsystems could be subjected to

**Possible Approaches to Constructing a Colonizer Cell**

- **Apply Lessons Learned from Systems Engineering**
- **Search for ancient metabolic pathways in diverse organisms (especially in Archaea)**
- **After elucidating these, focus on essentials**
- **Experimentally isolate subsystems or small groups of subsystems for development and test, before integrating later, e.g.,**
  - **membrane + transport system**
  - **energy capture and incorporation**
  - **syntheses of simple amino acids**
  - **streamlined DNA replication**
  - **primitive ribosomes + translation**
  - **simple cell division**
  - **and so forth**

**FIGURE 14.15** An example project plan for the application of systems engineering to the development and hardware implementation of functions of putative colonizer cells.

*Source:* Original art by author

improvements which would enhance the overall system in one or more ways. This is normal. However, there is too often neither time nor funding available for such improvements.

Furthermore, technologies advance indepdently of one another. It can happen that a portion or even all of a subsystem being implemented could be replaced by a superior newer technology. For example, early cameras used on spacecraft missions to the planets were a type of imager known as a vidicon. This used an electron beam to scan out an image produced on a vacuum tube. Eventually, the "CCD" semi-conductor-based, inherently digital imaging device replaced the bulky and less accurate vidicon. Originally, the CCD was too susceptible to damage by the ionizing radiation particles that permeate space, but this was overcome by improved processes and designs. Each imaging system has successfully made major new discoveries in planetary research, and if the CCD had never been invented, there would be vidicons still being flown today.

In the case of a biological cell, it does have most, if not all, of the subsystems analogous to engineering subsystems. As it reproduces, the cell is overall evolving but only typically one part of one portion of one sub-subsystem of one subsystem. The reality is that incremental improvements generally happen at the lower levels, and only occassionally at the highest level (the system). This is true for engineered systems and also of living entities.

### 14.6.2 Lesson 2: Simplicity Is Good Enough (KISS Principle)

On the other hand, it is paramount during initial development that subsystems or the system itself not be the victim of improvement-obsession. "Better is the Enemy of Good-Enough" is a mantra sometimes heard as a project wrestles with why it is behind schedule, over cost, or suffering from fatigue and low morale.

Another adversary is the goal of "perfection" and its attendant complexities. Rather, the KISS ("Keep It Simple, Stupid") principle is an additional rallying cry. When something works sufficiently well, but obviously could be better, it behooves attention to the larger goal of making the overall system perform its functions, rather than needlessly complexifying and delaying a particular subsystem in order to exceed its needed level of performance.

There will always be time later to pursue subsystem improvements. Sometimes, these can be retrofit into an existing system. The B-52 aircraft has, for example, been subjected to extensive replacement of various subsystems, from its avionics to its engines, over the 70 years since it was first built. The remaining aircraft, each a full-up system of its own, may still be flying at the time of their 100th year anniversary since their original manufacture!

The key lesson learned here is that in order for a system to come together in a timely manner, it is necessary to develop each subsystem only to the level of sophistication or other attributes (weight, cost, power consumption, etc.) necessary to fulfill the requirements levied upon it by the needs of the overall system. For the development of a prototype colonizer cell, it may be necessary to utilize metabolic pathways that are less efficient or more energy dependent or in some other way less optimum, as long as they do the job.

### 14.6.3 Lesson 3: Think "Out of the Box"

In contrast to Lesson 2, the alternatives to KISS should be recognized and evaluated. As evolution occurs in subsystems, the overall system may have the potential for major benefit. Whether it is technological advancement or cost reduction or improved reliability and robustness or other favorable factors, they should be recognized. Many systems engineering projects cannot necessariliy devote a significant amount of effort to such considerations, because any changes from a proven approach can negatively affect the cost, schedule, resilience, or effectiveness of the system implementation.

On the other hand, a systems-level project involves so many subsystems and components (all the way down to individual piece-parts) that there will always be new innovations coming on line. Furthermore, there are "new ways of doing things" that need to be considered.

Thus, beyond the tried and true applications, there should always be some attention to new and innovative solutions to problems and challenges. If every system is like its own predecssor, there can be no advancement. Biological systems inherently follow this process because DNA replication and chromosome integrity are never 100% perfect.

Likewise, in a continual development of simulations of the colonizer cell, it behooves the investigators to evaluate new results and unconventional ideas that may provide a better path to the goal of understanding this stage in the transition of early life to the establishment of a planet-wide biosphere.

### 14.6.4 Lesson 4: Develop and Test Subgroups Separately

A system is made up of a number of subsystems, and when system engineers define and derive a number of nested requirements, they allocate them to the subsystems or the parts which make them up. Because of this discipline, it is possible to set separate performance objectives for development and also the tests to be performed to verify that the requirements are being met by each subsystem.

This is of enormous benefit because if the requirements' deconvolution is done correctly, the system has the best chance that it will function as intended when all of its parts are made properly (although it will be necessary later to fully test the assembled system for unintended consequences, fragilities, etc.).

This quasi-independence of subsystems is never absolute. For example, when one subsystem operates, it dissipates thermal energy radiatively and into the structure, which then causes a temperature rise of nearby subsystems. In biological systems, the subsystems are often intermingled. For example, the human tongue has at least a half-dozen different functions (tasting, swallowing, speech, teeth cleaning, sucking, mashing, etc.), which are applicable to several different subsystems. Likewise, the ATP–ADP couple is involved in a wide variety of biochemical reactions that affect many subsystems in the metabolic scheme, over all forms of life.

And, as electronic circuits are operated, they also emit parasitic electromagnetic emissions and ground currents which could interfere with neighboring circuits in other units. To prevent this, engineered systems have specific requirements placed on both the level of emissions permitted by electronics units as well as on their ability to not be affected by certain levels of such potential interference.

Metabolic chemical systems are not as amenable to such limitations, in the sense that any one chemical compound may have some affect on numerous other chemical reactions that are underway. However, with the control of a multitude of enzymes, it is possible to orchestrate the overall activities and functions of the organism.

### 14.6.5 Lesson 5: Enforce Rigorous Quality Control (QC)

Verifying that a subsystem or subcomponent has been properly produced is the job of quality control, which can involve inspection, witnessing, or even periodic tests. In laboratory studies, it is imperative that procedures are sufficiently explicit (or known) that if the same experiments are repeated, they will yield the same outcomes. Part of this is simply by documenting what was done and what were the results.

In the event of some failure to produce the desired result, or an unexpected effect, consequence, or product, it is especially important to capture the details of what led to that anomaly. In critical major projects, if a failure occurs, all further activities are halted and the failure documented, while remedial actions are planned to probe the "root cause" of the failure so that it can be avoided in the future. In general, an unexpected failure is a cause to halt all further progress until the actual fundamental flaw can be determined, and remedial action to prevent it in the future is determined and approved.

In lab research settings, such unexpected results or "dead-ends" are not always documented so well, in part because of the human reaction to avoiding failures and focusing on making new achievements. This is not efficient for the science community at large and a project group in particular. In competitive markets, a company may keep such failures confidential, as part of their "trade secrets." For cooperative projects, such as advancements in science, all results should be documented and published (at the very least in the Supplementary Material section).

Proofing of a procedure also means that it can be later repeated, on the basis of the existing documentation of the prior success. It is also very useful to determine the robustness and limits of validity of a previous successful procedure. Athough resource-consuming with lesser reward, this helps determine the conditions under which the new result is valid and those under which it is less robust or not valid. This not only guarantees proper representation of achievements, but also saves colleagues from needless "rediscovery" and wasteful dead-ends.

### 14.6.6 Lesson 6: Simulate the System and Subsytems in Software

Modeling of all sorts of processes has now become much more common, given the advances in computer technology (processing speed, memory resources); software aids (advanced, user-friendly compiled languages); and data interpretation (trends and statistical analysis, graphing visualizations). Machine learning algorithms are also already being applied to a wide variety of data analysis problems.

For critical systems, there is thermal analysis, reliability analysis, and stability analysis of electronic circuits that have long been mainstay activities. Metabolic systems are intrinsically highly complex because so many of the molecular ingredients can interact in multiple ways. Enzyme soft-switching helps simplify the problem, although more exact simulations need to take into account the entire ensemble of ingredients that can interact.

As computer power, knowledge accumulation, and the number of human modelers with biological backgrounds increase, it will become possible to exponentially increase the capabilty to model the complex activities within a biological system. The colonizer cell is one of the more ideal targets for such activity because, as we have shown in Section 14.5, several subsystems will be primitive (i.e., simple) or absent.

There is also power and benefit in higher-level simulation of the activities of a CZ cell or various conceptual proto-CZ cells. In engineering endeavours, the simulations often simplify the subsystems or their internal sub-subsystems as "black boxes", each characterized by specific outputs as functions of their inputs. Such simulations are especially valuable for an early, higher-level understanding of system performance and predicted responses to varying conditions, as well as identification of any need for revisions to some subsysstem requirements.

### 14.6.7 Lesson 7: Assemble-Test-Assemble-Test-etc.

The way that engineers construct a complex system is gradual and piecemeal but according to master plans. Enough personnel are marshaled to enable everything to proceed in parallel at the subsystem level. For a spacecraft or a planetary rover, the process of assembling the final article proceeds from the lowest to highest level. This is done in parallel!

Each step of the way, there is testing to verify performance. For an electronics assembly, it begins at the piece-part level (microchips and a few individually discrete parts). A circuit is manufactured and tested at the "board level." The circuit boards are then assembled one at a time into the final unit, with testing generally performed at each assembly step.

Similarly, the planetary rover is assembled piece by piece in a many-months-long process known as ATLO (Assemby, Test, and Launch Operations). The bulk of the structure may be assembled in one location and then transferred to the final assembly location for population by various "black boxes" (these days, they are not actually always black in color) one by one. Typically, with each unit that is added, there is a system-level test of some sort. Early on, the main computer may not be part of the assembly, but separate units that emulate the flight computer's commands and data transfer are used to test the partially assembled system. Often delivered late in the process are the various science instruments, because they generally require the most development time for incorporating the latest, most advanced (and least proven) technologies.

Once fully assembled, the entire system is subjected to vibration and shock environments that simulate the launch and landing adversities. It is also typically exposed to the expected maximum extremes in temperature in suitable thermal-vacuum test chambers. If any anomalies in performace are detected as a result of the purposeful exposure to these environmental insults, remedial action must be taken (or a waiver documented).

Not all expected environments are exposed to the newly minted hardware. A prime example is that of the dust in the planetary environments. The reason is that even small grains of contaminants are a threat to the proper operation of the moving parts in various mechanisms, and especially the valves that are used to control propellants and hence the rocket engines needed to safely land on the target planet. Thus, construction of all components in the system is generally conducted under locally clean conditions or in "Clean Rooms" where the final assemblies occur. The designs are such that once the system is totally integrated, it should be much less suceptible to damage from exposure to the dust and rocks expected for the mission.

### 14.6.8 Lesson 8: When Indicated, Repair or Replace

Any unit that cannot perform its function to satisify its so-called "Derived Requirements" must undergo analysis, with one of three consequences: (1) Repair, (2) replace with similar unit, or (3) replace with different technology. One other possibility is to reexamine requirements in the light of new-established performance of other subsystems or units to see if the "failed" capability is truly needed.

In terms of this task of prototyping, it is anticipated that there will be many such "failures" either because the involved capability is not adequate or because of unexpected counterproductive characteristics. This is part of the "cut and try" approach that will be necessary in such an intrinsically complicated system. Decision-making, when such impediments become obvious, is very difficult, and it may be necessary for different groups to pursue different options in order to reveal the best choices to be made. To an outside observer, the main task may seem to be intractable at best, or even unachievable. Persistence in the face of sudden adversities is a hallmark of a dedicated and determined team with the end goal always in mind.

### 14.6.9 Lesson 9: Test Environmental Robustness

As with engineering systems, which test at the piece-part level to the total system level, and every significant stage of assembly in-between, it is important to determine how robust a CZ cell could be. For dispersal by eolian activity, the extremes in weather, such as temperature and humidity, are important. Most specifically, can the primitive CZ cell, which first escapes the macrobiont habitat, survive short-term desiccation on the time scales of hours, days, or weeks?

A large number of today's microbial species do have this tolerance. One means of their long-term preservation for laboratory use is via the technique of lyophilization. This process does typically entail placing the active bacteria in a nutrient medium, then freezing before drying. However, it is quite clear from the genetic assays of soils worldwide that an enormous variety of microbes easily spread around our globe in just a matter of years or decades at most. Although the global ocean may be one vector of transmission and the transport by migrating animals, especially birds, may be another, there are also global-scale transportation of dust particles and their attached microbes through levitation by storms and the global atmospheric circulation system. Thus, a resistance to temporary desiccation would seem to be one of the properties of high value for rapid spread of the CZ organisms to establish the global biosphere.

### 14.6.10 Lesson 10: Safety, Safety, Safety

In major team-engineered projects, there are a variety of opportunities for unintended accidents to occur. More than once has an organization accidentally "dropped" a spacecraft during fabrication and assembly. Personnel are sometimes injured, occassionally resulting in death. Something as simple as a nitrogen-filled compartment was fatal to personnel who walked in without oxygen masks. Rockets involve numerous explodable components. High voltages are sometimes present. For these reasons, separate safety responsibilities are assigned to specialist personnel for the review and remediation of practices that will avoid or control all identified potential safety risks.

Prebiotic chemistry research often employs far UV illumination, and there have been all too many incidents where laboratory personnel have suffered exposure to their unprotected eyes. Combustible species, such as $H_2$ and $CH_4$, are often involved in the chemical studies (given that the early Earth was anoxic, and these were not hazards then). Some reagents are chemical toxins, carcinogens, or irritants or pose other hazards. But this research invokes additional new hazards.

For any engineered organism, it is important to assure that it has multiple vulnerabilities if it were to become inadvertently released to the natural environment. This can include dependencies on nutrients rare in the environment, but, most importantly, its reproductive subsystem must have, in addition, a stand-down response to certain signaling molecules or physical conditions (a built-in "kill switch" function). It is also important to provide a "watermark" sequence in the organism's DNA to identify the laboratory in which it was invented or last modified.

An additional, newer concern is whether machine learning and Artificial Intelligence (AI) analysis techniques are employed to aid the research. Some, or possibly all, portions of AI that is used should be compartmentalized to avoid the release of critical information to the worldwide knowledge base. In the wrong hands of irresponsible humans, the capabilities achieved could be put to wrong uses. Similarly, in engineering systems organizations, there is a strict compartmentalization of sensitive information as "classified" or "proprietary" data.

In addition, if machine learning techniques are employed, they should not be shared with agents of AI without safeguards. Would it be warranted or wise, under any circumstance, for AI agents to be privy to techniques for creating biological entities that could spread and be potentially harmful to our precious biosphere or its underpinnings?

## 14.7 SOME APPROACHES TOWARD CONSTRUCTION OF A COLONIZER CELL

In this section, we will attempt to apply some of the techniques, lessons, experience, and disciplines from systems engineering to applied research which could illuminate the transition from the origin of life to the establishment of a biosphere. Specifically, the task here is to explore the conceptual parameter space in which the properties of a CZ cell might exist.

The two separate methodologies of bottom-up and top-down are part of a divide-and-conquer approach until they later can be integrated into the most meaningful conceptualization of what CZ cells might have been, and their potential universality across the most common habitable zones in the universe.

### 14.7.1 Types of Testbeds

The synthetic colonizer cell that is our goal need not be so physically small as microbes today and could be more sluggish and inefficient. To begin, we may construct physical parts which may not appear to be realistic relative to a cell. All the parameters of composition, activities, and sensitivities of each subsystem, or groups of subsystems, may be more primitive than today.

This can lead to the use of multiple testbeds, which simulate different portions of different subsystems. For example, a testbed could be just to test membranes or to just test certain metabolic pathways and their properties or to investigate simpler methods for DNA replication or for the transcription of DNA to RNA. The ultra-primitive ribosome itself and its attendant functions could be a separate object of intense experimental investigation.

However, in our quest to create a reasonable CZ cell, we must construct something that has all the correct underpinnings to allow it to evolve to something akin to the earliest characterized biology, i.e., the inferred original LUCA cell.

In doing so, we can also let biology help find the solutions – for example, by knocking out a particular capability and then drive test-tube evolution to let the organism find new solutions (or, hopefully, rediscover older simpler, although less optimum, ones).

Or synthetic chemistry and analysis can be used to find new unknown (or undiscovered) solutions. Then, each new solution can be compared with biological plausibility and the possibility it may have once existed but was later superseded.

As in any practical situation, we do not need to necessarily attempt to include all features of a subsystem at once. In addition to leaving out some functional features, in order to simplify and focus on core capabilities, there are issues in size and power requirements. Engineers working with a development testbed may postpone any needs for miniaturization until later when they can address the "packaging" issues. Thus, for our CZ cell testbeds, there may be no real size constraints – they could be bulky, clumsy cells compared to modern realities.

How may a *prototype* colonizer cell be constructed to provide a testbed for the various options for subsystems? This will need to be able to be modifiable and provide options for a variety of subsystem concepts and particularly their inner workings.

### 14.7.2 Cell Boundary in Prototype Testbed

Taking a lead from spacecraft ATLO, perhaps the first step could be the construction of the basic structure, namely the boundary. In a chemical system, such as our cell, this boundary needs to be able to contain key ingredients while also providing import and export capabilities that are selective. Thus, key metabolites must be acquired from the environmental milieu, while essential components must remain inside the boundary, and waste products must be exported through the boundary. In contemporary microbes, the boundary is a membrane composed of phospholipids, proteins, and other organic molecules. This membrane has an array of functional capabilities. However, a colonizer cell may have had a far simpler membrane than even the simplest contemporary microbial cell. It may not have included all the exquisite capabilities of even the modern minimal cells.

One line of research undoubtedly needs to continue the investigations into the means by which cell boundaries can form, ranging from coacervates to primitive amphiphiles to synthesized phospholipids mimicking real cellular membranes. In this regard, however, it may be particularly difficult to *a priori* construct a membrane as complex as LUCA or modern cells, because these membranes are decorated with many inserted elements which help the cell boundary perform more proactively than could a simple passive boundary. As seen in Figure 14.16, the cellular boundary can be rather complex. Certain ribosomes can also be attached to the membrane, facilitating the manufacture of membrane proteins "onsite".

One approach is to use artificial cellular-like envelopes. As reviewed by Llopis-Lorente et al. (2023), the use of special liposomes or coacervates has allowed the assembly of cell-like functions within a boundary that is readily available to studies, in contrast to the relatively fragile, tiny, and difficult-to-modify phospholipid layer(s).

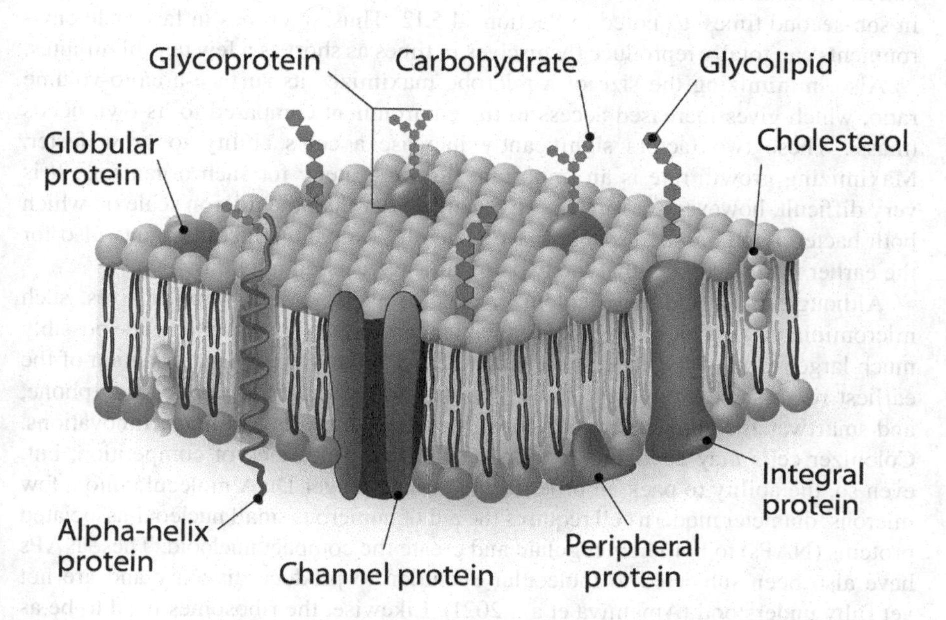


**FIGURE 14.16** Schematic portraying some of the complexity of the cellular membrane.
*Photo credit:* Shutterstock

This is a promising area of research because it allows an early implementation of a boundary that has many of the equivalent functions but is more amenable to investigations, especially where the emphasis is on the activities of the contents of the cell.

Or, as envisioned by Gordon and Gordon (2023), the diurnal transitioning of shaped droplets or vesicles, involving flattening-to-swelling cycling, may have been partially responsible for producing peptides from primitive amino acids, thus intimately combining internal metabolism with membrane evolution. They also speculate that it could provide the drive toward a genetic code. This is an example of a highly integrated scenario for the development of the cell as a reproductive system.

However, using the approach of systems engineering, the individual subsystems can also be developed and tested completely separately before attempting to assemble the entire system. Thus, the boundary of the cell might initially be totally artificial using hard boundaries but with the ability to program the injection or removal of substances, using microfluidics, filtration, sequestration, digestion, and other means. This allows the "system" to focus on what minimum capabilities are needed, while concentrating on exploring various conditions and reaction pathways *within* the boundary to learn what therein can be simplified.

### 14.7.3 Size of CZ Prototype Cell

The CZ cell may initially have been a much larger size than today's most common micron-scale microbes. Contemporary microbes live in competitive environments and as such must have optimized rates of acquisition and processing of available nutrients. Small molecules can diffuse across the entire diameter of a 1-micron cell in sub-second times, as noted in Section 14.5.12. Thus, microbes in favorable environments can totally reproduce themselves in times as short as a few tens of minutes.

Also, minimizing the size of a microbe maximizes its surface-area-to-volume ratio, which gives increased access to the environment compared to its own needs (mass). These two factors significantly increase a cell's ability to grow faster. Maximizing growth rate is an important fitness strategy for such organisms. It is very difficult, however, for a cell to be much smaller than the micron scale on which both bacteria and archaea species have generally evolved (which accounts also for the earlier misconception that the two domains were essentially only one).

Although this could drive colonizer cells to evolve to ever-smaller sizes, such microminiaturizing requires many refinements from the simplest and hence possibly much larger form of the earliest colonizer cell. Similarly, the miniaturization of the earliest room-sized computers to the desktop computer, then mobile smartphone, and smartwatch computers required long development time and many innovations. Colonizer cells may need smaller genomes because of a lack of competition, but, even so, the ability to pack an order-of-magnitude-longer DNA molecule into a few microns' diameter modern cell requires the aid of numerous small nucleoid-associated proteins (NAPs) to both help regulate and create the compact nucleoid. These NAPs have also been subjected to molecular evolution to produce diversity and are not yet fully understood (Amemiya et al., 2021). Likewise, the ribosomes need to be as compact as possible in order to pack tens of thousands of these independent protein manufacturing factories into the fastest-dividing cells.

In contrast to today's harsh realities, the colonizer cells face a world without competition, except with one another – which has the outcome that is most favorable for the overall success in establishing a biosphere at the planetary scale. Thus, in a laboratory simulation of the construction and metabolic activities of an example colonizer cell, scales of size and time need not be commensurate with those of the most successful contemporary organisms.

Perhaps the prototype CZ cell can have shorter DNA molecules and few, if any, NAPs. And perhaps it could have a simplified DNA polymerase and no multiple origins or bidirectional replications.

### 14.7.4 Energy and Metabolism

A key objective for understanding how simple or plausible could be a primitive colonizer cell is to focus on metabolic pathways that provide for both energy production and key anabolic syntheses. Along these lines, it is generally presumed that the first cells were chemolithoautotrophic, rather than phototrophic (because of the complexities of converting solar energy into useful chemical energy) or heterotrophic (because of the likely low levels of available environmental organic substrates).

There could be exceptions to the latter, if the organics from carbonaceous meteorites or comets were especially abundant such as the comet pond scenario (Clark and Kolb, 2018) or the carbonaceous meteorite pond scenario (Pearce et al., 2022). Otherwise, the average abundance of useful organics on a planet-wide basis would be sparse because post-accretion bombardment is spread over hundreds of millions $km^2$ and years, during which time many organic molecules would be dispersed, buried, and/or degraded into forms not useful for prebiotic chemical evolution. Local production of organics by lightning discharges may also be sporadic and limited because these concentrations depend on weather.

The metabolic functions required of the CZ cell depend, of course, on the environment in which it must flourish – noting that this may not be the same in which it was "born" (the macrobiont setting) but rather some class of environments that are common and reachable across the surface of the planet. One major class is the global ocean. Assuming its composition of solutes was similar to that of today, it can be rather poor in certain key nutrients, especially bioavailable forms of phosphorus, nitrogen, iron, and carbon. Lakes provide another widespread opportunity and can contain higher concentrations of useful nutrients and lower concentrations of nuisance $Na^+$. If the colonizer cell developed originally in the periphery of a macrobiont pond, it may be especially suited to small, stagnant lakes.

On the early Earth, there also may have been a variety of potential energy and carbon sources in widespread environments. We take here, as an exemplar, one promising set of favorable circumstances. Namely, we consider $H_2$ in a $CO_2$-rich atmosphere. Both $CH_4$ and HCN provide carbon sources, and HCN can be especially highly favorable to the cyanosulfidic concept of PCE for the origin of life (Sutherland, 2016, 2017; Ritson, 2021) and may have been present in an early, highly reducing atmosphere from a variety of sources (e.g., Pearce et al., 2022; Rimmer, 2023). We focus here on hydrogen gas because it has been invoked as a necessary greenhouse warming gas for both early Mars (Ramirez et al., 2014) and Earth (Wordsworth and Pierrehumbert, 2013) to counteract the lower luminosity of the new-born sun. Hydrogen can be produced by the serpentinization reaction between $H_2O$ and ultramafic rock and also by the radiolysis of $H_2O$ by mineral radioactivity, by galactic cosmic rays (GCRs), and by the ionizing energetic particles from solar outbursts.

Biological systems across all three domains of life and a multitude of phyla can utilize the reaction of $H_2$ with $CO_2$ as a fundamental metabolic resource to produce not only organic compounds but also a net energy gain which can be manifested as ATP. Alternatively, $H_2$ could provide metabolic energy via oxidation reactions with sulfate, to the extent oxidized forms of sulfur were available.

These metabolites can also be produced without enzymes, using just $Fe^0$ and $Ni^0$ (both common in meteorites) via selective reduction of $CO_2$ with $H_2$ to the acetate and pyruvate intermediates for the ancient reductive acetyl-CoA biosynthesis pathway (Varma et al., 2018). Previously, Cody et al. (2004) showed that minerals such as Cu- and Fe-sulfides could replace enzymes to promote several of the reactions in this pathway. Peters et al. (2023) demonstrated $CO_2$ fixation via catalysis by meteoritic material and also by Fe-rich volcanic ash.

Another proposed primitive pathway (Hartman, 1975) is the autocatalytic reverse Tricarboxylic Acid Cycle (rTCA, aka citric acid or Krebs cycle), which can utilize $CO_2$ such that several of its intermediates are the initiating molecules for biosynthesis of amino acids, lipids, the pyrimidine nucleotides, as well as sugars and cofactors. It has been shown that three important steps can be driven without enzymes through the use of $Ni^0$ or ground up meteorites or Pt-group metals (Rauscher and Moran, 2022). Morowitz et al. (2000) applied a set of plausible selection rules for rudimentary metabolic processes to a database of several million organic molecules and found 153 candidate compounds and that all 11 intermediate compounds in the rTCA cycle are within this group. From this, they suggested that our terrestrial biology's central metabolism via the TCA cycle may be a common natural consequence "and would likely characterize any aqueous carbon-based life anywhere it is found in this universe."

Energy consumption during protein synthesis is quite high, with an average of 4 ATPs utilized per amino acid incorporated into a nascent protein (Milo and Phillips, 2015, p. 203). Fortunately, enzymes are reusable and many degrade slowly unless purposely inactivated by regulatory proteases. And structural proteins generally need to be synthesized only once. In contrast, most of the pool of nonprotein organic molecules that are conducting the steps of energy acquisition and management, through the processes of anabolism and catabolism, are constantly being reacted away and regenerated anew. These can be considered two separate types of subsubsystems and, thus, to some extent, their formation separately studied, modeled, and synthetically emulated.

### 14.7.5 Growth

Microbes generally contain several thousand ribosomes. As the factories which produce all the proteins needed for many different critical functions, the cell cannot live without them. Given that they all work in parallel, these microminiature factories not only are key to survival but also are the engine of growth for producing the next organisms. Although the DNA and RNA molecules are not manufactured by the ribosomes, there are certain key proteins made by ribosomes which include the enzymes that enable DNA replication and RNA production from DNA. Since ribosomes are needed to make more ribosomes, they are, in some sense, the essence of reproductive life at the sub-cell scale. And perhaps because ribosomes have such an extraordinary effect on cellular health and vigor, their design has changed only slightly since the earliest incarnations in the cells closest to the LUCA.

In Figure 14.17, the ribosome for a bacterium is shown schematically (a) and also as the actual molecular configurations for the small subunit (b). For microbes, the 21 proteins that make up the 16S rRNA complex in the small subunit of the ribosome

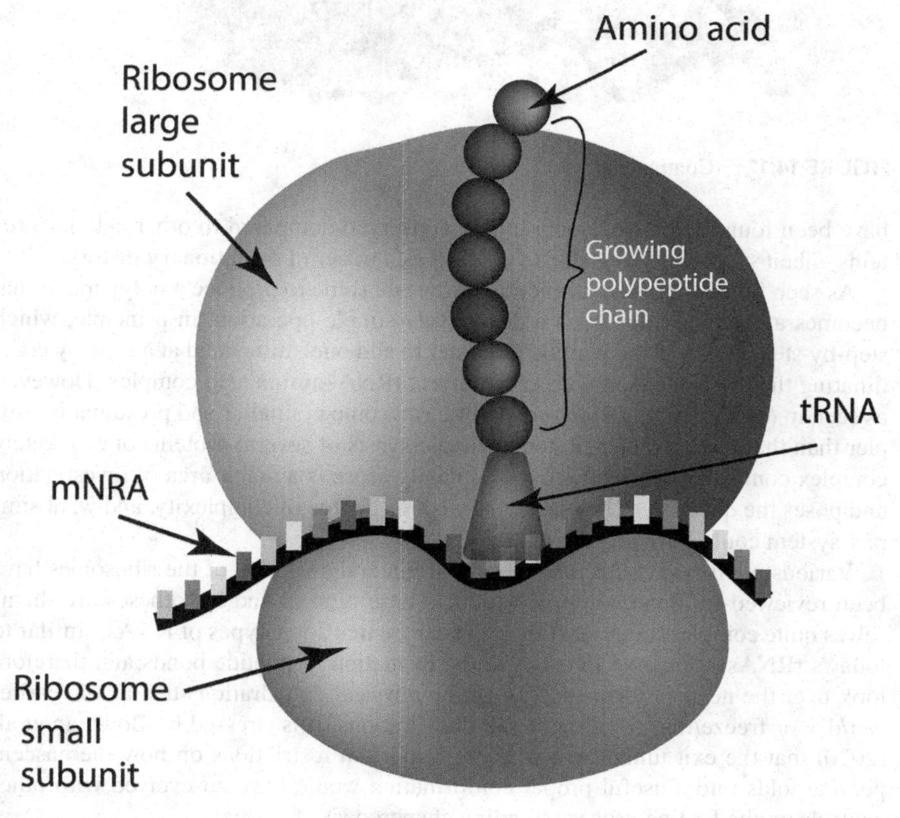


**FIGURE 14.17** (a) Idealized configuration of the ribosomal machinery. (b) Actual complex protein configurations in the small subunit (30S ribosome with three tRNA at top). *Photo courtesy: Shutterstock*

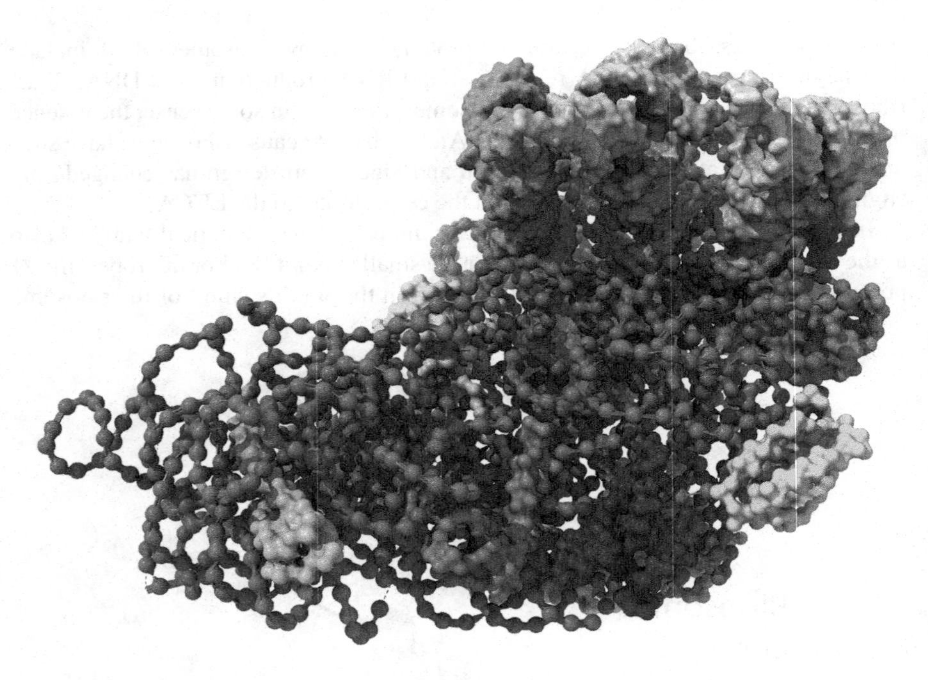

**FIGURE 14.17** (Continued)

have been found to be extremely highly conserved compared to other cellular proteins, albeit sufficiently variable to be used as a tracer of evolutionary history.

As seen in this figure, the function of the ribosome to produce a polypeptide that becomes a protein molecule is a deceptively simple operation, in principle, which step-by-step advances the peptide oligomer to add one amino acid at a time by coordinating the mRNA codon with each correct tRNA–amino acid complex. However, as seen in (b), even though the procaryotic ribosome is smaller and presumably simpler than that in eucaryotes, it nonetheless consists of several proteins of exquisitely complex conformations. How this complexity arose is a major area of investigation and poses the question of how necessary is this degree of complexity, and what simpler system could have proceeded it.

Various published conjectures on the origin and evolution of the ribosomes have been reviewed by Bowman et al. (2020). These models and hypotheses are themselves quite complex, but a starting point can be that some types of RNAs, similar to today's rRNAs, were able to catalyze the formation of peptide bonds and therefore took over the need to form such bonds by physical dehydration such as the earlier wet/dry or freeze/thaw cycling or salt dehydrations. It is stressed by Bowman et al. (2020) that the exit tunnel which exerts important restrictions on how the nascent peptide folds into a useful proper conformation would have co-evolved with other controls on the folding process (binding chaperones).

Lewandowski et al. (2013) have constructed artificial "small-molecule machines" that can synthesize peptides according to a prescribed sequence. Although akin to the biological ribosome, this is a "primitive analog" to a ribosome in function only and may not shed light on the origin and evolution of the ribosome in nature. Further investigations attempting to discover primitive versions, which could, in principle, evolve into modern ribosomes through the use or action of plausible metabolic ingredients, could provide likely candidates for simulating or providing "proof-of-principle" progress toward the evolution to the sophisticated ribosomal assemblages of LUCA and beyond.

### 14.7.6 Early Enzymes

As noted previously, certain mineral surfaces and solvated elements can perform as catalysts. Indeed, somewhere between 30% and 40% of all enzymes are metalloproteins, where a co-factor includes inorganic ions, especially transition element cations. Many early catalyzed reactions may have depended on metal ions from the environment and later led to the emergence of metalloproteins (Aithal et al., 2023; Muchowska et al., 2019).

But how might proteins have become enzymes in the earliest stages of life? Assume a bare-bones CZ cell, equipped with only a limited set of amino acids. What proteins might be constructed? It would seem that earliest life might not have had available or used all 20 (or 22) amino acids spread throughout the domain of biology. On the other hand, there are hundreds of different amino acids (a.a.'s) found in nature, and the carbonaceous meteorites include many dozens as well as thousands of other organic compounds (Naraoka et al., 2019; Martins, 2019), albeit most of them in very low concentrations. Also, the meteorite a.a.'s include many at similar concentrations that are *not* used by life.

Building a colonizer cell that "works", whether fully realistic or reasonable, will give a better chance of back-engineering, conceptually at least, the MB cell. These Gedanken and actual experiments can be fruitful in both directions. For example, if essential a.a.'s were uniquely available to the MB cell because of its special environment, but were rare in the general environment, then the CZ cell must include the functionality to synthesize them.

Many authors have considered the question of which of the modern proteinogenic amino acids would have been available in earliest times. Kirschning (2022) discusses five different suggested models. In a review by Gordon and Gordon (2023) of 16 different published papers which consider the likelihood of various amino acids appearing early in PCE, they find that some a.a.'s are highly expected (e.g., alanine and glycine). Others popular in these published studies include arginine and serine, followed by less popular Asp, Asn, Ile, Thr, and Val.

Judged highly *unlikely* are Lys, Phe, Tyr, Typ, and Cys. Yet, in the most ancient archaeal thermophiles, their heat-resistant proteins are generally stabilized by the strong, covalent disulfide bridges formed by pairs of advantageously located cysteine (Cys) amino acids that help maintain critical protein's 3D configurations. However, as shown by Foden et al. (2020), cysteine may have been readily synthesized abiotically, at high yield, and furthermore have also provided a mechanism of early catalysis of peptide ligation.

### 14.7.7 Gene Duplication and Diversification

A key path to the relatively rapid development of new enzymes is by simply making a modified copy of an existing protein. For an organism, this repurposing process is far more efficient than attempting to derive a new function by exploring the near-infinite space of arbitrary amino acid sequences to make some arbitrary new protein, which may be useless.

Duplicating a gene with minor changes increases the probability that the new protein will fold into a useful 3D construct that is compact yet has high integrity and protection from degradation. If the critical function of a new catalysis is not yielded the first time, there can be many additional attempts. Once some additional catalytic activity is achieved, there can be further evolution and enhancement. Analogously, in an electronic circuit, one can replace a preamplifier stage with an improved transistor with lower inherent noise or higher gain to obtain better overall performance, with only minor or no changes to other parts of the circuit.

Kauffman's famous example of the multi-functioned screwdriver (Kauffman, 2013) can allude to the repurposing of a given gene product to a different function by simple modifications. Although the design of a screwdriver embodies many different "requirements", one aspect of special importance is the blade at the end of the shaft. The thickness and width of the blade are set for being able to perform its primary function for a given range of screw sizes. Yet, independent of this, the shape of the blade is such that it is somewhat chisel-like, although the blunt end would be better as a chisel if replaced by a shaved portion with a resulting sharp edge.

However, for chopping a block of ice or anything needing to be pulverized or penetrated, the screwdriver can simply be tilted at an angle such that one of the two corners at the wide portion is used to impact the target. Thus, when striking a blow, the energy is concentrated in the smallest possible area and hence overcomes the impact strength of the target. Or, one could hold the screwdriver by its slender shaft and use the heavier handle to pound ice or nails or whatever.

Likewise, by small modifications, it may be possible to repurpose a given protein to accomplish some other task, in comparison with its original duty. Thus, for example, there are certain metalloenzymes with similar end-product functionality but using different key ions – for example, the nitrogenases which variously specialize in Fe-, V-, or Mo-sulfide cofactors and the various forms of which arose from gene duplications (Boyd et al., 2011). This approach of gene duplication with changed function that may not be optimum but is "good enough for now" is acceptable when the overall goal is performance of the system, not excellence in subsystems.

Another form of repurposing is where a single enzyme has more than one catalytic activity. Such multifunctional uses for catalysis are recognized as "enzyme promiscuity", a property that many elements and minerals in nature seem to have as well. Multifunction enzymes have been inferred for LUCA and related organisms and identified for their value in enabling early life to function with a lower gene content (Xavier et al., 2021).

Also, when simple gene duplication occurs, there still is the original gene performing its primary function so that the overall system functionality is preserved.

Thus, gene duplication has many obvious ways to bootstrap to improved capabilities without necessarily upsetting existing functions. In electronic circuitry, certain very fundamental circuit configurations (operational amplifiers, oscillators, analog-to-digital converters, binary scalers, etc.) are used over and over again, with evolutionary changes that improve performance, efficiency, or reliability without departing from the overall scheme or design principles.

### 14.7.8 Gene Sharing

Another very important aspect of how CZ cells could gain a foothold that would guarantee the long-term survival of life itself is horizontal gene transfer (HGT). Gene sharing is extraordinarily important to early life because some fortunate cell will have the opportunity to choose the best-of-the-best genes across a spectrum of functions and then supersede those overall less superior cells when environments become challenging.

### 14.7.9 Degradation as a Feature

Over time, things happen and not all good. Changes occur within a subsystem, and/or its interactions, or even at the system level due to wear, tear, failures, new challenges, etc. These can be categorized as temporary versus permanent versus stochastic, which may invoke different counter actions or solutions. Wear-out or deterioration is particularly devastating, but a system can have a built-in capability for repair or rejuvenation.

Are there repair possibilities? Can non-useful or deleterious molecules be identified, then exported, inactivated, converted, or destroyed? How does a system that is undergoing synthesis for reproduction deal with excess useless or detrimental components?

One answer seems to be that all components are either being routinely destroyed (e.g., by hydrolysis, denaturing, or targeting) and therefore have a finite lifetime. If there are incorrect or nonfunctional versions created, they will be replaced soon anyway, generally with correct versions.

The programmed protease enzymes routinely and systematically degrade proteins all the way down to their original amino acids, with little or no incomplete degradation which could leave ineffective and possibly deleterious oligopeptides behind. In the CZ cell, the proteases may be unfettered, but in archaea, they are sequestered in proteasome nanostructures under regulatory control (Maupin-Furlow et al., 2000).

Degradation of RNA molecules is also routine. Various forms of ribonuclease (RNase) are found in all organisms and is an ancient and key process. The master plan for the organisms resides, however, mainly in the DNA genome (plus epigenetic and related influences). For this reason, the genome must be inherently stable in the aqueous cytosol environment in which it resides, which explains the double helix which confers chemical stability upon the molecule (in spite of it adding impediments for the replication that is essential to create the next generation).

### 14.7.10 Dead-End Evolution

Primitive designs may work beautifully at the time they arise. But as those designs are enhanced, modified, and "improved", they can invoke additional complexity that adds new vulnerabilities or pitfalls. For example, today's advanced cells maintain critical control on detection and import or export of critical minor or trace elements in their surroundings. This is because they have highly tuned metalloenzymes which require specific elements as co-factors. Thus, for example, the human physiological system requires just a certain amount of copper to function optimally, because some enzymes require it, while for other enzymes it may be toxic because it replaces other key transition elements in some critical enzyme.

Such exquisite sensitivity is developed by the cell over generations of selection for the most optimum reproduction, as an advantage over competitors for the nutrients in the same milieu. But this elegant optimization renders the organism more susceptible to environmental variabilities, which necessitates enhanced regulatory sophistication.

### 14.7.11 Back-Ups and Redundancies

For rare engineered systems, such as a planetary rover, the system created is considered exceptionally valuable. Such systems therefore have certain back-up subsystems and functional redundancies that provide them with excellent survivability, if the environmental conditions they face are within the capabilities created by design and verified by test or validation analyses. With cell populations, the "back up" is the existence of a huge number of representatives of the basic design, but, with some variations, that can either be more or be less adapted to a future environment. This multiplicity in systems, with variation, provides a degree of robustness that is difficult to achieve in any other way.

The colonizers as a whole can have many missteps, but because of the enormous numbers involved and the quasi-independence of the multitude of natural environments, those with cells with appropriate capabilities will survive and propagate. Once CZ cells gain a foothold, their eradication becomes unlikely to happen before their evolution radiates to a multitude of diverse habitats and metabolic specialties, including those which are extreme. This guarantees survival of the biosphere. Eventually, catastrophic planetary-scale upheaval may occur (e.g., the red giant phase of main sequence stars such as our sun).

### 14.7.12 Knowledge Gaps

What we do not know, we cannot understand. This truism applies to much about biology, where the detailed and intertwined metabolic processes mask many of the underlying mechanisms and sensitivities. This does not prevent the formulation of questions about the inner workings of various subsystems, although the questions themselves often require revision with each new major discovery.

Nonetheless, the clear identification of significant gaps in knowledge or mechanisms or sensitivities of various portions of the metabolic machinery or cellular physiology can be one way to chart future progress. The practical realization of a functional prototype CZ cell could nonetheless inspire and stimulate the portfolio of research that will be needed to close those knowledge gaps and undoubtedly open further avenues of advanced research and investigation.

## 14.8 APPLICABILITY TO OUR SOLAR SYSTEM AND EXOPLANETS

Mercury and the Earth's moon may have deposits of $H_2O$ ice in permanently shadowed locations in their polar regions, but their lack of a sufficient atmosphere and the very high daily temperatures prevent the stable occurrence of $H_2O$ in the liquid phase.

Venus is Earth's twin in size and does have a dense atmosphere containing huge amounts of carbon, as $CO_2$, but the runaway greenhouse by this thick atmosphere also prevents aqueous environments, even at its poles. Being at greater solar distance, Mars is a more likely candidate for the rise of early life.

### 14.8.1 MARS

Mars is today too arid and too cold to support liquid water at its surface, except transiently, but its ancient climate was obviously more hospitable in terms of temperature and pressure. This is manifestly evident from the dry riverbeds and other features carved by flowing water, as well as a chemical sedimentary record of extenive and complex deposits of various salts.

From the extensive space exploration of the Red Planet, early Mars is revealed to have been an eminently urable planet (Clark et al., 2021), where life could have not only started but then should also have propagated to widespread regions. Although conditions are not as clement today, there may still be a biosphere lurking beneath the surface – perhaps shallow or perhaps at a few killometers of depth where the natural geotherms reach the melting point of ice.

### 14.8.2 OTHER SOLAR SYSTEM PLANETARY OBJECTS

The evidence for aqueous alteration in a number of types of meteorites is evidence for previous, albeit ancient, $H_2O$ in the liquid form. This includes carbonaceous chondrite meteorites, which therefore have the special circumstance of the combination of organic molecules of prebiotic interest and liquid water. However, although there are many intriguing constituents, including a large variety of amino acids (some of which are members of the canonical 20), there is no evidence of major prebiotic chemical evolution toward the key ingredients of life as we know it.

Furthermore, the asteroids from which these meteorites come are too small to have gravitationally or geomagnetically protected any nascent atmosphere for a geologically significant time. Thus, PCE and the potential for life are precluded.

An interesting special case is the asteroid Ceres, which had aqueous $H_2O$ activity in conjunction with a nitrogen-rich environment to form ammoniated phyllosilicates. In addition, its composition includes primary rock-forming minerals as well as carbonates and, most importantly, organics and even subsurface moisture (De Sanctis and Ammannito, 2021).

However, doses of ionizing radiation due to galactic cosmic rays and energetic solar events can reach population-sterilizing levels with exposure times of less than 10 Myr, even down to several meters depth in asteroidal material (Clark et al., 1999).

Frustration of life in icy bodies on cometary trajectories is a different situation. As the trajectory of a Kuiper belt object or other icy body is perturbed to bring its perihelion into the inner solar system, its surface layers are thermally eroded and dispersed. Prior to that, long-term GCR radiation may have sterilized the outer several meters. But after that, deeper regions are exposed. Some models have indicated the possibility of some melting in deep interiors. However, the lack of a sensible atmosphere precludes the possiblity of stable liquid $H_2O$ at the surface.

Several planetary bodies in the outer solar system, beyond the conventional "Goldilocks region", have been found to have subsurface water kept as liquid by tidally driven internal heating. These "ocean worlds", such as Enceladus, Titan, Europa, and other Jovian moons, have been postulated to potentially host living organisms. As with the other cases, there will be some location where a convergence of conditions, such as thermal processing of native organics and minerals, may have conspired to create the first primitive organisms. To inhabit the major portion of the ocean will require, however, the transition from primitive cells to colonizer cells. Once again, the research that characterizes such CZ cell's metabolic, reproductive, and evolutionary capabilities corresponds in a major way with Earth's early widespread, aqueously dominated environments.

### 14.8.3 Exoplanets

Main sequence stars are relatively short-lived (~ 10 Gyr), but M-dwarf stars and their planets may have far longer lives by several orders of magnitude, and given their higher abundance as well as tendency to host multiple rocky planets within their theoretically habitable zones, the prospects for life originating and then persisting along with the universe seem possible. Such planets may generally be tidally locked with only small regions or special circumstances allowing liquid water, although special conditions such as subglacial environments may provide ample conditions to support a restricted biosphere (Shields et al., 2016). However, detecting biosignatures on bodies around other stars depends on signals that can only result from a widespread and vigorous biosphere. Otherwise, the signals could be too weak.

Whereas the origin of life may be difficult to occur because it may generally depend on a fortuitous convergence of environmental conditions, there will be no widespread biosphere unless the next step is taken to turn the spark of life into a bonfire of habitation. The essential link in this process will be the advent of the equivalent to our colonizer cell.

Thus, the research about what it takes to convert the very first cells in their original home to acquire the greater capabilities that allow them to become distributed in the wider-spread environments of the home planet can be applied more generally. In particular, if it can be shown that these evolutionary steps are straightforward and expected, then the probability of a global biosphere taking hold will hinge mainly only upon the probability that the first protocell and its early progeny will arise from some rare, primordial circumstance.

## 14.9 SUMMARY AND DISCUSSION

Although there are various hypotheses of the special environments in which life arose, the earliest cells would need to spawn off a cell type which was more compatible with the widespread natural environments on the early Earth. From deep-sea hydrothermal vents, to hot springs, or to warm little ponds spiked with starter organics produced locally (lightning, UV) or imported from exogenous sources (meteoritic or cometary constituents), these conditions are relatively unique. But the CZ "colonizer" cells, which subsequently evolved in the fringes of the macrobiont where conditions were more like freshwater lakes, rivers, and ponds, would become the pioneers which spread globally and established the much more robust biosphere.

Efforts to derive the characteristics and features of the earliest, most primitive cells have centered around two distinct approaches. In the phylogenetic approach, the implied metabolic needs and tolerances of the LUCA cell embrace a chemolithoautotrophic lifestyle as a thermophilic and obligatory anaerobic methanogen living off atmospheric carbon dioxide and hydrogen gas. However, the extant organisms implied to be representative of those at the root of the implied Tree of Life have numerous features of metabolic complexity, which are difficult to conceive as arising full-blown early in the development of life (e.g., Peretó, 2019; Goldford and Segrè, 2018).

The second approach is that of discovering and then re-synthesizing a "minimalist cell." Through the use of the gene modification techniques, a version of the small microbe from the domain of Bacteria has been shown to need only a few hundred genes in order to grow and reproduce itself. This cell, however, also has a high degree of sophistication compared to what may have been possible in earliest times. The interactions between all the enzymes, reactants, and products that this entails would be difficult to imagine as the product of one small setting in which life occurred.

Here, it is proposed that a bottoms-up approach to intellectually conceiving and experimentally exploring the properties and internal workings of the colonizer cell could benefit from the methods of systems analysis, construction, and testing used by engineers. Given that the degree of fiscal support for the development of spacecraft and rovers in the name of astrobiological investigations far outweighs the investments in laboratory research into the origins of life and the biosphere, it only seems fitting that astrobiology researchers also benefit from the engineering approaches and experience for constructing systems with specialized functional goals. To the extent that a testbed can spawn prototype versions of cells that emulate the CZ cells, safety precautions must also be assured.

Central to this challenge are several "chicken or egg – which first?" conundrums. For example, DNA creates the rRNA that allows the ribosome to create the DNA synthase that is needed to replicate the DNA to create its progeny. Furthermore, the ribosome also creates the proteins which make up the next ribosome. And cell membrane is created by the synthesis machinery which the membrane protects from dilution and degradation from outside forces. Nonprotein reactions derive energy from the environment, which powers the chemical reactions needed for generation of protein enzymes, which direct the reactions which enable this extraction of energy. The solution to these various conundrums has been the concept of progressive coevolution of mutually reinforcing capabilities (Fry, 2019; Kirschning, 2022), which is also pointed out explicitly by Bartlett and Wong (2023), who refer to this conundrum as the "Avian Circularity."

Although these "which first?" conundrums needed to be at least partially solved by the protocells and MB cells and their progeny, much of the refinements of each of these subsystems may well have occurred after the CZ cell left the setting of its origin and began propagating in enormous numbers to facilitate biological evolution on timescales short compared to major geological processes.

The CZ cell testbed could help explore solutions to this challenge. By partially isolating portions of the overall system, it can be studied how to reduce them individually to their most primitive forms. Then, integrating those forms into a whole prototype system, insights into the plausibility and perhaps the reality of the emerging biosphere in the pre-LUCA era can be investigated.

Fundamental questions of architecture, of both the system and ecosystem of its peers, await us. Is the CZ cell an independent do-all cell type or is it part of a mini-ecosystem of mutualistic organisms with different specialties? Are its informational macromolecules a collection of plasmids, or an integrated master genome, or some of each?

Although an effort to develop working prototypes of plausible colonizer cell functionality will involve the consumption of science resources, the passion to explore our universe in search for life could hinge on how straightforward is the transition from the earliest protocells, locked in their unique environment, to evolve into the colonizers of the planet. Only then is the existence of life ensured on timescales competitive with geological transformations.

Research is the answer to this question of the likelihood of discovering life on various exoplanets. Of course, if we do happen to discover conclusive evidence of extinct or extant life on Mars or ocean worlds in our own solar system, there will be remarkable new evidence that can help further guide our research and conjectures of how, and how likely, protocells become biospheres.

Evolution allows a biosphere to span of time in not only one system's longevity, but in the connectivity from one given limited-life system to its progeny, and them to theirs. In that sense, the biosphere encompasses one continuous system, from the first reproductive entity to the vast array of biological diversity in existence today, spanning literally billions of years. All the evidence is consistent with that span being uninterrupted back to that first living entity down to the biosphere we observe today.

## ACKNOWLEDGMENTS

The author is deeply indebted to Professor Vera Kolb for many insightful suggestions, including those on tipping points and massive cellular transport, as well as the inspiration for this contribution.

## REFERENCES

Aithal, A; Dagar, S; Rajamani, S. Metals in prebiotic catalysis: A possible evolutionary pathway for the emergence of metalloproteins. *ACS Omega* **2023**, *8*(6), 5197–5208. https://doi.org/10.1021/acsomega.2c07635

Amemiya, HM; Schroeder, J; Freddolino, PL. Nucleoid-associated proteins shape chromatin structure and transcriptional regulation across the bacterial kingdom. *Transcription* **2021**, *12*(4), 182–218. https://doi.org/10.1080/21541264.2021.1973865

Ausiannikava, D; Allers, T. Diversity of DNA replication in the archaea. *Genes* **2017**, *8*(2), 56. https://doi.org/10.3390/genes8020056

Bartlett, S; Wong, ML. Emergence, construction, or unlikely? Navigating the space of questions regarding life's origins. In *Conflicting Models for the Origin of Life*. Scrivener Publishing: Beverly, MA, and John Wiley & Sons: Hoboken, NJ, **2023**, pp. 53–64.

Bowman, JC; Petrov, AS; Frenkel-Pinter, M; Penev, PI, Williams LD. Root of the tree: The significance, evolution, and origins of the ribosome. *Chemical Reviews* **2020**, *120*(11), 4848–4878. https://doi.org/10.1021/acs.chemrev.9b00742

Boyd, ES; Anbar, AD; Miller, S; Hamilton, TL; Lavin, M; Peters, JW. A late methanogen origin for molybdenum-dependent nitrogenase. *Geobiology* **2011**, *9*(3), 221–232. http://doi.org/10.1111/j.1472-4669.2011.00278.x

Cael, BB; Heathcote, AJ; Seekell, DA. The volume and mean depth of Earth's lakes. *Geophysical Research Letters* **2017**, *44*(1), 209–218. https://doi.org/10.1002/2016GL071378

Clark, BC; Baker, AL; Cheng; AF; Clemett, SJ; McKay, D; McSween, HY; Pieters, CM; Thomas, P; Zolensky, M. Survival of life on asteroids, comets and other small bodies. *Origins of Life and Evolution of the Biosphere* **1999**, *29*, 521–545. https://doi.org/10.1023/A:1006589213075

Clark, BC; Kolb, VM. Comet pond II: Synergistic intersection of concentrated extraterrestrial materials and planetary environments to form procreative Darwinian ponds. *Life* **2018**, *8*(2), 12. https://doi.org/10.3390/life8020012

Clark, BC; Kolb, VM. Macrobiont: Cradle for the origin of life and creation of a biosphere. *Life* **2020**, *10*(11), 278. https://doi.org/10.3390/life10110278

Clark, BC; Kolb, VM; Steele, A; House, CH; Lanza, NL; Gasda, PJ; VanBommel, SJ; Newsom HE; Martínez-Frías, J. Origin of life on Mars: Suitability and opportunities. *Life* **2021**, *11*(6), 539. https://doi.org/10.3390/life11060539

Cody, GD; Boctor, NZ; Brandes, JE; Filley, TR; Hazen, RM; Yoder Jr, HS. Assaying the catalytic potential of transition metal sulfides for abiotic carbon fixation. *Geochimica et Cosmochimica Acta* **2004**, *68*(10), 2185–2196. https://doi.org/10.1016/j.gca.2003.11.020

Cole, JJ; Pace, ML; Caraco, NF; Steinhart, GS. Bacterial biomass and cell size distributions in lakes: More and larger cells in anoxic waters. *Limnology and Oceanography* **1993**, *38*(8), 1627–1632. https://doi.org/10.4319/lo.1993.38.8.1627

Deamer, DW; Cary, F; Damer, B. Urability: A property of planetary bodies that can support an origin of life. *Astrobiology* **2022**, *22*(7), 889–900. https://doi.org/10.1089/ast.2021.0173

Deamer, DW; Dworkin, JP; Sandford, SA; Bernstein, MP; Allamandola, LJ. The first cell membranes. *Astrobiology* **2002**, *2*(4), 371–381. https://doi.org/10.1089/153110702762470482

Deamer, DW; Pashley RM. Amphiphilic components of the Murchison carbonaceous chondrite: Surface properties and membrane formation. *Origins of Life and Evolution of the Biosphere*. **1989**, *19*, 21–38. https://doi.org/10.1007/BF01808285

de Duve, C. Clues from present-day biology: The thioester world. In *The Molecular Origins of Life: Assembling Pieces of the Puzzle*, Brack, A, Ed. Cambridge University Press: Cambridge, **2000**, pp. 219–236.

De Sanctis, MC; Ammannito, E. Organic matter and associated minerals on the dwarf planet Ceres. *Minerals* **2021**, *11*(8), 799. https://doi.org/10.3390/min11080799

Downing, JA; Prairie, YT; Cole, JJ; Duarte, CM; Tranvik, LJ; Striegl, RG; McDowell, WH; Kortelainen, P; Caraco, NF, Melack, JM; Middelburg, JJ. The global abundance and size distribution of lakes, ponds, and impoundments. *Limnology and Oceanography* **2006**, *51*(5), 2388–2397. https://doi.org/10.4319/lo.2006.51.5.2388

Eto, S; Matsumura, R; Shimane, Y; Fujimi, M; Berhanu, S; Kasama, T; Kuruma Y. Phospholipid synthesis inside phospholipid membrane vesicles. *Communications Biology* **2022**, *5*(1), 1016. https://doi.org/10.1038/s42003-022-03999-1

Foden, CS; Islam, S; Fernández-García, C; Maugeri L; Sheppard, TD; Powner, MW. Prebiotic synthesis of cysteine peptides that catalyze peptide ligation in neutral water. *Science* **2020**, *370*(6518), 865–869. https://doi-org/10.1126/science.abd5680

Fry, I. The origin of life as an evolutionary process: Representative case studies. In *Handbook of Astrobiology*, Kolb, VM, Ed. CRC Press: Boca Raton, FL, **2019**, pp. 437–462.

Gibson, DG; Glass, JI; Lartigue, C; Noskov, VN, Chuang, RY; Algire, MA; Benders, GA; Montague, MG; Ma, L; Moodie, MM; Merryman, C. Creation of a bacterial cell controlled by a chemically synthesized genome. *Science* **2010**, *329*(5987), 52–56. https://doi.org/10.1126/science.1190719

Gibson, DG; Hutchison III, CA; Smith; HO; Venter JC. Synthetic cells and minimal life. In *Handbook of Astrobiology*, Kolb, VM, Ed. CRC Press: Boca Raton, FL, **2019**, pp. 75–90.

Gladwell, M. *The Tipping Point*, Back Bay Books/Brown, Little, and Company: New York, **2002**.

Glass, JI; Merryman, C; Wise, KS; Hutchinson III, CA; Smith, HO. Minimal cells real and imagined. *Cold Spring Harbor Perspectives on Biology* **2017**, *9*, a023861. http://doi.org/10.1101/cshperspect.a023861

Gogarten, JP; Deamer, D. Is LUCA a thermophilic progenote? *Nature Microbiology* **2016**, *1*, 16229. https://doi.org/10.1038/nmicrobiol.2016.229

Gogarten-Boekels, M; Hilario, E; Gogarten, JP. The effects of heavy meteorite bombardment on the early evolution – The emergence of the three Domains of life. *Origins of Life and Evolution of Biosph*eres **1995**, *25*, 251–264. https://doi.org/10.1007/BF01581588

Goldford, JE; Segrè, D. Modern views of ancient metabolic networks. *Current Opinion in Systems Biology* **2018**, *8*, 117–124. https://doi.org/10.1016/j.coisb.2018.01.004

Gordon, R; Gordon, NK. How to make a transmembrane domain at the origin of life: A possible origin of proteins. In *Conflicting Models for the Origin of Life*. Scrivener Publishing: Beverly, MA, and John Wiley & Sons: Hoboken, NJ, **2023,** pp. 131–173.

Gull, M; Pasek, M. Chapter 6. *Systems Geochemistry in Astrobiological Context*, this Guidebook **2024**.

Hartman, H. Speculations on the origin and evolution of metabolism. *Journal of Molecular Evolution* **1975**, *4*, 359–370. https://doi.org/10.1007/BF01732537

Hu, B; Tu, Y. Behaviors and strategies of bacterial navigation in chemical and nonchemical gradients. *PLoS Computational Biology* **2014**, *10*(6), e1003672. https://doi.org/10.1371/journal.pcbi.1003672

Hutchison III, CA; Chuang, RY; Noskov, VN; Assad-Garcia, N; Deerinck, TJ; Ellisman, MH; Gill, J; Kannan, K; Karas, BJ; Ma, L; Pelletier, JF. Design and synthesis of a minimal bacterial genome. *Science* **2016**, *351*(6280), aad6253. http://doi.org/10.1126/science.aad6253

Joyce, GF; Orgel, LE. Prospects for understanding of RNA world. In *The RNA World: The Nature of Modern RNA Suggests a Prebiotic RNA World*, Gesteland, RF; Atkins, JF, Eds. Cold Spring Harbor University Press: Cold Spring Harbor, NY, **1993**, pp. 1–25.

Kasting, JF. Atmosphere on early Earth and its evolution as it impacted life. In *Handbook of Astrobiology*, Kolb, VM, Ed. CRC Press: Boca Raton, FL, **2019**, pp. 207–218.

Kauffman, SA. Evolution beyond Newton, Darwin, and entailing law: The origin of complexity in the evolving biosphere. In *Complexity and the Arrow of Time*, Lineweaver, CG; Davies, PCW; Ruse, M, Eds. Cambridge University Press: Cambridge, **2013**, pp. 162–190, see specifically p. 173.

Kirschning, A. On the evolutionary history of the twenty encoded amino acids. *Chemistry – A European Journal* **2022**, *28*(55), e202201419. https://doi.org/10.1002/chem.202201419

Kohyama, S; Merino-Salomón, A; Schwille, P. In vitro assembly, positioning and contraction of a division ring in minimal cells. *Nature Communications*. **2022**, *13*(1), 6098. https://doi.org/10.1038/s41467-022-33679-x

Kolb, VM. Astrobiology: definition, scope, and a brief overview. In *Handbook of Astrobiology*, Kolb, VM, Ed., CRC Press: Boca Raton, FL, **2019**, pp. 3–14.

Kolb, VM; Clark, BC. *Systems Approach to Astrobiology*, CRC Press: Boca Raton, FL, **2023**.

Koonin, EV. How many genes can make a cell: the minimal-gene-set concept. *Annual Review of Genomics and Human Genetics* **2000**, *1*(1), 99–116. https://doi.org/10.1146/annurev.genom.1.1.99

Lambert, JB. Silicon and life. In *Handbook of Astrobiology*, Kolb, VM, Ed., CRC Press: Boca Raton, FL, **2019**, pp. 371–378.

Lewandowski, B; De Bo, G; Ward, JW; Papmeyer, M; Kuschel, S; Aldegunde, MJ; Gramlich, PM; Heckmann, D; Goldup, SM; D'Souza, DM; Fernandes, AE. Sequence-specific peptide synthesis by an artificial small-molecule machine. *Science* **2013**, *339*(6116), 189–193. https://doi-org/10.1126/science.1229753

Lindås, AC; Bernander, R. The cell cycle of archaea. *Nature Reviews Microbiology* **2013**, *11*(9), 627–638. https://doi.org/10.1038/nrmicro3077

Liu, R; Ochman, H. Origins of flagellar gene operons and secondary flagellar systems. *Journal of Bacteriology* **2007**, *189*(19), 7098–7104. https://doi.org/10.1128/jb.00643-07

Llopis-Lorente, A; Yewdall, NA; Mason, AF; Abdelmohsen, LK: van Hest, JC. Synthetic cells: A route toward assembling life. In *Conflicting Models for the Origin of Life*. Scrivener Publishing: Beverly, MA, and John Wiley & Sons: Hoboken, NJ, **2023**, pp. 279–302.

Martins, Z. Organic molecules in meteorites and their astrobiological significance. In *Handbook of Astrobiology*, Kolb, VM, Ed., CRC Press: Boca Raton, FL, **2019**, pp. 177–194.

Maupin-Furlow, JA; Wilson, HL; Kaczowka, SJ; Ou, MS. Proteasomes in the archaea: From structure to function. *Frontiers in Bioscience* **2000**, *5*(3), 837–865. https//doi.org/10.2741/furlow.

Milkoreit, M; Hodbod, J; Baggio, J; Benessaiah, K; Calderón-Contreras, R; Donges, JF; Mathias, JD; Rocha, JC; Schoon, M; Werners, SE. Defining tipping points for social-ecological systems scholarship – an interdisciplinary literature review. *Environmental Research Letters* **2018**, *13*(3), 033005. https://doi.org/10.1088/1748-9326/aaaa75

Milo, R; Phillips, R. *Cell Biology by the Numbers*. Garland Science, Taylor & Francis Group, LLC: New York, **2015**.

Morales, GP; Delaye, L. Was LUCA a hyperthermophilic prokaryote? The impact-bottleneck hypothesis revisited. In *Astrobiology and Cuatro Ciénegas Basin as an Analog of Early Earth Cuatro Ciénegas Basin: An Endangered Hyperdiverse Oasis*, Souza, V; Segura, A; Foster, J, Eds. Springer: Cham, Switzerland, **2020**.

Morowitz, HJ; Kostelnik, JD; Yang, J; Cody GD. The origin of intermediary metabolism. *Proceedings of the National Academy of Sciences* **2000**, *97*(14), 7704–7708. https://doi.org/10.1073/pnas.110153997

Muchowska, KB; Chevallot-Beroux, E; Moran, J. Recreating ancient metabolic pathways before enzymes. *Bioorganic & Medicinal Chemistry* **2019**, *27*(12), 2292–2297. https://doi.org/10.1016/j.bmc.2019.03.012

Naraoka, H; Hashiguchi, M; Sato, Y; Hamase K. New applications of high-resolution analytical methods to study trace organic compounds in extraterrestrial materials. *Life* **2019**, *9*(3), 62. https://doi.org/10.3390/life9030062

Nascimento, FF; Reis, MD; Yang, Z. A biologist's guide to Bayesian phylogenetic analysis. *Nature Ecology & Evolution* **2017**, *1*(10), 1446–1454. https://doi.org/10.1038/s41559-017-0280-x

Ohland, CL; Jobin, C. Microbial activities and intestinal homeostasis: a delicate balance between health and disease. *Cellular and Molecular Gastroenterology and Hepatology* **2015**, *1*(1), 28–40. https://doi.org/10.1016/j.jcmgh.2014.11.004

Ouzounis, CA; Kunin, V; Darzentas, N; Goldovsky, L. A minimal estimate for the gene content of the last universal common ancestor – exobiology from a terrestrial perspective. *Research in Microbiology* **2006**, *157*(1), 57–68. https://doi.org/10.1016/j.resmic.2005.06.015

Pearce, BK; Molaverdikhani, K; Pudritz, RE; Henning, T; Cerrillo, KE. Toward RNA life on early Earth: From atmospheric HCN to biomolecule production in warm little ponds. *The Astrophysical Journal* **2022**, *932*(1), 9. https://doi.org/10.3847/1538-4357/ac47a1

Peretó, J. Prebiotic chemistry that led to life. In *Handbook of Astrobiology*, Kolb, VM, Ed. CRC Press: Boca Raton, FL, **2019**, pp. 219–234.

Peters, S; Semenov, DA; Hochleitner, R; Trapp, O. Synthesis of prebiotic organics from CO2 by catalysis with meteoritic and volcanic particles. *Scientific Reports* **2023**, *13*(1), 6843. https://doi.org/10.1038/s41598-023-33741-8

Prangishvili D. Archaeal viruses: Living fossils of the ancient virosphere? *Annals of the New York Academy of Sciences* **2015**, *1341*(1), 35–40.

Ramirez, RM; Kopparapu, R; Zugger, ME.; Robinson, TD; Freedman, R; Kasting, JF. Warming early Mars with $CO_2$ and $H_2$. *Nature Geoscience* **2014**, *7*, 59–63. https://doi.org/10.1038/ngeo2000

Rauscher, SA; Moran, J. Hydrogen drives part of the reverse Krebs cycle under metal or meteorite catalysis. *Angewandte Chemie International Edition* **2022**, e202212932. https://doi.org/10.1002/anie.202212932

Repko, AF. *Interdisciplinary Research: Process and Theory*, 2nd ed. Sage: Thousand Oaks, CA, **2012**, pp. 20–22, 73, 94–96.

Rimmer, PB. Origins of life on exoplanets. In *Conflicting Models for the Origin of Life*. Scrivener Publishing: Beverly, MA, and John Wiley & Sons: Hoboken, NJ, **2023**, pp. 407–424.

Ritson, DJ. A cyanosulfidic origin of the Krebs cycle. *Science Advances* **2021**, *7*(33), eabh3981. https://doi.org/10.1126/sciadv.abh3981

Robertson, MP; Joyce, GF. The origins of the RNA world. *Cold Spring Harbor Perspectives in Biology*. **2012**, *4*, a003608.

Schulz, HN; Jørgensen, BB. Big bacteria. *Annual Reviews in Microbiology* **2001**, *55*(1), 105–137. https://doi.org/10.1146/annurev.micro.55.1.105

Shields, AL; Ballard, S; Johnson, JA. The habitability of planets orbiting M-dwarf stars. *Physics Reports* **2016**, *663*, 1–38. https://doi.org/10.1016/j.physrep.2016.10.003

Smoukov, SK; Seckbach, J; Gordon, R, eds. *Conflicting Models for the Origin of Life*. Scrivener Publishing: Beverly, MA, and John Wiley & Sons: Hoboken, NJ, **2023**.

Soucy, SM; Huang, J; Gogarten, JP. Horizontal gene transfer: Building the web of life. *Nature Reviews Genetics* **2015**, *16*(8), 472–482. https://doi.org/10.1038/nrg3962

Strazewski, P. Prebiotic chemical pathways to RNA and the importance of its compartmentation. In *Handbook of Astrobiology*, Kolb, VM, Ed., CRC Press: Boca Raton, FL, **2019**, pp. 235–264.

Sutherland, JD. The origin of life – out of the blue. *Angewandte Chemie International Edition* **2016**, *55*(1), 104–121. https://doi.org/10.1002/anie.201506585

Sutherland, JD. Studies on the origin of life: The end of the beginning. *Nature Reviews Chemistry* **2017**, *1*, 1–7. https://doi.org/10.1038/s41570-016-0012

Todd, ZR. Sources of Nitrogen-, Sulfur-, and phosphorus-containing feedstocks for prebiotic chemistry in the planetary environment. *Life* **2022**, *12*, 1268. https://doi.org/10.3390/life12081268

Varma, SJ; Muchowska, KB; Chatelain, P; Moran, J. Native iron reduces $CO_2$ to intermediates and end-products of the acetyl-CoA pathway. *Nature Ecology and Evolution* **2018**, *2*, 1019–1024. https://doi.org/10.1038/s41559-018-0542-2

Walton, CR; Ewens, S; Coates, JD; Blake, RE; Planavskyy, NJ; Reinhard, C; Ju, P; Hao, J; Pasek MA. Phosphorus availability on the early Earth and the impacts of life. *Nature Geoscience* **2023**, *16*, 399–409. https://doi.org/10.1038/s41561-023-01167-6

Weiss, MC; Sousa, FL; Mrnjavac, N; Neukirchen, S; Roettger, M; Nelson-Sathi S, et al. The physiology and habitat of the last universal common ancestor. *Nature Microbiology* **2016**, *1*, 16116. pmid:27562259 https://doi.org/10.1038/nmicrobiol.2016.116

Witzany, G. Communication as the main characteristic of life. In *Handbook of Astrobiology*, Kolb, VM, Ed. CRC Press: Boca Raton, FL, **2019**, pp. 91–108.

Wordsworth, R; Pierrehumbert, R. Hydrogen-nitrogen Greenhouse warming in Earth's early atmosphere. *Science* **2013**, *339*, 64–67. https://doi-org/10.1126/science.1225759

Xavier, JC; Gerhards, RE; Wimmer, JL; Brueckner, J; Tria, FD; Martin, WF. The metabolic network of the last bacterial common ancestor. *Communications Biology* **2021**, *4*(1), 413. https://doi.org/10.1038/s42003-021-01918-4

# Index

Note: Page numbers in *italic* indicate a figure and page numbers in **bold** indicate a table on the corresponding page.

## N



## O

## P



## Q



## R



## S

## T

## U



## V



## W

Printed in the United States
by Baker & Taylor Publisher Services

Printed in the United States
by Baker & Taylor Publisher Services